AF389441

Big Data Analytics and Information Science for Business and Biomedical Applications II

About the Editors

S. Ejaz Ahmed

S. Ejaz Ahmed is a Professor of Statistics and Dean of the Faculty of Math and Science at Brock University, Canada. Previously, he was Professor and Head of the Mathematics and Statistics Department at the University of Windsor, Canada, and University of Regina, Canada, as well as Assistant Professor at the University of the Western Ontario, Canada. He holds adjunct professorship positions at many Canadian and International universities. He has supervised more than 20 Ph.D. students and organized several international workshops and conferences around the globe. He is a Fellow of the American Statistical Association and has held the prestigious ASEAN Chair Professorship position. His areas of expertise include big data analysis, statistical learning, and shrinkage estimation strategy. He has authored several books and edited and co-edited several volumes and special issues of scientific journals. He has been a Technometrics Review Editor for the past ten years. Further, he is editor and associate editor of many statistical journals. Overall, he has published more than 200 articles in scientific journals and reviewed more than 100 books. Having been among the Board of Directors of the Statistical Society of Canada, he was also Chairman of its Education Committee. Moreover, he was Vice President of Communications for The International Society for Business and Industrial Statistics (ISBIS) as well as a member of the "Discovery Grants Evaluation Group" and the "Grant Selection Committee" of the Natural Sciences and Engineering Research Council of Canada.

Farouk Nathoo

Farouk Nathoo received his B.Sc. in Mathematics and Statistics (combined honours) from the University of British Columbia in 1998, his M.Math. from the University of Waterloo in 2000 and his Ph.D. in statistics at Simon Fraser University in 2006. He joined the Department of Mathematics and Statistics at the University of Victoria in 2006, and became a Full Professor in 2021. He currently holds the Tier 2 Canada Research Chair in Biostatistics for Spatial and High-Dimensional Data. His research interests focus on Bayesian Methods, High-Dimensional Data, Statistical Computation, Neuroimaging Statistics, and Machine Learning. He is President-Elect, Business and Industrial Statistics Section of the Statistical Society of Canada (2022–2023), and a Member of the Board of Directors of the Canadian Statistical Sciences Institute, (2018–2024).

Preface to "Big Data Analytics and Information Science for Business and Biomedical Applications II"

This book is the second volume of *Big Data Analytics and Information Science for Business and Biomedical Applications*. As with the first volume, it provides a venue for the presentation of cutting-edge research and discussion of powerful statistical methods developed for the analysis of Big Data in these areas. This second volume comprises nine papers showcasing both theoretical and applied developments.

In the first article, Shahhosseini and Miranda present a review article discussing techniques for the estimation of functional brain connectivity with an emphasis on functional magnetic resonance imaging (fMRI) data. In the second article, Chakraborty and Shojaie discuss the problem of learning the structure of directed acyclic graphs within the setting of non-Gaussian data. They develop nonparametric methods and associated algorithms to learning causal structure at high dimensions. In the third article, Fan and Bu discuss the diagnosis of lung disease from X-ray imaging using deep neural networks and transfer learning to incorporate existing pretrained networks to handle small sample sizes. The fourth article considers the examination of associations between longitudinal gestational weight gain and infant birth weight. Pietrosanu et al. develop a Bayesian joint modelling approach where parameters representing trajectories in a longitudinal model for gestational weight gain are incorporated as predictors of infant birthweight. Naiman and Song consider high-frequency data collected by mobile devices and develop semiparametric kernel machine regression with variable selection for functional predictors. In the sixth article, Liu et al. present a study examining the differences in social support communication among people with different types of cancers in online health communities using a network analysis. In the seventh article, Söderbäck et al. develop an improved estimation strategy for financial quantities by accounting for the high resolution and heteroscedastic nature of intraday data from liquid financial markets. In the eight article, Opoku et al. consider the estimation of fixed effects in the high-dimensional linear mixed model in settings where there is some prior information in the form of linear restrictions on the parameters. Shrinkage estimators are developed based on a full ridge regression estimator as a base model. The final contribution focuses on denoising image sequences such as those that arise from satellite imaging or fMRI. Yi and Qiu develop an edge-preserving image denoising procedure based on a jump-preserving local smoothing procedure. The procedure incorporates tunning parameters representing the bandwidths chosen to account for spatio-temporal correlation.

We hope that this second volume will continue to generate new ideas and research focussed on the many modern problems involving big data and high-dimensional inference.

S. Ejaz Ahmed and Farouk Nathoo

Editors

Review

Functional Connectivity Methods and Their Applications in fMRI Data

Yasaman Shahhosseini and Michelle F. Miranda *

Department of Mathematics and Statistics, University of Victoria, Victoria, BC V8W 2Y2, Canada;
yshahhosseini@uvic.ca
* Correspondence: michellemiranda@uvic.ca

Abstract: The availability of powerful non-invasive neuroimaging techniques has given rise to various studies that aim to map the human brain. These studies focus on not only finding brain activation signatures but also on understanding the overall organization of functional communication in the brain network. Based on the principle that distinct brain regions are functionally connected and continuously share information with each other, various approaches to finding these functional networks have been proposed in the literature. In this paper, we present an overview of the most common methods to estimate and characterize functional connectivity in fMRI data. We illustrate these methodologies with resting-state functional MRI data from the Human Connectome Project, providing details of their implementation and insights on the interpretations of the results. We aim to guide researchers that are new to the field of neuroimaging by providing the necessary tools to estimate and characterize brain circuitry.

Keywords: fMRI; functional connectivity; brain network; Human Connectome Project; statistics

Citation: Shahhosseini, Y.; Miranda, M.F. Functional Connectivity Methods and Their Applications in fMRI Data. *Entropy* 2022, 24, 390. https://doi.org/10.3390/e24030390

Academic Editors: S. Ejaz Ahmed and Farouk Nathoo

Received: 15 January 2022
Accepted: 8 March 2022
Published: 11 March 2022

Publisher's Note: MDPI stays neutral with regard to jurisdictional claims in published maps and institutional affiliations.

1. Introduction

Functional magnetic resonance imaging (fMRI) techniques have emerged as a powerful tool for the characterization of human brain connectivity and its relationship to health, behavior, and lifestyle [1]. The fMRI measurements comprise of an indirect and non-invasive measurement of brain activity based on the blood oxygen level dependent (BOLD) contrast [2]. Compared to alternative brain imaging modalities such as positron emission tomography (PET) and eletroencephalography (EEG), fMRIs are non-invasive and have a high spatial resolution, which makes them a popular choice in large brain imaging studies. An example of such studies is the Human Connectome Project that aims at understanding the underlying function of the brain by describing the patterns of connectivity in the healthy adult human brain [3].

There are mainly two goals in such studies: first, to identify location signatures in the brain that respond to external stimuli, and second, to identify brain space–time association patterns that emerge when the brain is either at rest or performing a task. These association patterns are measures of co-activation in functionally connected time series of anatomically different brain regions, known as functional connectivity [4,5]. There is evidence that individual differences in these connectivity patterns are responsible for important differences in cognitive and behavioral functions. Therefore, understanding these patterns can play an important role in predicting the early onset of neurodegenerative diseases and in monitoring disease care and treatment [6,7].

Functional MRI data is often high-dimensional and consists of images of 3D brain volumes collected over a period of time. In a typical study, the number of voxels N_v is in the hundred of thousands, and the number of time points T is in the hundreds. Therefore, estimating the $N_v \times N_v$ correlation matrix of voxelwise spatial connectivities is challenging and requires a few strategies and assumptions. A simple technique is to first pre-specify regions called *seeds* and then compute the cross-correlation of seeds and the functional time series of every other voxel in the brain. This *seed-based approach* became popular due to its straightforward calculation and interpretation. Seeds can vary in size and be as small as a single voxel. If the seed is a region, it is customary to average the time courses of the region and use that as the reference time course to be correlated with all the other voxels. In order to improve scalability, it is also common to first parcellate the brain into small regions and use the average time series of these regions to estimate the networks. The seed-based method can be a helpful resource when comparing patterns of neuropathologies and the normal brain. For example, ref. [8] uses this method to show that connectivity between the hippocampus and other brain regions change with the early signs of Alzheimer's disease when compared to control subjects. Despite the popularity of these approaches, there are various criticisms to the method. First, by focusing on pre-determined seeds, potential patterns that emerge in different brain regions are ignored [9]. Second, the method neglects the variability across voxels by averaging the time series in the ROIs. Third, the approach computes correlations between pairs of nodes and ignores the potential influence of other nodes in the entire network.

In contrast to region pre-specification, dimension reduction approaches characterize the spatial and/or temporal connectivity patterns by representing the data through a small number of latent components [10]. Principal component analysis (PCA) and independent component analysis (ICA) are the two most common of these methods. Both methods project the high-dimensional imaging data into a low-dimensional subspace. In PCA, this projection consists of orthogonal components that maximize the variance of the data projected into the low-dimensional subspace [11]. In ICA, the projection consists of components that are as spatially independent as possible [12]. Each of these components is then assembled into brain maps with the value in each voxel representing the relative amount of that particular voxel, which is modulated by the activation of that component. Compared to the seed-based approach, both PCA and ICA have the advantage of providing automated components with no need for the pre-specification of a seed region, i.e., these methods are data-driven. The authors in reference [13] used ICA to decompose brain networks into spatial sub-networks with similar functions in both the resting state and task fMRI data.

Other methods combine the brain parcellation strategy used in seed-based methods with dimension reduction approaches to characterize brain circuitry. Reference [10] uses an anatomical atlas to pre-determine clusters (ROIs) and then extract features from each cluster via principal components. Multiple extracted components were then used to estimate connectivity between these ROIs using the RV coefficient, a measure that summarizes the correlation among sets of features.

In addition to the methods utilized to estimate connectivity, it is common to characterize the functional networks by using the tools of *graph theory*. In a graph, brain networks are treated as a collection of nodes connected by edges. Commonly, the edges are defined by an estimated connectivity. Following the specification of the nodes, a binary matrix is obtained by thresholding the connectivity matrix. The binary graph is then used to compute various graph parameters that describe the nature of the brain network. These parameters express key characteristics of the network and usually include quantities that help determine if the graph nodes are connected in a random or small-world order. Random networks have a more globally connected pattern while small-world networks show a high level of local ordering [14]. Statistical network models take these graph measures as inputs for the prediction of global networks that characterize multiple individuals.

The goal of this paper is to provide an overview of the most commonly used methods to estimate and characterize functional connectivity in resting-state fMRI data. We illustrate these methods by analyzing data from a single-subject in the Human Connectome Project. Although we do not attempt to offer an exhaustive presentation of the rapidly evolving methods in the field, we expect that the information provided here will guide researchers that are new to the field of neuroimaging in exploring these data.

The remainder of the paper is organized as follows: In Section 2, we describe the different methods of estimating functional connectivity, focusing on data from a single subject. In Section 3, we estimate functional connectivity for a single-subject resting-state data from the Human Connectome Project, using the methods described in Section 2. In Section 4, we present a few multiple-subject estimation methods. In Section 5, we describe statistical network models. Finally, in Section 6, we present some final remarks on the topic.

2. Methods for Functional Connectivity

In this section, we review the different methods to estimate functional connectivity for single-subject data. We focus on data for a single subject and discuss group connectivity in Section 4. For all calculations, let the matrix Y be a matrix of size $T \times N_v$, consisting of N_v time courses representing the BOLD signal at each voxel $v = 1, \ldots, N_v$ [2] for a single subject. Here, for simplicity, we centralized the matrix Y by subtracting each voxel data (column) by the average of its corresponding time course. The goal of a connectivity-based analysis is to describe how various brain regions interact, either when the brain is resting or performing a task [15].

2.1. Seed-Based Analysis

It is computationally expensive to compute all pairwise correlations among a large number of voxels as it would require N_v^2 operations. Seed-based analysis (SBA) relies on estimating pairwise correlations between regions of interest (ROIs) or between a seed region and all the other voxels across the brain.

To estimate correlations between ROIs, a common approach is to divide the brain volume according to anatomical templates, usually called *brain atlases* [16]. There are several human brain atlases available, including the popular *Automated Anatomical Labelling (AAL)*, *Tailarach Atlas*, and the *MN1-152 atlas* [16,17]. To estimate correlations between a seed region and the other voxels, a seed is usually selected either by expert opinion or by choosing the voxel that shows greater activation during the fMRI experiment as the seed. The latter is more common during experiments involving tasks. After selection, the connectivity is estimated through a measure that quantifies correlation. Various measures were proposed in the literature. We provide more details about these measures in the Appendix A. We can summarize the seed-based approach the following way.

(i)   Choose a seed region or voxel;
(ii) Correlate the time series of the region or voxel with all other voxels in the brain. If the seed is a region, average the time series of the region prior to correlating that with all other voxels in the brain. Use one of the measures described in Appendix A;
(iii) Display the 3D volumes of the correlation measure or display the thresholded correlations (just the ones that are significant). **Note:** To determine significance, we need to account for multiple comparisons. Bonferroni and FDR are widely used procedures.

Alternatively, after dividing the brain into various ROIs using an atlas, we can summarize the time series of that region, either by averaging across voxels or by calculating the first principal component [15]. Next, we use those summary time series to be correlated between all regions. We illustrate both options in Section 3.

2.2. Decomposition Methods

Although seed-based methods have a straightforward interpretation, they are biased to the seed selection technique [18] and, therefore, should be used with caution. Principal component analysis and independent component analysis aim at solving the issue by providing a data-driven approach to functional connectivity. These decomposition methods play many roles in functional neuroimaging applications. They are used in the pre-processing steps to remove data artifacts and to reduce data dimensionality, and they will likely appear in at least one step of estimating functional connectivity in various populations. In this section, we will focus on their role as a method to describe functional connectivity in single-subject fMRI data, while in Section 4, we explore their contribution in multi-subject analysis.

As an alternative to seed-based analysis, the goal of the decomposition methods is to represent the voxel domain as a smaller subset of spatial components. Each spatial component has a separate time course and represents simultaneous changes in the fMRI signals of many voxels [12]. In this section, we assume that for each column of Y the average was subtracted from the data.

2.2.1. Principal Component Analysis (PCA)

PCA is a common method to reduce data dimensionality while minimizing the loss of data information and maximizing data variability [11]. The principal components are obtained either by the eigendecomposition of the sample covariance matrix $Y^T Y$ or by finding the eigenvectors of the data matrix Y using the theory of singular value decomposition (SVD). The rank of the data matrix is $r = min\{T, N_v\}$ (usually $T < N_v$ and $r = T$) and therefore we can find r principal components through the decomposition

$$Y = U\Sigma W^T = \sum_{k=1}^{r} \sigma_k u_k w_k^T, \tag{1}$$

where the $T \times r$ matrix U contains an orthonormal left singular vector $u_k \in \Re^T$, the $r \times N_v$ matrix W contains orthonormal right singular vectors $w_k \in \Re^{N_v}$, and the $r \times r$ diagonal matrix Σ contains the ordered singular values [11,15,19]. Notice that the eigendecomposition of $Y^T Y$ is defined as $W^T \Sigma^2 W$. The orthonormal rows of the $r \times N_v$ matrix W are referred to as eigenimages and can be assembled into brain maps, each representing the relative amount from a given voxel that is modulated by the activation of that component.

A different approach is to project the original fMRI data into the space spanned by the first p principal components, where the choice of p is based on the amount of data variability explained by the component. The projected data matrix, YW, consists of the time series of regions in this new subspace. The authors in reference [20] used this idea to reduce the dimensionality of the fMRI data in certain ROIs and then applied a Granger causality analysis on the block time series of two brain regions to infer directional connections. Although it is possible to compute correlations using the time series of these projected data points, the results have no clear interpretation since each of these spatial regions in the new subspace represent a linear combination of different voxels in the original data space.

2.2.2. Independent Component Analysis (ICA)

ICA aims at representing the brain data using a latent representation of independent factors. Differently from PCA, the goal is to decompose Y as a product of a mixing matrix and a combination of spatially independent components (ICs).

$$Y = MC + E = \sum_{k=1}^{K} m_k c_k + E, \tag{2}$$

where M is the $T \times K$ mixing matrix with columns m_k, and the $K \times N_v$ matrix C is the matrix of independent components with rows c_k, where each c_k contains brain networks corresponding to component k for a total of K independent components. These components represent the networks of various functions, such as motor, vision, auditory, etc. The elements of the matrix E are independent Gaussian noises.

It is assumed that the component maps, $c_k, k = 1, \ldots, K$ represent possible overlapping and statistically dependent signals, but the individual component map distributions are independent, i.e., if $P(c_k)$ represents the probability distribution of the voxels values in the kth component map, we have

$$P(c_1, c_2, \ldots, c_K) = \prod_{k=1}^{K} P(c_k). \tag{3}$$

Each independent component c_k is a vector of size N_v and can be assembled into brain maps. As in PCA, these maps represent the relative amount of a given voxel that is modulated by the activation of that component.

2.3. Computational Aspects

In imaging applications, estimating the principal components requires the decomposition of the $N_v \times N_v$ matrix $Y^T Y$, which is usually unfeasible. Many algorithms were proposed in the literature to efficiently estimate the components in such high-dimensional settings. Ref. [21] develops the sparse PCA (SPCA), which is based on a regression optimization problem using a lasso penalty, [22] a multilevel functional principal component for high-dimensional settings, and [23] estimate a sparse set of principal components through an iterative thresholding algorithm. Some of these toolboxes are easy to access and available for downloading at the authors' website.

Similarly, estimating the independent components is not straightforward, and ranking the components is challenging because the ICs are usually not orthogonal, and the sum of the variances explained by each component will not sum to the variance of the original data. One of the first algorithms was the Infomax, which aims at maximizing the joint entropy of suitably transformed component maps [12,24]. Recently, more modern algorithms focus on extracting a sparse set of features from data matrices containing a very large number of features. Examples are the ICA with a reconstruction cost (RICA) proposed by [25], which is available as a Matlab toolbox.

2.4. A Hybrid Method

A different approach to estimate functional connectivity is given by reference [10]. The authors propose a multi-scale model based on networks at multiple topological scales, from voxel level to regions consisting of clusters of voxels, and larger networks consisting of collections of those regions. In practice, these collections of voxels are pre-specified and usually taken as anatomical ROIs. These anatomical ROIs can be then combined to form clusters of ROIs. Their approach consists of a dimension reduction step through to a factor model within each ROI. Let r represent the r-th ROI and Y_r be a $T \times p_r$ data matrix consisting of the time series of voxels belonging to the r-th ROI (containing a total of p_r voxels, where $\sum_{r=1}^{R} p_r = N_v$ and R is the total number of ROIs). Then, we write

$$Y_r(t) = Q_r f_r(t) + E_r(t), \tag{4}$$

where $Y_r(t)$ is a column vector of size p_r, $f_r(t)$ is a $m_r \times 1$ vector of latent common factors with a number of factors $m_r \ll p_r$, Q_r is a $p_r \times m_r$ factor-loading matrix that defines the dependence between the p_r voxels through the mixing of f_r, and $E_r(t) = [e_{r1}(t), \ldots, e_{rp_r}(t)]'$ is a $p_r \times 1$ vector of white noise with $E(E_r(t)) = 0$ and $\Sigma_{E_r, E_r} = Cov(E_r(t)) = \text{diag}(\sigma_{e_{r1}}^2, \ldots, \sigma_{e_{rp_r}}^2)$.

These factor models are then concatenated to define

$$Y(t) = Q f(t) + E(t), \tag{5}$$

where $Y(t)$ is a column vector of size $\sum_{r=1}^{R} p_r = N_v$, $Q = \text{diag}(Q_1, \ldots, Q_R)$ is a $\sum_{r=1}^{R} p_r \times \sum_{r=1}^{R} m_r$ block-diagonal mixing matrix and $f(t) = [f_r(t), \ldots, f_R(t)]'$ is a $\sum_{r=1}^{R} m_r \times 1$ vector of aggregated latent factors.

Network covariance matrices in these different topological scales are estimated using the low-rank matrix in the following way. Let $\Sigma_{Y_r Y_r}$ be the covariance matrix within ROI r. Model (4) implies the following decomposition

$$\Sigma_{Y_r Y_r} = Q_r \Sigma_{f_r f_r} Q_r' + \Sigma_{E_r E_r}. \tag{6}$$

Similarly, from Model (5) we have

$$\Sigma_{YY} = Q\Sigma_{ff}Q' + \Sigma_{EE}.$$ (7)

The low-dimensional factor covariance matrix Σ_{ff} is a block matrix used to estimate the lag-zero dependency between ROIs as follows.

$$\Sigma_{ff} = \begin{pmatrix} \Sigma_{f_1 f_1} & & \Sigma_{f_1 f_R} \\ & \ddots & \\ \Sigma_{f_R f_1} & & \Sigma_{f_R f_R} \end{pmatrix}$$

The diagonal blocks $\Sigma_{f_r f_r}, r = 1, \ldots, R$ are diagonal covariance matrices, capturing the total variance of factors within each ROI. The off-diagonal blocks $\Sigma_{f_k f_j}, j \neq j$ are cross-covariance matrices between factors and summarize the dependence between clusters j and k.

The authors summarize the dependence between ROIs using the RV coefficient, a multivariate generalization of the squared correlation coefficient. The RV coefficient between factors in clusters j and k is defined by

$$RV = \frac{\text{tr}(C_{f_k f_j}, C_{f_j f_k})}{\sqrt{\text{tr}(C_{j_k f_j}, C_{f_j f_j})\text{tr}(C_{f_k f_k}, C_{f_k f_k})}},$$ (8)

where $C_{f_j f_k} = (\Sigma_{f_j f_j})^{-\frac{1}{2}}\Sigma_{f_j f_k}(\Sigma_{f_k f_k})^{-\frac{1}{2}}$.

In practice, the authors apply this model to estimate resting-state networks. They estimate the factors f_r and matrices Q_r using PCA and pre-specify the ROIs based on an anatomical atlas. The authors in reference [26] use this approach to estimate background functional connectivity between ROIs using data from the Working Memory Task in the Human Connectome Project.

2.5. Brain Networks

It is common to represent the brain using tools from *graph theory*. In this framework, we can think of functional connectivity as a network represented by a graph, where the spatial units are the nodes and the connection between them are the edges. Networks are treated as a collection of nodes (vertices) connected by links (edges). The graph (network) is represented as the pair $G = (V, E)$, where V and E are the sets of vertices and edges, respectively. In addition, graphs may be weighted and, in such cases, will be denoted by the triple $G = (V, E, W)$, with $W(E)$ indicating the weight for each edge.

The first decision to make is the selection of the nodes of the network. Similar to the seed-based connectivity, these nodes are defined by either voxels or the ROI parcellations given by anatomical atlases. Following the specification of the nodes, their edges (links) must be determined. These edges quantify the strength of association between these different nodes, i.e., they are the functional connectivity. The same measures discussed previously for functional connectivity and described in Appendix A are used to quantify the strength of the edges.

Most of the standard tools of graph theory have been developed for binary networks, where each edge is either present or not. A binary matrix, usually called an *adjacency matrix*, is obtained by thresholding the connectivity matrix. Although it is convenient to threshold the weighted graphs to apply the standard graph theoretical machinery, information about the original signal is discarded in the process. Moreover, in most situations, the choice of a threshold is not unique, and such a decision may be difficult to justify. One strategy is the use of a mass-univariate approach, in which a statistical test is performed for every possible edge in the network and then corrected for multiple comparisons using standard techniques, such as the Bonferroni correction or the false discovery rate (FDR) [27,28].

After the network is estimated, some descriptive measures are calculated as means to describe the topological graph properties. In brain networks, the popular metrics are the characteristic path length, the clustering coefficient, and the degree distribution. For a list of the comprehensive topological measures used in neuroimaging, see reference [29].

Characteristic path length. Paths are the sequences of distinct nodes that represent the potential flow of information between pairs of brain regions with shorter paths, implying stronger potential for integration. The length of a path estimates the potential for functional integration between brain regions. One of the most commonly used measures of functional integration is the average shortest path length between all pairs of nodes in the network, which is defined as the characteristic path length [15]. Paths between disconnected nodes are defined to have infinite length, which is a problem when calculating this measure, especially in sparse networks such as in functional connectivity. In practice, we take the average only between the existing paths, which can be a problem. For a discussion on this issue please refer to reference [29].

Degree distribution. A measure of centrality, the degree of an individual node is equal to the number of links connected to that node, i.e., the number of neighbors of the node. The degree distribution is, therefore, the distribution of the degrees of all the nodes in the network. In functional connectivity, nodes with a high degree are interacting functionally with many other nodes in the network [29] and are referred to as *hubs*.

Clustering coefficient. A measure of segregation, the clustering coefficient is the fraction of the node's neighbors that are also neighbors of each other, which in graph theory is the fraction of triangles around an individual node. The presence of clusters in functional networks suggests an organization of statistical dependencies indicative of segregated functional neural processing, which is the ability for specialized processing to occur within densely interconnected groups of brain regions. The mean clustering coefficient for the network reflects, on average, the prevalence of clustered connectivity around individual nodes. The mean clustering coefficient is normalized individually for each node and can disproportionately be influenced by nodes with a low degree.

Many other network measures are greatly influenced by basic network characteristics, such as the number of nodes and links and the degree of distribution presented in this section.

3. Real Data Example

We analyzed the resting-state data from the Human Connectome Project (HCP). We chose to work with the data that had been previously denoised using the FIX pipeline [30]. This pipeline uses a gentle high-pass temporal filter, performs motion regression (i.e., the regression of 24 movement parameters: six rigid-body motion parameters, their backward temporal derivatives, and squares of those 12 time series), and applies a regression based on ICA to remove the variance in noise components that was orthogonal to signal components [31]. For the single-subject analysis, we considered the volumes collected from the right–left phase of the example, Subject 100307. Volumes of fMRI were obtained every 720 ms. Each volume consisted of images of size $91 \times 109 \times 91$ for a total of 1200 time frames.

3.1. Single-Subject Examples

3.1.1. ROI-Based Connectivity

We partitioned the brain into ROIs using the AAL Atlas version that was registered into the MNI152 space. We considered a total of 166 ROIs and estimated the connectivity using the following methods:

(a)   Cross correlation of the average time series in each ROI;
(b)   Partial correlation of the average time series in each ROI;
(c)   Cross correlation of the time series of the ROI data projected into the space of its first principal component;

(d)   Partial correlation of the time series of the ROI data projected into the space of its first principal component;

(e)   For each ROI, we consider the principal components that account for 20% of the ROI variability and calculate the RV coefficient as described in Equation (8).

The results for the estimated connectivity values are shown in Figure 1. Inspecting Figure 1 reveals that cross-correlation measures in panels (a) and (c) capture larger correlations than their corresponding partial cross-correlation measures (panels (b) and (d)). The RV coefficient from the method described in (e) seems to be able to capture a large number of small correlations among ROIs. Before drawing any conclusions from the figure, we should first test whether these values are significant. For the first four matrices, the test is done by first transforming these values to z-scores and then thresholding them to identify important correlations. For the RV coefficient in panel (e), significance is based on the asymptotic distribution of the coefficient as detailed in reference [10].

Next, we used these connectivity matrices to obtain a binary graph with the edges determined based on the p-values obtained from the z-scores of the correlation coefficient, as described in Appendix A, Equation (A2). The p-values were thresholded based on the Bonferroni correction and a significance of 5%. For the RV coefficient in panel (e), we use the asymptotic distribution of the coefficients to convert the values to z-scores and thresholded based on the Bonferroni correction to find the quantile of the standard normal distribution with a significance of 5%. Considering this criteria, we compute the adjacency matrices shown in Figure 2.

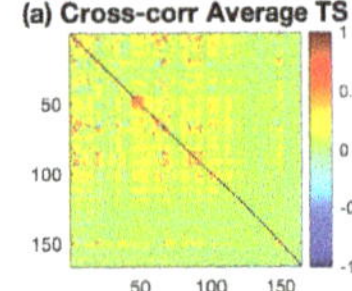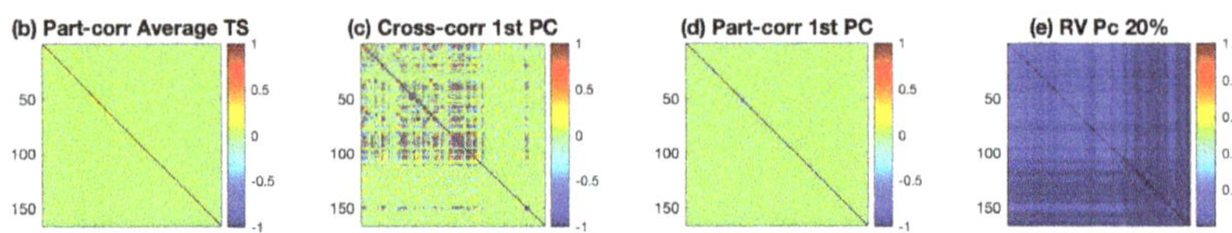

Figure 1. Estimated connectivity for the ROIs based on the AAL parcellation. Panel (**a**) depicts the cross-correlation for the average time series of the ROIs, panel (**b**) depicts the partial cross correlation for the average time series of the ROIs, panel (**c**) depicts the cross correlation for the time series of the ROI data projected with the first PC, panel (**d**) depicts the partial cross correlation for the time series of the ROI data projected with the first PC, and panel (**e**) represents the RV coefficient with each ROI retaining the principal components that explain 20% of its variability.

Inspecting Figure 2 reveals the presence of a large number of edges for both (a) and (c) graphs. This indicates a high level of interaction between the different anatomical regions. This high-interaction level was not found in graphs (b) and (d). In panel (e), we observe a moderate level of interaction with a few ROIs connecting with many others, while some regions are quiet during the resting-state experiment.

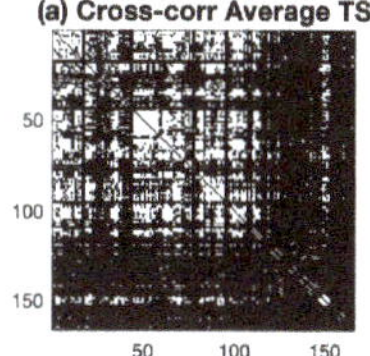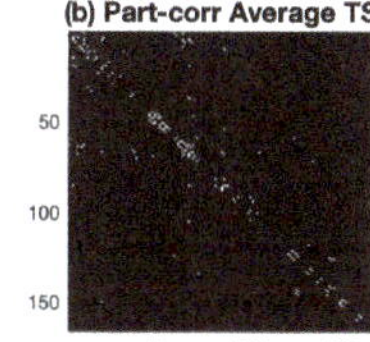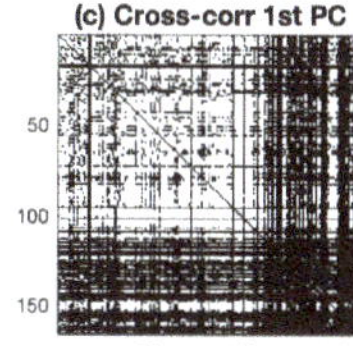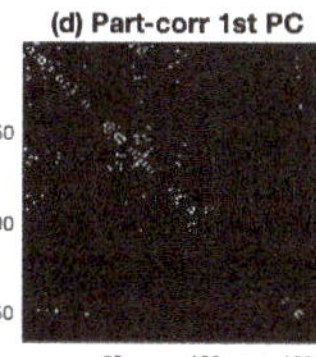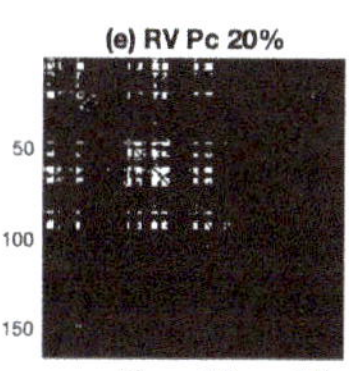

Figure 2. Binary Graphs obtained from the thresholded connectivities matrices of Figure 1. For all panels, the white color indicates an edge between the ROIs. Panel (**a**) is the graph obtained by thresholding the cross correlation of the average time series of the ROIs, panel (**b**) depicts the graph from the thresholded partial cross correlation for the average time series of the ROIs, panel (**c**) depicts the graph obtained by thresholding the cross correlation for the time series of the ROI data projected with the first PC, panel (**d**) depicts the graph obtained by thresholding the partial cross correlation for the time series of the ROI data projected with the first PC, and panel (**e**) represents the graph obtained by thresholding the RV coefficient.

3.1.2. Network Summary Measures

We used the binary graphs obtained above to estimate a few summary measures, using graph theory as described in Section 2.5. The formulas used in each calculation are detailed in Appendix B. Table 1 shows the results. *CPL* is the characteristic path length excluding all infinity paths from the network, *DG* is the average degree of the network, where the degree indicates the number of links in each node, *CC* is the average clustering coefficient of the network, and *Inf* is the number of infinity paths in the network. The quantities in Table 1 reflect what we observe in Figure 2. The degree indicates the number of connections between regions. As noticed before, the graphs in panels (a) and (c) indicate a high degree, with many interactions between ROIs. The characteristic path length (CPL) of the RV coefficient indicates that on average the network has a short path length, with a value that is comparable to the networks in panels (a) and (c) of Figure 2. This indicates that despite few regions being connected, the ones that are connected are near each other.

Table 1. Network summary measures.

	(a) Av. CCorr	(b) Av. Pcorr	(c) PC1 Ccorr	(d) PC1 Pcorr	(e) RV
CPL	2.137	5.029	1.508	3.9685	1.8789
DG	34.193	1.313	69.518	1.386	4.217
CC	0.640	0.072	0.825	0.179	0.820
Inf	2963	11,239	2964	11,789	12,366

3.1.3. Volume-Based Connectivity

Seed-Based Analysis. For seed-based analysis, we chose the left pars opercularis (left interior frontal gyrus) as the seed [32]. We take the average time series for this region and compute the cross correlation with the remaining voxels. We perform a Bonferroni correction considering $\alpha = 0.05$ to threshold the correlation values. Figure 3 shows the resulting brain map. We display clusters bigger than 125 as significant voxels, and their mask is overlaid on a template brain consisting of the average time points of the example subject data used here.

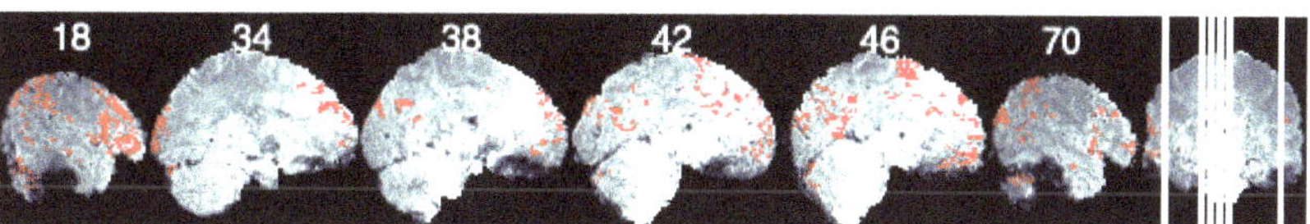

Figure 3. Seed- based connectivity of the left pars opercularis. Figure shows sagittal slices with voxels that have a significant connection with the seed ROI depicted in red.

Decomposition Methods. We first obtain the principal components of the data matrix Y. It is important to notice that a large number of principal components is needed to represent data variability and that traditional principal components have the issues discussed in Section 2.3. For this particular data, 150 components are needed to represent 20% of the data variability and 463 are needed to represent 50%. We illustrate the first five components scaled by their eigenvalues (i.e., the loadings) in Figure 4.

Next, to estimate the independent components, we use the probabilistic independent components analysis proposed in reference [33] and implemented in the MELODIC (multivariate exploratory linear optimized decomposition into independent components) function in FSL. Figure 5 depicts the results.

For illustration purposes, we present the components without thresholding their values. It is more common to use the individual components' maps as inputs in a multi-subject approach and then perform thresholding in the group components to identify a group network. We comment more on the topic in the next section.

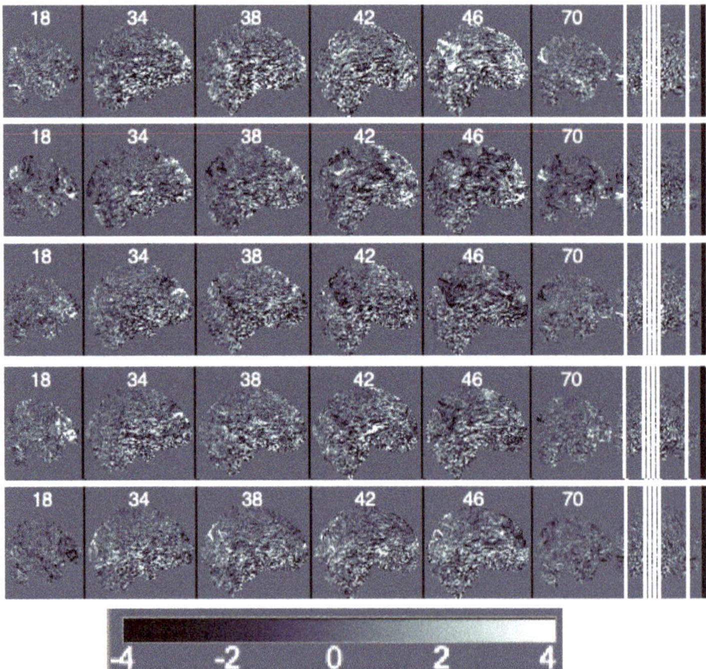

Figure 4. Sagittal view of the ordered principal components' maps from first (**top**) to fifth (**bottom**).

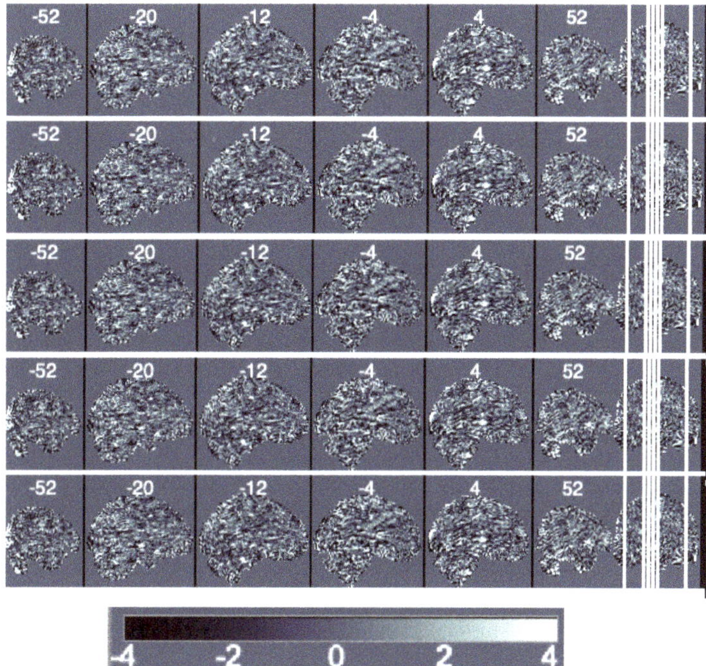

Figure 5. Sagittal view of the independent components' maps ordered based on increasing amounts of uniquely explained variance from first (**top**) to fifth (**bottom**).

4. Multiple-Subject Functional Connectivity

When modeling functional MRI, an important goal is to identify the functional connectivity structure in multi-subject data by leveraging a shared structure across subjects. Multi-subject functional connectivity models can range from constrained tensor decomposition models, e.g., PARAFAC, to more flexible approaches where subject-specific connectivity matrices or PCA and ICA models are estimated first, and their concatenated results are used as inputs on a group-based estimation. The optimal model will depend on which level of flexibility best captures the functional connectivity features within the group [34].

In multi-subject ICA models, a simple procedure is to estimate the single-subject connectivity matrix using pre-specified ROIs, as in the seed-based approach described in Section 2, and then aggregate those results into a single matrix, subsequently further decomposing this matrix using principal components. The principal components can then be mapped to estimate a group-based functional connectivity. Ref. [35] used this idea to estimate a dynamical group-based resting-state connectivity of minimally disabled relapsing–remitting patients.

A multi-stage approach is implemented in reference [36] to compare functional connectivity between subjects at a high familial risk for Alzheimer's disease that are clinically asymptomatic versus matched controls. The method follows four steps, including subject-specific SVD, a population-level decomposition of aggregated subject-specific eigenvectors, a projection of the subject-level data onto the population eigenvectors to obtain subject-specific loadings, and the use of the subject-specific loadings in a functional logistic regression model.

A group of methods propose a *group ICA* approach, where fMRI data is either temporally concatenated across subjects or taken as a multi-dimensional array. The FMRIB Software Library (FSL), a software library containing image analysis and statistical tools for various imaging data, provides group ICA and tensorial ICA in its MELODIC function. This section will focus on these two approaches.

4.1. Group ICA

Ref. [37] proposed for the first time an approach to perform ICA on fMRI data from a group of subjects. Suppose we observe fMRI data from n subjects. Let Y_i be a matrix of size $T \times N_v$ consisting of N_v time courses representing the BOLD signal at each voxel $v = 1, \ldots, N_v$ for subject $i = 1, \ldots, n$. Their model involves a multi-stage approach as follows.

1. Subject-level data reduction. In this step, reduction is applied in the temporal domain. For each subject $i = 1, \ldots .n$, the reduced data is given by

$$X_i = F_i^{-1} Y_i,$$

   where F_i^{-1} is a $L \times T$ reducing matrix and X_i is a $L \times N_v$ matrix representing the reduced data. In practice, F^{-1} is obtained by PCA decomposition;

2. Data reduction of the aggregated subject-level data. Data reduction is applied to the $NL \times N_v$ matrix $[X_1^T, \ldots, X_N^T]^T$ to obtain a $K \times N_v$ matrix $X = G^{-1}[X_1^T, \ldots, X_N^T]^T$, where K is the number of components to be obtained and G^{-1} is a $K \times NL$-reducing matrix that is in practice obtained by principal components;

3. Estimation of independent sources. An ICA decomposition is applied to the matrix X, as described in Section 2.2.2.

$$X = MC,$$

   where M is a $K \times K$-mixing matrix and C is a $K \times N_v$ component map matrix. The resulting group ICA components can be thresholded by first converting them into Z-scores.

Individual maps can be obtained by partitioning GM (where $G = (G^{-1})^T$) by subject and going back along the previous steps as follows.

$$GX = GMC = \begin{bmatrix} F_1^{-1} Y_1 \\ \ddots \\ F_N^{-1} Y_N \end{bmatrix}.$$

Based on these steps, the matrix GMC is a matrix of size $NL \times N_v$ of individual maps and can be partitioned such that $G_i M_i C_i = F_i^{-1} Y_i$, and C_i contains the individual maps.

4.2. Tensorial ICA

The tensor ICA is based on tensor decomposition, which obtains a low-rank representation of a multi-dimensional array. PARAFAC is a common decomposition method [38]. Let $X \in \mathbb{R}^{T \times N_v \times N}$ be an array with fMRI data and dimension times, voxels, and subjects, respectively. The three-dimensional array X can be decomposed as a sum of R outer products in the following way

$$X = \sum_{r=1}^{R} a_r \circ b_r \circ c_r,$$

where $a_r \in \mathbb{R}^T$, $B_r \in \mathbb{R}^{N_V}$, and $c_r \in \mathbb{R}^N$. This decomposition implies that each element in the array X can be written as

$$\{x_{ijk}\} = \sum_{r=1}^{R} a_{ir} b_{jr} c_{kr}.$$

The vectors in the decomposition can be represented in matrices, e.g., $A = [a_1 a_2 \ldots a_R]$, and likewise to obtain matrices B and C. The three-dimensional array can be unfolded into matrices in a process called matricization. The unfolding can happen in any of the three dimensions. On the second dimension, $X_{(2)} \in \mathbb{R}^{N_v \times NT}$ is the mode-two matricization of X. Similarly, we can use the unfolding to generate mode-two and mode-three matrices. For details on the PARAFAC decomposition and matricization, please refer to reference [38]. Using these definitions, we can write:

$$X_{(2)} = B(C \odot A)^T,$$

where $\odot$ denotes the Katri–Rao product. In reference [39], the authors propose an ICA decomposition of the form

$$X^* = (C \odot A)B^T + E,$$

where $X^* = X_{(2)}^T$ and the mixing matrix $M = (C \odot A)$ and component matrix B^T are estimated as in reference [33].

5. Statistical Network Models

In this section, we follow the notation in reference [40] to describe statistical network models with the purpose of characterizing brain circuitry. In these models, individual functional connectivity is estimated first, using the techniques described in Section 2. After individual estimation, the effects of multiple variables of interest and topological network features are taken into account on the overall network structure.

Let $(\mathcal{Y}_i, \mathcal{X}_i)$ represent the network and covariates for subject i, respectively. The probability density function of the network given the covariates is denoted by $P(\mathcal{Y}_i | \mathcal{X}_i, \theta_i)$, where θ_i describes the relationship of $\mathcal{Y}_i$ and $\mathcal{X}_i$. These covariates can be node-specific covariates, such as brain location and also functions of the network $\mathcal{Y}_i$, such as the path length or other metrics described in Section 2.5. Popular ways of modeling the density function include exponential random graph models (ERGMs) and mixed models [40].

In ERGMs, we consider binary graphs and the models are fitted for each subject individually as follows. Let $\mathcal{Y}_i$ be a network consisting of $R \times R$ nodes. Then, $\mathcal{Y}_{ijk} = 1$ if a link exists between nodes j and k, and $\mathcal{Y}_{ijk} = 0$ otherwise. The probability mass function has the form of a regular exponential family:

$$P(\mathcal{Y}_i = y_i | \mathcal{X}_i) = \kappa(\theta)^{-1} \exp\left\{ \theta^T g(y_i, \mathcal{X}_i) \right\}.$$

The estimation of the parameters θ is done by MCMC MLE. In reference [41], they identify the most important explanatory metrics $g(y_i)$ for each subject's network. Next, the authors create a group-based summary measure of the fitted parameter values θ for all subjects. They use these group-based explanatory metrics and parameters to fit a group-based representative network via ERGMs.

One limitation of the current estimation methods for ERGMs is scalability. The major issue is not the number of ROIs per se but the edge structure of the network, which can cause convergence problems. Moreover, most models were developed for binary graphs and are not well-suited for link-level examination [40].

As an alternative to ERGMs, mixed models allow for both link-level examination and multiple-subject comparisons. The framework defines a two-part mixed effect that models both the probability of a connection being present or absent and the strength of a connection if it exists. Let $\mathcal{Y}_i$ be a representation of the functional connectivity strength given by one of the correlation measures listed in Appendix A, and let $\mathcal{R}_{ijk}$ be an indicator of whether a connection between j and k is present. Then the conditional probabilities are

$$P(\mathcal{R}_{ijk} = r_{ijk} | \beta_r; b_{ri}) = \begin{cases} 1 - p_{ijk}(\beta_r; b_{ri}), & \text{if } r_{ijk} = 0 \\ p_{ijk}(\beta_r; b_{ri}), & \text{if } r_{ijk} = 1, \end{cases}$$

where β_r are the vector of fixed effects that relate the covariates $\mathcal{X}_{ijk}$ for each participant and pair of nodes, and b_{ri} are random effects representing subject-specific and node-specific parameters.

Let Z_{ijk} be the design matrix associated with the random effects b_{ri}; the models are divided into two parts. The first part of the model uses a logit link function to relate the probability of connection between nodes j and k to the covariates as follows.

$$\text{logit}(p_{ijk}) = \mathcal{X}'_{ijk}\beta_r + Z'_{ijk}b_{ri}.$$

The second part models the strength of the connection between nodes j and k given that the connection is present, by linearly linking the Fisher's Z-transform of the correlation coefficient between nodes i and j and the covariates. Let $S_{ijk} = \mathcal{Y}_{ijk}|R_{ijk} = 1$, then

$$\text{Fisher's Z-transform}(S_{ijk}) = \mathcal{X}'_{ijk}\beta_s + Z'_{ijk}b_{si} + e_{ijk},$$

where β_r is a vector of population parameters that related the strength of connection to the same set of covariates $\mathcal{X}_{ijk}$ for each participant and pair of nodes, b_{si} is a vector of subject and node-specific parameters that capture how this relationship varies about the population average β_s, and e_{ijk} is the random noise for subject i and nodes j and k. Details of the two-parts modeling approach is presented in reference [42].

One issue that arises from these models is that thresholding choices based on the connectivity weights will impact the network topology, even if multiple comparisons are taken into account. The authors in reference [40] argue that persistent homology provides a multi-scale hierarchical framework that addresses the threshold issue. The method is a technique of computational topology that provides a coherent mathematical framework for comparing networks. Instead of looking at the networks at a fixed threshold, persistent homology records the changes in topological network features over multiple resolutions and scales. By doing so, it reveals the features that are robust to noise, i.e., the most 'persistent' topological features.

6. Summary

In this paper, we have reviewed the most common methods to estimate functional connectivity in fMRI data. For single-subject data, estimation can be done by directly quantifying correlations across regions of interest and/or seed regions, or by finding a set of latent components that represent simultaneous activity, and while interpretation is straightforward for the former approach, it is not as clear for the later. In the example provided, the number of component maps needed to represent the data variability is very high and, therefore, the investigation of only a few components might not reflect the whole picture of the brain network.

The results obtained in Section 2 indicate that even if the regions are defined in an equivalent way, different estimation procedures of connectivity will lead to different interpretations of the networks. Therefore, it is of great importance to be aware of the limitations of each approach, especially when interpreting results from individual datum.

Despite the challenges with the single-subject analysis, a consistent procedure, applied to various subjects, might translate into a successful representation of multiple-subject networks. This is specially true if the method does not require a multi-stage approach and performs, instead, a joint estimation as in the tensorial ICA framework. Other emerging multi-subject network methods, such as persistent homology, are a promising way to estimate brain circuitry, especially if scalability can be achieved.

Funding: This research was funded by Natural Sciences and Engineering Research Council grant number RGPIN-2020-06941.

Conflicts of Interest: The authors declare no conflict of interest.

Appendix A. Methods to Quantify Correlation

Cross Correlation. Cross correlation measures the (lagged) temporal dependencies between two signals, and it was first proposed by reference [43] as an effective way to describe functional connectivity. Suppose we want to calculate the correlation between the BOLD time series for a given voxel v, i.e., $y_v(t), t = 1, \ldots, T$ and a reference time series $r_{v'}(t), t = 1, \ldots, T$ for $v \neq v'$. Let μ_y and μ_r be the average value of the vectors $\boldsymbol{y}_v$ and $\boldsymbol{r}_{v'}$, respectively. Then, the cross correlation between the vectors $\boldsymbol{y}_v$ and $\boldsymbol{r}_{v'}$ is defined as

$$
cc_{y,r} = \frac{\sum_{t=1}^{T}(\boldsymbol{y}_v(t) - \mu_y)(\boldsymbol{r}_{v'}(t) - \mu_r)}{\sqrt{\sum_{t=1}^{T}(\boldsymbol{y}_v(t) - \mu_y)^2}\sqrt{\sum_{t=1}^{T}(\boldsymbol{r}_{v'}(t) - \mu_r)^2}}
\tag{A1}
$$

The reference vector can be a pre-selected voxel, the seed, or it can be an average of time series in a certain region. For cross correlation between ROIs, both $y(t), t = 1, \ldots, T$ and $r(t), t = 1, \ldots, T$ can be the average time series in the pre-determined regions y and r, respectively.

It is common to transform the correlation coefficient obtained in (A1) using a Fisher's Z-transformation for each correlation coefficient as follows

$$
\text{z-score} = \frac{ln(1 + cc_{y,r}) - ln(1 - cc_{y,r})}{2}.
\tag{A2}
$$

These coefficients are approximately normally distributed, and cutoff values are obtained from the standard normal distribution.

Partial Cross Correlation. Cross correlation quantifies only the marginal linear dependence between two signals and does not consider the effect of a third signal [15,44]. To remove the linear influence of a third signal $k(t)$ we define the partial correlation as follows.

$$
PCC_{y,r|k} = \frac{cc_{y,r} - cc_{y,k}cc_{r,k}}{\sqrt{1 - cc_{y,k}^2}\sqrt{1 - cc_{r,k}^2}}.
\tag{A3}
$$

Partial cross correlation is a valuable metric for estimating brain networks because it can estimate the direct relationship between two signals [15].

The calculation of cross correlation and partial cross-correlation measures assumes the signals to be stationary. When this assumption is not satisfied, detrended cross correlation and detrended partial cross correlation should be used instead [45].

Time-varying connectivity. It is possible to obtain a dynamical functional connectivity to understand its pattern over time. Both static measures mentioned in this section have a natural time-varying analogue in conjunction with a sliding window [15].

The concept of the sliding window is simple. Starting from the first time point, a window (a fixed number of time points) is selected, and all data points within the window are used to estimate the FC. This window is then shifted a certain number of time points, and the FC is estimated on the new set of data points. The process is repeated until the end of the time course. The series of estimated FC over time is the time-varying FC.

Appendix B. Calculation of Network Measures

For completeness, we present the mathematical definitions of the network measures presented in Section 2.5. For a complete list of network measures, please refer to reference [29].

We use the graph notation as defined in Section 2.5. Let n be the number of nodes in the network and N be the set of all nodes. Let l be the number of links in the network and L be the set of all links. Then, (i, j) is a link between nodes i and j and $a_{ij} = 1$ when there is a link (i, j). We define $l = \sum_{i,j} a_{ij}$ (counting each indirect link twice).

Degree of a node. The degree of a node i is the sum of all the links connected to the node and is defined as

$$
k_i = \sum_{j \in N} a_{ij}.
$$

Shortest path length. The shortest path length measures the shortest distance between nodes i and j and is defined as:

$$d_{ij} = \sum_{a_{uv} \in g_{i \leftrightarrow j}} a_{uv},$$

where $g_{i \leftrightarrow j}$ is the shortest distance between i and j. For all disconnected pairs (i, j), $d_{ij} = \infty$.

Characteristic path length. Let L_i be the average distance between node i and all other nodes. The characteristics path length is defined as

$$L = \frac{1}{n} \sum_{i \in N} L_i = \frac{1}{n} \sum_{i \in N} \frac{\sum_{j \in N, j \neq i} d_{ij}}{n - 1}.$$

Number of triangles. The number of triangles of a node i is defined as

$$t_i = \frac{1}{2} \sum_{j, h \in N} a_{ij} a_{ih} a_{jh}.$$

Clustering coefficient. The clustering coefficient of the network is defined as

$$C = \frac{1}{n} \sum_{i \in N} C_i = \frac{1}{n} \sum_{i \in N} \frac{2t_i}{k_i(k_i - 1)}.$$

References

1. Elam, J.S.; Glasser, M.F.; Harms, M.P.; Sotiropoulos, S.N.; Andersson, J.L.; Burgess, G.C.; Curtiss, S.W.; Oostenveld, R.; Larson-Prior, L.J.; Schoffelen, J.M.; et al. The Human Connectome Project: A retrospective. *NeuroImage* **2021**, *244*, 118543, doi: [CrossRef] [PubMed]
2. Ogawa, S.; Lee, T.M.; Kay, A.R.; Tank, D.W. Brain magnetic resonance imaging with contrast dependent on blood oxygenation. *Proc. Natl. Acad. Sci. USA* **1990**, *87*, 9868–9872. doi: [CrossRef] [PubMed]
3. Barch, D.M.; Burgess, G.C.; Harms, M.P.; Petersen, S.E.; Schlaggar, B.L.; Corbetta, M.; Glasser, M.F.; Curtiss, S.; Dixit, S.; Feldt, C.; et al. Function in the human connectome: Task-fMRI and individual differences in behavior. *NeuroImage* **2013**, *80*, 169–189. doi: [CrossRef] [PubMed]
4. Van den Heuvel, M.P.; Hulshoff Pol, H.E. Exploring the brain network: A review on resting-state fMRI functional connectivity. *Eur. Neuropsychopharmacol.* **2010**, *20*, 519–534. doi: [CrossRef] [PubMed]
5. Biswal, B.B.; Kylen, J.V.; Hyde, J.S. Simultaneous assessment of flow and BOLD signals in resting-state functional connectivity maps. *NMR Biomed.* **1997**, *10*, 165–170. doi: [CrossRef]
6. Belliveau, J.W.; Cohen, M.S.; Weisskoff, R.M.; Buchbinder, B.R.; Rosen, B.R. Functional studies of the human brain using high-speed magnetic resonance imaging. *J. Neuroimaging* **1991**, *1*, 36–41. [CrossRef] [PubMed]
7. Glover, G.H. Overview of functional magnetic resonance imaging. *Neurosurg. Clin.* **2011**, *22*, 133–139. [CrossRef]
8. Wang, L.; Zang, Y.; He, Y.; Liang, M.; Zhang, X.; Tian, L.; Wu, T.; Jiang, T.; Li, K. Changes in hippocampal connectivity in the early stages of Alzheimer's disease: Evidence from resting state fMRI. *NeuroImage* **2006**, *31*, 496–504. doi: [CrossRef]
9. Rajamanickam, K. A Mini Review on Different Methods of Functional-MRI Data Analysis. *Arch. Intern. Med. Res.* **2020**, *3*, 44–60. [CrossRef]
10. Ting, C.; Ombao, H.; Salleh, S.; Latif, A.Z.A. Multi-Scale Factor Analysis of High-Dimensional Functional Connectivity in Brain Networks. *IEEE Trans. Netw. Sci. Eng.* **2020**, *7*, 449–465. [CrossRef]
11. Jolliffe, I.T.; Cadima, J. Principal component analysis: A review and recent developments. *Philos. Trans. R. Soc. A Math. Phys. Eng. Sci.* **2016**, *374*, 20150202. [CrossRef] [PubMed]
12. Mckeown, M.J.; Makeig, S.; Brown, G.G.; Jung, T.P.; Kindermann, S.S.; Bell, A.J.; Sejnowski, T.J. Analysis of fMRI data by blind separation into independent spatial components. *Hum. Brain Mapp.* **1998**, *6*, 160–188. doi: [CrossRef]
13. Smith, S.M.; Fox, P.T.; Miller, K.L.; Glahn, D.C.; Fox, P.M.; Mackay, C.E.; Filippini, N.; Watkins, K.E.; Toro, R.; Laird, A.R.; et al. Correspondence of the brain's functional architecture during activation and rest. *Proc. Natl. Acad. Sci. USA* **2009**, *106*, 13040–13045. doi: [CrossRef] [PubMed]
14. Van den Heuvel, M.; Stam, C.; Boersma, M.; Hulshoff Pol, H. Small-world and scale-free organization of voxel-based resting-state functional connectivity in the human brain. *NeuroImage* **2008**, *43*, 528–539. doi: [CrossRef] [PubMed]
15. Ombao, H.; Lindquist, M.; Thompson, W.; Aston, J. *Handbook of Neuroimaging Data Analysis*; CRC Press: Boca Raton, FL, USA, 2016.

16. O'Reilly, J.X.; Woolrich, M.W.; Behrens, T.E.; Smith, S.M.; Johansen-Berg, H. Tools of the trade: Psychophysiological interactions and functional connectivity. *Soc. Cogn. Affect. Neurosci.* **2012**, *7*, 604–609. [CrossRef] [PubMed]

17. Tzourio-Mazoyer, N.; Landeau, B.; Papathanassiou, D.; Crivello, F.; Etard, O.; Delcroix, N.; Mazoyer, B.; Joliot, M. Automated Anatomical Labeling of Activations in SPM Using a Macroscopic Anatomical Parcellation of the MNI MRI Single-Subject Brain. *NeuroImage* **2002**, *15*, 273–289. doi: [CrossRef] [PubMed]

18. Wu, L.; Caprihan, A.; Bustillo, J.; Mayer, A.; Calhoun, V. An approach to directly link ICA and seed-based functional connectivity: Application to schizophrenia. *NeuroImage* **2018**, *179*, 448–470. [CrossRef] [PubMed]

19. Andersen, A.H.; Gash, D.M.; Avison, M.J. Principal component analysis of the dynamic response measured by fMRI: A generalized linear systems framework. *Magn. Reson. Imaging* **1999**, *17*, 795–815. [CrossRef]

20. Zhou, Z.; Ding, M.; Chen, Y.; Wright, P.; Lu, Z.; Liu, Y. Detecting directional influence in fMRI connectivity analysis using PCA based Granger causality. *Brain Res.* **2009**, *1289*, 22–29. doi: [CrossRef] [PubMed]

21. Zou, H.; Hastie, T.; Tibshirani, R. Sparse Principal Component Analysis. *J. Comput. Graph. Stat.* **2006**, *15*, 265–286. doi: [CrossRef]

22. Zipunnikov, V.; Caffo, B.; Yousem, D.M.; Davatzikos, C.; Schwartz, B.S.; Crainiceanu, C. Multilevel Functional Principal Component Analysis for High-Dimensional Data. *J. Comput. Graph. Stat.* **2011**, *20*, 852–873. doi: [CrossRef] [PubMed]

23. Ma, Z. Sparse principal component analysis and iterative thresholding. *Ann. Stat.* **2013**, *41*, 772–801. doi: [CrossRef]

24. Bell, A.J.; Sejnowski, T.J. An information-maximization approach to blind separation and blind deconvolution. *Neural Comput.* **1995**, *7*, 1129–1159. [CrossRef] [PubMed]

25. Le, Q.; Karpenko, A.; Ngiam, J.; Ng, A. ICA with Reconstruction Cost for Efficient Overcomplete Feature Learning. In *Advances in Neural Information Processing Systems*; Shawe-Taylor, J., Zemel, R., Bartlett, P., Pereira, F., Weinberger, K.Q., Eds.; Curran Associates, Inc.: New York, NY, USA, 2011; Volume 24.

26. Miranda, M.F.; Morris, J.S. Novel Bayesian method for simultaneous detection of activation signatures and background connectivity for task fMRI data. *arXiv* **2021**, arXiv:2109.00160.

27. He, Y.; Chen, Z.J.; Evans, A.C. Small-World Anatomical Networks in the Human Brain Revealed by Cortical Thickness from MRI. *Cereb. Cortex* **2007**, *17*, 2407–2419. doi: [CrossRef] [PubMed]

28. Achard, S.; Salvador, R.; Whitcher, B.; Suckling, J.; Bullmore, E. A Resilient, Low-Frequency, Small-World Human Brain Functional Network with Highly Connected Association Cortical Hubs. *J. Neurosci.* **2006**, *26*, 63–72. doi: 10.1523/JNEUROSCI.387 4-05.2006. [CrossRef] [PubMed]

29. Rubinov, M.; Sporns, O. Complex network measures of brain connectivity: Uses and interpretations. *NeuroImage* **2010**, *52*, 1059–1069. doi: [CrossRef] [PubMed]

30. Salimi-Khorshidi, G.; Douaud, G.; Beckmann, C.F.; Glasser, M.F.; Griffanti, L.; Smith, S.M. Automatic denoising of functional MRI data: Combining independent component analysis and hierarchical fusion of classifiers. *NeuroImage* **2014**, *90*, 449–468. doi: [CrossRef]

31. Burgess, G.C.; Kandala, S.; Nolan, D.; Laumann, T.O.; Power, J.D.; Adeyemo, B.; Harms, M.P.; Petersen, S.E.; Barch, D.M. Evaluation of Denoising Strategies to Address Motion-Correlated Artifacts in Resting-State Functional Magnetic Resonance Imaging Data from the Human Connectome Project. *Brain Connect.* **2016**, *6*, 669–680. doi: [CrossRef]

32. Smitha, K.; Raja, K.A.; Arun, K.; Rajesh, P.; Thomas, B.; Kapilamoorthy, T.; Kesavadas, C. Resting state fMRI: A review on methods in resting state connectivity analysis and resting state networks. *Neuroradiol. J.* **2017**, *30*, 305–317. doi: [CrossRef]

33. Beckmann, C.; Smith, S. Probabilistic independent component analysis for functional magnetic resonance imaging. *IEEE Trans. Med. Imaging* **2004**, *23*, 137–152. doi: [CrossRef] [PubMed]

34. Madsen, K.H.; Churchill, N.W.; Mørup, M. Quantifying functional connectivity in multi-subject fMRI data using component models. *Hum. Brain Mapp.* **2017**, *38*, 882–899. doi: [CrossRef] [PubMed]

35. Leonardi, N.; Richiardi, J.; Gschwind, M.; Simioni, S.; Annoni, J.M.; Schluep, M.; Vuilleumier, P.; Van De Ville, D. Principal components of functional connectivity: A new approach to study dynamic brain connectivity during rest. *NeuroImage* **2013**, *83*, 937–950. doi: [CrossRef] [PubMed]

36. Caffo, B.S.; Crainiceanu, C.M.; Verduzco, G.; Joel, S.; Mostofsky, S.H.; Bassett, S.S.; Pekar, J.J. Two-stage decompositions for the analysis of functional connectivity for fMRI with application to Alzheimer's disease risk. *NeuroImage* **2010**, *51*, 1140–1149. doi: [CrossRef] [PubMed]

37. Calhoun, V.; Adali, T.; Pearlson, G.; Pekar, J. A method for making group inferences from functional MRI data using independent component analysis. *Hum. Brain Mapp.* **2001**, *14*, 140–151. doi: [CrossRef] [PubMed]

38. Kolda, T.G.; Bader, B.W. Tensor decompositions and applications. *SIAM Rev.* **2009**, *51*, 455–500. doi: [CrossRef]

39. Beckmann, C.; Smith, S. Tensorial extensions of independent component analysis for multisubject FMRI analysis. *NeuroImage* **2005**, *25*, 294–311. doi: [CrossRef]

40. Solo, V.; Poline, J.B.; Lindquist, M.A.; Simpson, S.L.; Bowman, F.D.; Chung, M.K.; Cassidy, B. Connectivity in fMRI: Blind Spots and Breakthroughs. *IEEE Trans. Med. Imaging* **2018**, *37*, 1537–1550. doi: [CrossRef]

41. Simpson, S.L.; Moussa, M.N.; Laurienti, P.J. An exponential random graph modeling approach to creating group-based representative whole-brain connectivity networks. *NeuroImage* **2012**, *60*, 1117–1126. doi: [CrossRef]

42. Simpson, S.L.; Laurienti, P.J. A two-part mixed-effects modeling framework for analyzing whole-brain network data. *NeuroImage* **2015**, *113*, 310–319. doi: [CrossRef]

43. Bandettini, P.A.; Jesmanowicz, A.; Wong, E.C.; Hyde, J.S. Processing strategies for time-course data sets in functional MRI of the human brain. *Magn. Reson. Med.* **1993**, *30*, 161–173. [CrossRef] [PubMed]
44. Marrelec, G.; Krainik, A.; Duffau, H.; Pélégrini-Issac, M.; Lehéricy, S.; Doyon, J.; Benali, H. Partial correlation for functional brain interactivity investigation in functional MRI. *NeuroImage* **2006**, *32*, 228–237. doi: [CrossRef] [PubMed]
45. Podobnik, B.; Stanley, H.E. Detrended cross-correlation analysis: A new method for analyzing two nonstationary time series. *Phys. Rev. Lett.* **2008**, *100*, 084102. [CrossRef] [PubMed]

Article

Nonparametric Causal Structure Learning in High Dimensions

Shubhadeep Chakraborty and Ali Shojaie *

Department of Biostatistics, University of Washington, Seattle, WA 98195, USA; deep20@uw.edu
* Correspondence: ashojaie@uw.edu

Abstract: The PC and FCI algorithms are popular constraint-based methods for learning the structure of directed acyclic graphs (DAGs) in the absence and presence of latent and selection variables, respectively. These algorithms (and their order-independent variants, PC-stable and FCI-stable) have been shown to be consistent for learning sparse high-dimensional DAGs based on partial correlations. However, inferring conditional independences from partial correlations is valid if the data are jointly Gaussian or generated from a linear structural equation model—an assumption that may be violated in many applications. To broaden the scope of high-dimensional causal structure learning, we propose nonparametric variants of the PC-stable and FCI-stable algorithms that employ the conditional distance covariance (CdCov) to test for conditional independence relationships. As the key theoretical contribution, we prove that the high-dimensional consistency of the PC-stable and FCI-stable algorithms carry over to general distributions over DAGs when we implement CdCov-based nonparametric tests for conditional independence. Numerical studies demonstrate that our proposed algorithms perform nearly as good as the PC-stable and FCI-stable for Gaussian distributions, and offer advantages in non-Gaussian graphical models.

Keywords: causal structure learning; consistency; FCI algorithm; high dimensionality; nonparametric testing; PC algorithm

Citation: Chakraborty, S.; Shojaie, A. Nonparametric Causal Structure Learning in High Dimensions. *Entropy* **2022**, *24*, 351. https://doi.org/10.3390/e24030351

Academic Editors: S. Ejaz Ahmed and Farouk Nathoo

Received: 20 January 2022
Accepted: 25 February 2022
Published: 28 February 2022

Publisher's Note: MDPI stays neutral with regard to jurisdictional claims in published maps and institutional affiliations.

1. Introduction

Directed acyclic graphs (DAGs) are commonly used to represent causal relationships among random variables [1–3]. The PC algorithm [3] is the most popular constraint-based method for learning DAGs from observational data under the assumption of causal sufficiency, i.e., when there are no unmeasured common causes and no selection variables. It first estimates the skeleton of a DAG by recursively performing a sequence of conditional independence tests, and then uses the information from the conditional independence relations to partially orient the edges, resulting in a completed partially directed acyclic graph (CPDAG). In Section 2, we provide a review of these and other notions commonly used in the graphical modeling literature that are relevant to our work. In addition, we refer to estimating the CPDAG as structure learning of the underlying DAG throughout the rest of the paper.

Observational studies often involve latent and selection variables, which complicate the causal structure learning problem. Ignoring such unmeasured variables can make the causal inference based on the PC algorithm erroneous; see, e.g., Section 1.2 in [4] for some illustrations. The Fast Causal Inference (FCI) algorithm and its variants [3–6] utilize similar strategies as the PC algorithm to learn the DAG structure in the presence of latent and selection variables.

Both PC and FCI algorithms adopt a hierarchical search strategy—they recursively perform conditional independence tests given subsets of increasingly larger cardinalities in some appropriate search pool. The PC algorithm is usually order-dependent, in the sense that its output depends on the order in which pairs of adjacent vertices and subsets of their adjacency sets are considered. The FCI algorithm suffers from a similar limitation. To overcome this limitation, Ref. [7] proposed two variants of the PC and FCI algorithms, namely the PC-stable and FCI-stable algorithms that resolve the order dependence at different stages of the algorithms.

In general, testing for conditional independence is a problem of central importance in the causal structure learning. The literature on the PC and FCI algorithms predominantly uses partial correlations to infer conditional independence relations. It is well-known that the characterization of conditional independence by partial correlations, or, in other words, equivalence between conditional independence and zero partial correlations only holds for multivariate normal random variables. Therefore, the high-dimensional consistency results for the PC and FCI algorithms [4,8] are limited to Gaussian graphical models, where the nodes correspond to random variables with a joint Gaussian distribution. Although the Gaussian graphical model is the standard parametric model for continuous data, it may not hold in many real data applications. Although this limitation can be somewhat relaxed by considering linear structural equation models (SEMs) with general noise distributions [9], linear SEMs and joint Gaussianity are essentially equivalent [10]. Moreover, neither approach is appropriate when the observations are categorical, discrete, or are supported on a subset of the real line. In Section 4.3, for example, we present a real application where all the observed variables are categorical, and therefore far from being Gaussian. As an improvement, ref. [11] used rank-based partial correlations to test for conditional independence relations, showing that the high-dimensional consistency of the PC algorithm holds for a broader class of Gaussian copula models. Some nonparametric versions of the PC algorithm have been also proposed in the literature via kernel-based tests for conditional independence [12,13]; however, they lack theoretical justifications of the correctness of the algorithms, and are not studied in high dimensions.

This work aims to broaden the applicability of the PC-stable and FCI-stable algorithms to general distributions by employing a nonparametric test for conditional independence relationships. To this end, we utilize recent developments on dependence metrics that quantify nonlinear and non-monotone dependence between multivariate random variables. More specifically, our work builds on the idea of distance covariance (dCov) proposed by [14] and its extension to conditional distance covariance (CdCov) by [15] as a nonparametric measure of nonlinear and non-monotone conditional independence between two random vectors of arbitrary dimensions given a third. Utilizing this flexibility, we use the conditional distance covariance (CdCov) to test for conditional independence relationships in the sample versions of the PC-stable and FCI-stable algorithms. The resulting algorithms—which, for distinction, are termed *nonPC* and *nonFCI*—facilitate causal structure learning from general distributions over DAGs and are shown to be consistent in sparse high-dimensional settings. We establish the consistency of the proposed algorithms using some moment and tail conditions on the variables, without requiring strict distributional assumptions. To our knowledge, the proposed generalizations of PC/PC-stable or the FCI/FCI-stable algorithms provide the first general nonparametric framework for causal structure learning with theoretical guarantees in high dimensions.

The rest of the paper is organized as follows: In Section 2, we review the relevant background, including preliminaries on graphical modeling (Section 2.1), an outline of the PC-stable and FCI-stable algorithms (Section 2.2) and a brief overview of dCov and CdCov (Section 2.3). The nonparametric version of the PC-stable algorithm is presented in Section 3.1. As a key contribution of the paper, we establish that the algorithm consistently estimates the skeleton and the equivalence class of the underlying sparse high-dimensional DAG in a general nonparametric framework. We then present the nonparametric version of the FCI-stable algorithm in Section 3.2 and establish its consistency in sparse high-dimensional settings. As the FCI involves the adjacency search of the PC algorithm, any improvement on the PC/PC-stable directly carries over to the FCI/FCI-stable as well. In Section 4, we compare the performances of our algorithms with the PC-stable and FCI-stable using both simulated datasets (involving both Gaussian and non-Gaussian examples), as well as a real dataset. These numerical studies clearly demonstrate that nonPC and nonFCI algorithms are comparable with PC-stable and FCI-stable for Gaussian data and offer improvements for non-Gaussian data.

2. Background

2.1. Preliminaries on Graphical Modeling

We start with introducing some necessary terminologies and background information. Our notations and terminologies follow standard conventions in graphical modeling (see, e.g., [3]). A graph $\mathcal{G} = (V, E)$ consists of a vertex set $V = \{1, \ldots, p\}$ and an edge set $E \subseteq V \times V$. In a graphical model, the vertices or nodes are associated with random variables X_a for $1 \leq a \leq p$. Throughout, we index the nodes by the corresponding random variables. We also allow the edge set E of the graph $\mathcal{G}$ to contain (a subset of) the following six types of edges: $\rightarrow$ (*directed*), $\leftrightarrow$ (*bidirected*), $-$ (*undirected*), $\circ\!-\!\circ$ (*nondirected*), $\circ\!-$ (*partially undirected*) and $\circ\!\rightarrow$ (*partially directed*). The endpoints of an edge are called marks, which can be tails, arrowheads or circles. A "$\circ$" at the end of an edge indicates it is not known whether an arrowhead should occur at that place. We use the symbol '$\star$' to denote an arbitrary edge mark; for example, the symbol $\star\!\rightarrow$ represents an edge of the type $\rightarrow$, $\leftrightarrow$ or $\circ\!\rightarrow$ in the graph. A *mixed graph* is a graph containing directed, bidirected and undirected edges. A graph containing only directed edges $(\rightarrow)$ is called a *directed graph*, one containing only undirected edges $(-)$ is called an *undirected graph*, and one containing directed and undirected edges is called a *partially directed graph*.

The *adjacency set* of a vertex X_a in the graph $\mathcal{G} = (V, E)$, denoted adj$(\mathcal{G}, X_a)$, is the set of all vertices in V that are adjacent to X_a, or, in other words, are connected to X_a by an edge. The *degree* of a vertex X_a, $|\text{adj}(\mathcal{G}, X_a)|$, is defined as the number of vertices adjacent to it. A graph is *complete* if all pairs of vertices in the graph are adjacent. A vertex $X_b \in \text{adj}(\mathcal{G}, X_a)$ is called a *parent* of X_a if $X_b \rightarrow X_a$, a *child* of X_a if $X_a \rightarrow X_b$ and a *neighbor* of X_a if $X_a - X_b$. The *skeleton* of the graph $\mathcal{G}$ is the undirected graph obtained by replacing all the edges of $\mathcal{G}$ by undirected edges, in other words, ignoring all the edge orientations. Three vertices $\langle X_a, X_b, X_c \rangle$ are called an *unshielded triple* if X_a and X_b are adjacent, X_b and X_c are adjacent, but X_a and X_c are not adjacent. A *path* is a sequence of distinct adjacent vertices. A node X_a is an *ancestor* of its *descendent* X_b, if $\mathcal{G}$ contains a directed path $X_a \rightarrow \cdots \rightarrow X_b$. A non-endpoint vertex X_a on a path is called a *collider* on the path if both the edges preceding and succeeding it have an arrowhead at X_a, or, in other words, the path contains $\star\!\rightarrow X_a \leftarrow\!\star$. An unshielded triple $\langle X_a, X_b, X_c \rangle$ is called a *v-structure* if X_b is a collider on the path $\langle X_a, X_b, X_c \rangle$.

A *cycle* occurs in a graph when there is a path from X_a to X_b, and X_a and X_b are adjacent. A directed path from X_a to X_b forms a *directed cycle* together with the edge $X_b \rightarrow X_a$, and it forms an *almost directed cycle* together with the edge $X_b \leftrightarrow X_a$. Three vertices that form a cycle are called a *triangle*. A *directed acyclic graph* (DAG) is a directed graph that does not contain any cycle. A DAG entails conditional independence relationships via a graphical criterion called *d-separation* (Section 1.2.3 in [16]). Two vertices X_a and X_b that are not adjacent in a DAG $\mathcal{G}$ are d-separated in $\mathcal{G}$ by a subset $X_S \subseteq V \backslash \{X_a, X_b\}$. A probability distribution P on $\mathbb{R}^p$ is said to be *faithful* with respect to the DAG $\mathcal{G}$ if the conditional independence relationships in P can be inferred from $\mathcal{G}$ using d-separation and vice versa; in other words, $X_a \perp\!\!\!\perp X_b | X_S$ if and only if X_a and X_b are d-separated in $\mathcal{G}$ by X_S.

A graph that is both (partially) directed and acyclic is called a *partially directed acyclic graph (PDAG)*. DAGs that encode the same set of conditional independence relations form a Markov equivalence class [17]. Two DAGs belong to the same Markov equivalence class if and only if they have the same skeleton and the same v-structures. A Markov equivalence class of DAGs can be uniquely represented by a *completed partially directed acyclic graph (CPDAG)*, which is a PDAG that satisfies the following: (i) $X_a \rightarrow X_b$ in the CPDAG if $X_a \rightarrow X_b$ in every DAG in the Markov equivalence class, and (ii) $X_a - X_b$ in the CPDAG if the Markov equivalence class contains a DAG in which $X_a \rightarrow X_b$ as well as a DAG in which $X_a \leftarrow X_b$.

2.2. The PC-Stable and FCI-Stable Algorithms

In this section, we provide an outline of the PC/PC-stable and FCI/FCI-stable algorithms. Estimation of the CPDAG by the PC algorithm involves two steps: (1) estimation of the skeleton and separating sets (also called the adjacency search step); and (2) partial orientation of edges; see Algorithms 1 and 2 in [8] for details.

Intuitively, the PC algorithm works as follows. In the first step (the adjacency search step), the algorithm starts with a complete undirected graph Then, for conditioning sets of increasing cardinality, $k = 0, 1, \ldots$, the algorithm removed an edge $X_a - X_b$ if X_a and X_b are conditionally independent given a subset S of size k chosen among the current neighbors of nodes a and b. This process continues up to the order $q - 1$, where q is the maximum degree of the underlying DAG. By searching over the neighboring nodes, the algorithm is adaptive and can efficiently infer sparse high-dimensional DAGs, where the sparsity is characterized by the maximum node degree, q.

In the presence of latent and selection variables, one needs a generalization of an DAG, called a *maximal ancestral graph* (MAG). A mixed graph is called an *ancestral graph* if it contains no directed or almost directed cycles and no subgraph of the type $X_a - X_b \leftarrow \star X_c$. DAGs form a subset of ancestral graphs. A MAG is an ancestral graph in which every missing edge corresponds to a conditional independence relationship via the m-separation criterion [18], a generalization of the notion of d-separation. Multiple MAGs may represent the same set of conditional independence relations. Such MAGs form a Markov equivalence class which can be represented by a *partial ancestral graph* (PAG) [19]; see [18] for additional details.

Under the faithfulness assumption, the Markov equivalence class of a DAG with latent and selection variables can be learned using the FCI algorithm (e.g., Algorithm 3.1 in [4]), which is a modification of the PC algorithm. The FCI algorithm first employs the adjacency search of the PC algorithm, and then performs additional conditional independence queries because of the presence of latent variables followed by partial orientation of the edges, resulting in an estimated PAG. The FCI algorithm adopts the same hierarchical search strategy as the PC algorithm: It starts with a complete undirected graph and recursively removes edges via conditional independence queries given subsets of increasingly larger cardinalities in some appropriate search pool.

The PC algorithm is usually order-dependent, in the sense that its output depends on the order in which pairs of adjacent vertices and subsets of their adjacency sets are considered. The FCI algorithm suffers from a similar limitation, as it shares the adjacency search step of the PC algorithm as its first step. To overcome this limitation, ref. [7] proposed variants of the PC and FCI algorithms, namely the PC-stable and FCI-stable algorithms that resolve the order dependence at different stages of the algorithms. The basic difference between the PC algorithm and the PC-stable algorithm is that, in the adjacency search step, the latter computes and stores the adjacency sets of all the variables after each new cardinality, $k = 0, 1, \ldots$, of the conditioning sets. These stored adjacency sets are then used to search for conditioning sets of this given size k. As a consequence, the removal of an edge no longer affects which conditional independence relations need to be checked for other pairs of variables at this given size of the conditioning sets.

We would refer the reader to Appendix A, where we provide in full detail the pseudocodes of the *oracle* versions of the PC-stable and FCI-stable algorithms. In the *oracle* versions of the algorithms, it is assumed that perfect knowledge is available about all the necessary conditional independence relations. As such, conditional independence relations are not estimated from data. Of course, this perfect knowledge is not available in practice. *Sample* versions of the PC-stable and FCI-stable algorithms can be obtained by replacing the conditional independence queries by a suitable test for conditional independence at some pre-specified level. For example, if the variables are jointly Gaussian, one can test for zero partial correlations (see, e.g., [8]). The next subsection is devoted to discussions on nonparametric tests for independence and conditional independence.

2.3. Distance Covariance and Conditional Distance Covariance

We start by describing the notation used throughout the paper. We denote by $\| \cdot \|_p$ the Euclidean norm of $\mathbb{R}^p$ and use $\| \cdot \|$ when the dimension is clear from the context. We use $X \perp\!\!\!\perp Y$ to denote the independence of X and Y and use $\mathbb{E}_U$ to denote expectation with respect to the probability distribution of the random variable U. For any set S, we denote its cardinality by $|S|$.

We use the usual asymptotic notation, 'O' and 'o', as well as their probabilistic counterparts, O_p and o_p, which denote stochastic boundedness and convergence in probability, respectively. For two sequences of real numbers $\{a_n\}_{n=1}^{\infty}$ and $\{b_n\}_{n=1}^{\infty}$, $a_n \asymp b_n$ if and only if $a_n/b_n = O(1)$ and $b_n/a_n = O(1)$ as $n \to \infty$. We use the symbol "$a \lesssim b$" to indicate that $a \leq C b$ for some constant $C > 0$. For a matrix $A = (a_{kl})_{k,l=1}^{n} \in \mathbb{R}^{n \times n}$, we denote its determinant by $|A|$ and define its $\mathcal{U}$-centered version $\tilde{A} = (\tilde{a}_{kl})_{k,l=1}^{n}$ as

$$
\tilde{a}_{kl} = \begin{cases} a_{kl} - \dfrac{1}{n-2}\sum_{j=1}^{n} a_{kj} - \dfrac{1}{n-2}\sum_{i=1}^{n} a_{il} + \dfrac{1}{(n-1)(n-2)}\sum_{i,j=1}^{n} a_{ij}, & k \neq l, \\ 0, & k = l, \end{cases} \tag{1}
$$

for $k, l = 1, \ldots, n$. We denote the indicator function of any set A by $\mathbf{1}(A)$. Finally, we denote the integer part of $a \in \mathbb{R}$ by $\lfloor a \rfloor$.

Ref. [14], in their seminal paper, introduced the notion of distance covariance (dCov, henceforth) to quantify nonlinear and non-monotone dependence between two random vectors of arbitrary dimensions. Consider two random vectors $X \in \mathbb{R}^p$ and $Y \in \mathbb{R}^q$ with $\mathbb{E}\|X\|_p < \infty$ and $\mathbb{E}\|Y\|_q < \infty$. The distance covariance between X and Y is defined as the positive square root of

$$
\mathrm{dCov}^2(X,Y) = \frac{1}{c_p c_q} \int_{\mathbb{R}^{p+q}} \frac{|f_{X,Y}(t,s) - f_X(t)f_Y(s)|^2}{\|t\|_p^{1+p} \|s\|_q^{1+q}} dt ds
$$

where f_X, f_Y and $f_{X,Y}$ are the individual and joint characteristic functions of X and Y, respectively, and $c_p = \pi^{(1+p)/2}/\Gamma((1+p)/2)$ is a constant with $\Gamma(\cdot)$ being the complete gamma function.

The key feature of dCov is that it completely characterizes the independence between two random vectors, or in other words $\mathrm{dCov}(X,Y) = 0$ if and only if $X \perp\!\!\!\perp Y$. According to Remark 3 in [14], dCov can be equivalently expressed as

$$
\begin{aligned}
\mathrm{dCov}^2(X,Y) &= \mathbb{E}\|X - X'\|_p\|Y - Y'\|_q + \mathbb{E}\|X - X'\|_p\,\mathbb{E}\|Y - Y'\|_q \\
&\quad - 2\,\mathbb{E}\|X - X'\|_p\|Y - Y''\|_q .
\end{aligned}
$$

This alternate expression comes handy in constructing V or U-statistic type estimators for the quantity. For an observed random sample $(X_i, Y_i)_{i=1}^{n}$ from the joint distribution of X and Y, define the distance matrices $d^X = (d_{ij}^X)_{i,j=1}^{n}$ and $d^Y = (d_{ij}^Y)_{i,j=1}^{n} \in \mathbb{R}^{n \times n}$, where $d_{ij}^X := \|X_i - X_j\|_p$ and $d_{ij}^Y := \|Y_i - Y_j\|_q$. Following the $\mathcal{U}$-centering idea in [20], an unbiased U-statistic type estimator of $\mathrm{dCov}^2(X,Y)$ can be expressed as

$$
\mathrm{dCov}_n^2(X,Y) := (\tilde{d}^X \cdot \tilde{d}^Y) := \frac{1}{n(n-3)} \sum_{i \neq j} \tilde{d}_{ij}^X \tilde{d}_{ij}^Y , \tag{2}
$$

where $\tilde{d}^X = (\tilde{d}_{ij}^X)_{i,j=1}^{n}$ and $\tilde{d}^Y = (\tilde{d}_{ij}^Y)_{i,j=1}^{n}$ are the $\mathcal{U}$-centered versions of the matrices d^X and d^Y, respectively, as defined in (1).

Ref. [15] generalized the notion of dCov and introduced the conditional distance covariance (CdCov, henceforth) as a measure of conditional dependence between two random vectors of arbitrary dimensions given a third. CdCov essentially replaces the characteristic functions used in the definition of dCov by conditional characteristic functions. Consider a third random vector $Z \in \mathbb{R}^r$ with $\mathbb{E}(\|X\|_p + \|Y\|_q \mid Z) < \infty$. Denote by $f_{X,Y|Z}$ the conditional joint characteristic function of X and Y given Z, and by $f_{X|Z}$ and $f_{Y|Z}$ the conditional marginal characteristic functions of X and Y given Z, respectively. Then, CdCov between X and Y given Z is defined as the positive square root of

$$
\mathrm{CdCov}^2(X,Y|Z) = \frac{1}{c_p c_q} \int_{\mathbb{R}^{p+q}} \frac{|f_{X,Y|Z}(t,s) - f_{X|Z}(t)f_{Y|Z}(s)|^2}{\|t\|_p^{1+p} \|s\|_q^{1+q}} dt ds.
$$

The key feature of CdCov is that $\mathrm{CdCov}\,(X, Y|Z) = 0$ almost surely if and only if $X \perp\!\!\!\perp Y|Z$, which is quite straightforward to see from the definition.

Similar to dCov, an equivalent alternative expression can be established for CdCov that avoids complicated integrations involving conditional characteristic functions. Let $\{W_i = (X_i, Y_i, Z_i)\}_{i=1}^{n}$ be an i.i.d. sample from the joint distribution of $W := (X, Y, Z)$. Define $d_{ijkl} := \left(d_{ij}^X + d_{kl}^X - d_{ik}^X - d_{jl}^X\right)\left(d_{ij}^Y + d_{kl}^Y - d_{ik}^Y - d_{jl}^Y\right)$, which is not symmetric with respect to $\{i, j, k, l\}$, and therefore necessitates defining the following symmetric form: $d_{ijkl}^S := d_{ijkl} + d_{ijlk} + d_{ilkj}$. Lemma 1 in [15] establishes an equivalent representation of $\mathrm{CdCov}^2(X, Y|Z = z)$ as

$$\mathrm{CdCov}^2(X, Y|Z = z) \;=\; \frac{1}{12}\,\mathbb{E}\left[d_{1234}^S \,|\, Z_1 = z, Z_2 = z, Z_3 = z, Z_4 = z\right]. \qquad (3)$$

Remark 1. *In a recent work, [21] explore the connection between conditional independence measures induced by distances on a metric space and reproducing kernels associated with a reproducing kernel Hilbert space (RKHS). They generalize CdCov to arbitrary metric spaces of negative type— termed generalized CdCov (gCdCov)—and develop a kernel-based measure of conditional independence, namely the Hilbert–Schmidt conditional independence criterion (HSCIC). Theorem 1 in their paper establishes an equivalence between gCdCov and HSCIC, or, in other words, between distance and kernel-based measures of conditional independence.*

For $w \in \mathbb{R}^r$, let $K_H(w) := |H|^{-1} K(H^{-1}w)$ be a kernel function, where H is the diagonal matrix $\mathrm{diag}(h, \ldots, h)$ determined by a bandwidth parameter h. K_H is typically considered to be the Gaussian kernel $K_H(w) = (2\pi)^{-\frac{r}{2}} |H|^{-1} \exp\left(-\frac{1}{2}w^T H^{-2}w\right)$, where $w \in \mathbb{R}^r$.

Let $K_{iu} := K_H(Z_i - Z_u) = |H|^{-1} K(H^{-1}(Z_i - Z_u))$ and $K_i(Z) := K_H(Z - Z_i)$ for $1 \leq i, u \leq n$. Then, by virtue of the equivalent representation of CdCov in (3), a V-statistic type estimator of $\mathrm{CdCov}^2(X, Y|Z)$ can be constructed as

$$\mathrm{CdCov}_n^2(X, Y|Z) \;:=\; \sum_{i,j,k,l} \frac{K_i(Z)\,K_j(Z)\,K_k(Z)\,K_l(Z)}{12\left(\sum_{i=1}^{n} K_i(Z)\right)^4}\, d_{ijkl}^S. \qquad (4)$$

Under certain regularity conditions, Theorem 4 in [15] shows that, conditioned on Z, $\mathrm{CdCov}_n^2(X, Y|Z) \xrightarrow{P} \mathrm{CdCov}^2(X, Y|Z)$ as $n \to \infty$.

3. Methodology and Theory

3.1. The Nonparametric PC Algorithm in High Dimensions

To obtain a measure of conditional independence between X and Y given Z that is free of Z, we define

$$\rho_0^*(X, Y|Z) \;:=\; \mathbb{E}\left[\mathrm{CdCov}_n^2(X, Y|Z)\right]. \qquad (5)$$

Clearly, $\rho_0^*(X, Y|Z) = 0$ if and only if $X \perp\!\!\!\perp Y \,|\, Z$. Consider a plug-in estimate of $\rho_0^*(X, Y|Z)$ as

$$\hat{\rho}^*(X, Y|Z) \;:=\; \frac{1}{n}\sum_{u=1}^{n} \mathrm{CdCov}_n^2(X, Y|Z_u) \;=\; \frac{1}{n}\sum_{u=1}^{n} \Delta_{i,j,k,l;u}$$

$$\text{where} \qquad \Delta_{i,j,k,l;u} \;:=\; \sum_{i,j,k,l} \frac{K_{iu}\,K_{ju}\,K_{ku}\,K_{lu}}{12\left(\sum_{i=1}^{n} K_{iu}\right)^4}\, d_{ijkl}^S. \qquad (6)$$

We reject $H_0 : X \perp\!\!\!\perp Y|Z$ vs $H_A : X \not\perp\!\!\!\perp Y|Z$ at level $\alpha \in (0, 1)$ if $\hat{\rho}^*(X, Y|Z) > \xi_\alpha$, for a suitably chosen threshold ξ_α. In Appendix A, we present a local bootstrap procedure for choosing ξ_α in practice, which is also used in our numerical studies. Henceforth, we will often denote $\rho_0^*(X, Y|Z)$ and $\hat{\rho}^*(X, Y|Z)$ simply by ρ_0^* and $\hat{\rho}^*$ respectively for notational simplicity, whenever there is no confusion.

In view of the complete characterization of conditional independence by ρ_0^*, we propose testing for conditional independence relations nonparametrically in the sample version of the PC-stable algorithm based on ρ_0^*, rather than partial correlations. We coin the resulting algorithm the 'nonPC' algorithm, to emphasize that it is a nonparametric generalization of parametric PC-stable algorithms.

The *oracle version* of the first step of nonPC, or the skeleton estimation step, is exactly the same as that of the PC-stable algorithm (Algorithm A1 in Appendix A). The second step, which extends the skeleton estimated in the first step to a CPDAG (Algorithm A2 in Appendix A), is comprised of some purely deterministic rules for edge orientations, and is exactly the same for both the nonPC and PC-stable as well. The only difference lies in the implementation of the tests for conditional independence relationships in the *sample versions* of the first step. Specifically, we replace all the conditional independence queries in the first step by tests based on $\rho_0^*(X, Y|Z)$. At some pre-specified significance level α, we infer that $X_a \perp\!\!\!\perp X_b \mid X_S$ when $\widehat{\rho}^*(X_a, X_b|X_S) \leq \xi_{n,\alpha}$, where $a, b \in V$ and $S \subseteq V$, $|S| \neq \phi$. When $|S| = \phi$, $\widehat{\rho}^*(X_a, X_b|X_S) = \mathrm{dCov}_n^2(X_a, X_b)$ and $\rho_0^*(X, Y|Z) = \mathrm{dCov}^2(X, Y)$. The critical value $\xi_{n,\alpha}$ in this case is obtained by a bootstrap procedure (see, e.g., Section 4 in [22] with $d = 2$).

Given that the equivalence between conditional independence and zero partial correlations only holds for multivariate normal random variables, our generalization broadens the scope of applicability of causal structure learning by the PC/PC-stable algorithm to general distributions over DAGs. This nonparametric approach is thus a natural extension of Gaussian and Gaussian copula models. It enables capturing nonlinear and non-monotone conditional dependence relationships among the variables, which partial correlations fail to detect.

Next, we establish theoretical guarantees on the correctness of the nonPC algorithm in learning the true underlying causal structure in sparse high-dimensional settings. Our consistency results only require mild moment and tail conditions on the set of variables, without making any strict distributional assumptions. Denote by m_p the maximum cardinality of the conditioning sets considered in the adjacency search step of the PC-stable algorithm. Clearly, $m_p \leq q$, where $q := \max_{1 \leq a \leq p} |\mathrm{adj}(\mathcal{G}, a)|$ is the maximum degree of the DAG $\mathcal{G}$. For a fixed pair of nodes $a, b \in V$, the conditioning sets considered in the adjacency search step are elements of $J_{a,b}^{m_p} := \{S \subseteq V \backslash \{a, b\} : |S| \leq m_p\}$.

We first establish a concentration inequality that gives the rate at which the absolute difference of $\rho_0^*(X_a, X_b|X_S)$ and its plug-in estimate $\widehat{\rho}^*(X_a, X_b|X_S)$ decays to zero, for any fixed pair of nodes a and $b \in V$ and a fixed conditioning set S. Towards that, we impose the following regularity conditions.

(A1) There exists $s_0 > 0$ such that, for $0 \leq s < s_0$, $\sup_p \max_{1 \leq a \leq p} \mathbb{E} \exp(sX_a^2) < \infty$.

(A2) The kernel function $K(\cdot)$ is non-negative and uniformly bounded over its support.

Condition (A1) imposes a sub-exponential tail bound on the squares of the random variables. This is a quite commonly used condition, for example, in the high-dimensional feature screening literature (see, for example, [23]). Condition (A2) is a mild condition on the kernel function $K(\cdot)$ that is guaranteed by many commonly used kernels, including the Gaussian kernel. Under conditions (A1) and (A2), the next result shows that the plug-in estimate $\widehat{\rho}^*(X_a, X_b|X_S)$ converges in probability to its population counterpart $\rho_0^*(X_a, X_b|X_S)$ exponentially fast.

Theorem 1. *Under conditions (A1) and (A2), for any $\epsilon > 0$, there exist positive constants A, B and $\gamma \in (0, 1/4)$ such that*

$$\mathbb{P}(|\widehat{\rho}^*(X_a, X_b|X_S) - \rho_0^*(X_a, X_b|X_S)| > \epsilon) \leq O\left(2 \exp\left(-A\, n^{1-2\gamma} \epsilon^2\right) + n^4 \exp\left(-B\, n^\gamma\right)\right).$$

The proof of Theorem 1 is long and somewhat technical; it is thus relegated to Appendix B. Theorem 1 serves as the main building block towards establishing the consistency of the nonPC algorithm in sparse high-dimensional settings.

For notational convenience, henceforth, we denote $\rho_0^*\,(X_a, X_b|X_S)$ and $\widehat{\rho}\,^*(X_a, X_b|X_S)$ by $\rho_{0;a\,b|S}^*$ and $\widehat{\rho}_{ab|S}^*$, respectively. In Theorem 2 below, we establish a uniform bound for the errors in inferring conditional independence relationships using the ρ_0^*-based test in the skeleton estimation step of the sample version of the nonPC algorithm.

Theorem 2. *Under conditions (A1) and (A2), for any* $\epsilon > 0$, *there exist positive constants A, B and* $\gamma \in (0, 1/4)$ *such that*

$$\sup_{\substack{a,b\in V \\ S\in J_{a,b}^{m_p}}} \mathbb{P}(|\widehat{\rho}_{ab|S}^* - \rho_{0;ab|S}^*| > \epsilon) \;\le\; \mathbb{P}\left(\sup_{\substack{a,b\in V \\ S\in J_{a,b}^{m_p}}} |\widehat{\rho}_{ab|S}^* - \rho_{0;ab|S}^*| > \epsilon \right)$$

$$\le\; O\left(p^{m_p+2} \left[2\exp\left(-A\,n^{1-2\gamma}\,\epsilon^2\right) + n^4\exp\left(-B\,n^\gamma\right) \right] \right). \tag{7}$$

Next, we turn to proving the consistency of the nonPC algorithm in the high-dimensional setting where the dimension p can be much larger than the sample size n, but the DAG is considered to be sparse. We impose the following regularity conditions, which are similar to the assumptions imposed in Section 3.1 of [8] in order to prove the consistency of the PC algorithm for Gaussian graphical models. We let the number of variables p grow with the sample size n and consider $p = p_n$, and also the DAG $\mathcal{G} = \mathcal{G}_n := (V_n, E_n)$ and the distribution $P = P_n$.

(A3) The dimension p_n grows at a rate such that the right-hand side of (7) tends to zero as $n \to \infty$. In particular, this is satisfied when $p_n = O(n^r)$ for any $0 \le r < \infty$.
(A4) The maximum degree of the DAG $\mathcal{G}_n$, denoted by $q_n := \max_{1\le a\le p_n} |adj(\mathcal{G}_n, a)|$, grows at the rate of $O(n^{1-b})$, where $0 < b \le 1$.
(A5) The distribution P_n is faithful to the DAG $\mathcal{G}_n$ for all n. In other words, for any $a, b \in V_n$ and $S \in J_{a,b}^{m_{p_n}}$,

$$X_a \text{ and } X_b \text{ are d-separated by } X_S \iff X_a \perp\!\!\!\perp X_b \mid X_S \iff \rho_{0;a\,b|S}^* = 0.$$

Moreover, $\rho_{0;a\,b|S}^*$ values are uniformly bounded both from above and below. Formally,

$$C_{min} := \inf_{\substack{a,b\in V_n \\ S\in J_{a,b}^{m_{p_n}} \\ \rho_{0;ab|S}^*\neq 0}} \rho_{0;ab|S}^* \ge \lambda_{min}\,\lambda_{min}^{-1} = O(n^v)$$

$$\text{and} \quad C_{max} := \sup_{\substack{a,b\in V_n \\ S\in J_{a,b}^{m_{p_n}}}} \rho_{0;ab|S}^* \le \lambda_{max}$$

where λ_{max} is a positive constant and $0 < v < 1/4$.

Condition (A3) allows the dimension to grow at any arbitrary polynomial rate of the sample size. Condition (A4) is a sparsity assumption on the underlying true DAG, allowing the maximum degree of the DAG to also grow, but at a slower rate than n. Since $m_p \le q_n$, we also have $m_p = O(n^{1-b})$. Finally, Condition (A5) is the strong faithfulness assumption (Definition 1.3 in [24]) on P_n and is similar to condition (A4) in [8]. This essentially requires $\rho_{0;ab|S}^*$ to be bounded away from zero when the vertices X_a and X_b are not d-separated by X_S. It is worth noting that the faithfulness assumption alone is not enough to prove the consistency of the PC/PC-stable/nonPC algorithms in high-dimensional settings, and the more stringent strong faithfulness condition is required.

Remark 2. *For notational convenience, treat* X_a, X_b *and* X_S *as X, Y and Z, respectively, for any* $a, b \in V_n$ *and* $S \in J_{a,b}^{m_{p_n}}$. *From Equation* (3), *we have*

$$CdCov^2(X, Y|Z) = \frac{1}{12}\,\mathbb{E}\left[d_{1234}^S \mid Z_1 = Z, \ldots, Z_4 = Z \right],$$

which implies

$$\rho_0^* \;=\; \mathbb{E}\left[\mathrm{CdCov}^2(X, Y|Z)\right] \;=\; \frac{1}{12}\,\mathbb{E}\left[d_{1234}^S\right] \;=\; \frac{1}{12}\,\mathbb{E}\left[d_{1234} + d_{1243} + d_{1432}\right].$$

Condition (A1) implies $\displaystyle\sup_{p}\;\max_{1\le a\le p}\;\mathbb{E}\,X_a^2 < \infty$. *With this and the definition of* d_{ijkl} *in Section* 2.3, *it follows from some simple algebra and the Cauchy–Schwarz inequality that* $\rho_0^* < \infty$. *This provides a justification for the second part of Assumption (A5) that* $\displaystyle\sup_{\substack{a,b\in V_n \\ S\in J_{a,b}^{m_{pn}}}}\rho_{0;ab|S}^* \le \lambda_{max}$ *for some positive*

constant λ_{max}.

The next theorem establishes that the nonPC algorithm consistently estimates the skeleton of a sparse high-dimensional DAG, thereby providing the necessary theoretical guarantees to our proposed methodology. It is worth noting that, in the sample version of the PC-stable and hence the nonPC algorithm, all the inference is done during the skeleton estimation step. The second step that involves appropriately orienting the edges of the estimated skeleton is purely deterministic (see Sections 4.2 and 4.3 in [7]). Therefore, to prove the consistency of the nonPC algorithm in estimating the equivalence class of the underlying true DAG, it is enough to prove the consistency of the estimated skeleton. We include the detailed proof of Theorem 3 in Appendix B.

Theorem 3. *Assume that Conditions (A1)–(A5) hold. Let* $\mathcal{G}_{skel,n}$ *be the true skeleton of the graph* $\mathcal{G}_n$, *and* $\hat{\mathcal{G}}_{skel,n}$ *be the skeleton estimated by the nonPC algorithm. Then, as* $n \to \infty$, $\mathbb{P}\big(\hat{\mathcal{G}}_{skel,n} = \mathcal{G}_{skel,n}\big) \to 1$.

Remark 3. *In the proof of Theorem 3, we consider the threshold* ξ_α *to be of constant order. However, the proof continues to work as long as* ξ_α *is of the same order as* C_{min} *as* $n \to \infty$.

3.2. The Nonparametric FCI Algorithm in High Dimensions

The FCI is a modification of the PC algorithm that accounts for latent and selection variables. Thus, generalizations of the PC algorithm naturally extend to the FCI as well. Similar to nonPC, we propose testing for conditional independence relations nonparametrically in the *sample version* of the FCI-stable algorithm (Algorithm A3 in Appendix A) based on ρ_0^*, instead of partial correlations. We coin the resulting algorithm the 'nonFCI' algorithm, to emphasize that it is a generalization of parametric FCI-stable algorithms. Again, the *oracle version* of the nonFCI is exactly the same as that of the FCI-stable algorithm. The difference is in the implementation of the tests for conditional independence relationships in their *sample versions*. This broadens the scope of the FCI algorithm in causal structural learning for observational data in the presence of latent and selection variables when Gaussianity is not a viable assumption. More specifically, it enables capturing nonlinear and non-monotone conditional dependence relationships among the variables that partial correlations would fail to detect.

Equipped with the theoretical guarantees we established for the nonPC in Section 3.1, we establish below in Theorem 4 the consistency of the nonFCI algorithm for general distributions in sparse high-dimensional settings. Let $\mathcal{H} = (V, E)$ be a DAG with the vertex set partitioned as $V = V_X \cup V_L \cup V_T$, where V_X indexes the set of p observed variables, V_L denotes the set of latent variables and V_T stands for the set of selection variables. Let $\mathcal{M}$ be the unique MAG over V_X. We let p grow with n and consider $p = p_n$, $\mathcal{H} = \mathcal{H}_n$ and $Q = Q_n$, where Q is the distribution of $(U_1, \ldots, U_p) := (X_1 \mid V_T, \ldots, X_p \mid V_T)$. We provide below the definition of possible-D-SEP sets (Definition 3.3 in [4]).

Definition 1. *Let* C *be a graph with any of the following edge types :* $\circ\!\!-\!\!\circ$, $\circ\!\!\rightarrow$ *and* $\leftrightarrow$. *A possible-D-SEP* (X_a, X_b) *in* C, *denoted* $\mathrm{pds}(C, X_a, X_b)$, *is defined as follows:* $X_c \in \mathrm{pds}(C, X_a, X_b)$ *if and only if there is a path* π *between* X_a *and* X_c *in* C *such that, for every subpath* $\langle X_e, X_f, X_g \rangle$ *of* π, X_f *is a collider on the subpath in* C *or* $\langle X_e, X_f, X_g \rangle$ *is a triangle in* C.

To prove the consistency of the nonFCI algorithm in sparse high-dimensional settings, we impose the following regularity conditions, which are similar to the assumptions imposed in Section 4 in [4].

(C3) The distribution Q_n is faithful to the underlying MAG $\mathcal{M}_n$ for all n.

(C4) The maximum size of the possible-D-SEP sets for finding the final skeleton in the FCI-stable algorithm (Algorithm A6 in Appendix A), q'_n, grows at the rate of $O(n^{1-b})$, where $0 < b \leq 1$.

(C5) For any $U_i, U_j \in \{U_1, \ldots, U_{p_n}\}$ and $U_S \subseteq \{U_1, \ldots, U_{p_n}\}\backslash\{U_i, U_j\}$ with $|U_S| \leq q'_n$, assume

$$\inf\left\{|\rho_0^*(U_i, U_j | U_S)| : \rho_0^*(U_i, U_j | U_S) \neq 0\right\} \geq \lambda'_{min} \quad (\lambda'_{min})^{-1} = O(n^v)$$

$$\text{and} \quad \sup |\rho_0^*(U_i, U_j | U_S)| \leq \lambda'_{max}$$

where λ'_{max} is a positive constant and $0 < v < 1/4$.

Theorem 4. *Suppose conditions (A1)–(A3) and (C3)–(C5) hold. Denote by $\mathcal{C}_n$ and $\mathcal{C}_n^*$ the true underlying FCI-PAG and the output of the nonFCI algorithm, respectively. Then, as $n \to \infty$, $\mathbb{P}(\mathcal{C}_n^* = \mathcal{C}_n) \to 1$.*

4. Numerical Studies

4.1. Performance of the NonPC Algorithm

In this subsection, we compare the performances of the nonPC and the PC-stable algorithms in finding the skeleton and the CPDAG for various simulated datasets. We simulate random DAGs in the following examples and sample from probability distributions faithful to them.

Example 1 (Linear SEM). *We first fix a sparsity parameter $s \in (0,1)$ and enumerate the vertices as $V = \{1, \ldots, p\}$. We then construct a $p \times p$ adjacency matrix Λ as follows. First, initialize Λ as a zero matrix. Next, fill every entry in the lower triangle (below the diagonal) of Λ by independent realizations of Bernoulli random variables with success probability s. Finally, replace each nonzero entry in Λ by independent realizations of a Uniform$(0.1, 1)$ random variable.*

In this scheme, each node has the same expected degree $\mathbb{E}(m) = (p - 1)s$, where m is the degree of a node and follows a Binomial $(p - 1, s)$ distribution. Using the adjacency matrix Λ, the data are then generated from the following linear structural equation model (SEM) :

$$X = \Lambda X + \epsilon$$

where $\epsilon = (\epsilon_1, \ldots, \epsilon_p)$ and $\epsilon_1, \ldots, \epsilon_p$ are jointly independent. To obtain samples $\{X_1^k, \ldots, X_p^k\}_{k=1}^n$ on $\{X_1, \ldots, X_p\}$, we first sample $\{\epsilon_1^k, \ldots, \epsilon_p^k\}_{k=1}^n$ from the three following data-generating schemes. For $1 \leq k \leq n$ and $1 \leq i \leq p$,

1. Normal: Generate ϵ_i^k's independently from a standard normal distribution.
2. Copula: Generate ϵ_i^k's as in (1) and then transform the marginals to a $F_{1,1}$ distribution.
3. Mixture: Generate ϵ_i^k's independently from a 50–50 mixture of a standard normal and a standard Cauchy distribution.

Example 2 (Nonlinear SEM). *In this example, we first generate a $p \times p$ adjacency matrix Λ in the similar way as in Example 1 and then generate the data from the following nonlinear SEM (similar to [10]) : $X_i = \sum_{j \,:\, \Lambda_{ij} \neq 0} f_{ij}(X_j) + \epsilon_i$ with $\epsilon_i \overset{i.i.d.}{\sim} N(0,1)$, where $1 \leq j < i \leq p$. If the functions f_{ij}'s are chosen to be nonlinear, then the data will typically not correspond to a well-known multivariate distribution. We consider $f_{ij}(x_j) = b_{ij1} x_j + b_{ij2} x_j^2$, where b_{ij1} and b_{ij1} are independently sampled from $N(0,1)$ and $N(0,0.5)$ distributions, respectively.*

With the exception of Example 1.1, the above examples are all non-Gaussian graphical models. We would thus expect the nonPC to perform better than the PC-stable in learning the unknown causal structure in these examples. For each of the four data generating methods considered above, we compare the Structural Hamming Distance (SHD) [25] between the estimated and the true skeletons of the underlying DAGs using the nonPC and PC-stable algorithms. The SHD between two undirected graphs is the number of edge additions or deletions necessary to make the two graphs match. Therefore, larger SHD values between the estimated and the true skeleton correspond to worse estimates.

We consider 199 bootstrap replicates for the CdCov-based conditional independence tests in the implementation of our nonPC algorithm and the significance level $\alpha = 0.05$. Table 1 presents the average SHD for the different data generating schemes over 20 simulation runs, for different choices of n, p and $\mathbb{E}(m)$.

Table 1. Comparison of the average structural Hamming distances (SHD) of nonPC and PC-stable algorithms across simulation studies.

			Normal		Copula	
n	p	$\mathbb{E}(m)$	nonPC	PC-stable	nonPC	PC-stable
50	9	1.4	3.35	3.05	5.55	5.75
100	27	2.0	14.55	11.00	25.6	28.6
150	81	2.4	53.70	43.45	97.3	121.3
200	243	2.8	186.2	183.4	331.00	471.45
			Mixture		Nonlinear SEM	
n	p	$\mathbb{E}(m)$	nonPC	PC-stable	nonPC	PC-stable
50	9	1.4	3.8	3.5	2.9	3.7
100	27	2.0	17.75	18.00	15.05	20.05
150	81	2.4	69.05	77.75	62.583	95.083
200	243	2.8	250.3	336.1	213.70	375.45

The results in Table 1 demonstrate that the nonPC performs nearly as good as the PC-stable for the Gaussian data example, in terms of the average SHD. However, for each of the non-Gaussian data examples, the nonPC performs better than the PC-stable in estimating the true skeleton of the underlying DAGs. The improvement in SHD becomes more substantial as the dimension grows. The superior performance of the nonPC over PC-stable for the non-Gaussian graphical models is expected, as the characterization of conditional independence by partial correlations is only valid under the assumption of joint Gaussianity.

4.2. Performance of the NonFCI Algorithm

In this subsection, we compare the performances of the nonFCI and the FCI-stable algorithms over various simulated datasets. We first generate random DAGs as in Examples 1 and 2. To assess the impact of latent variables, we randomly define half of the variables with no parents and at least one child as latent. We do not consider selection variables. We run both the nonFCI and the FCI-stable algorithms on the above data examples with $n = 200$, $p = \{10, 20, 30, 100, 200\}$ and $\alpha = 0.01$, using 199 bootstrap replicates for the CdCov-based conditional independence tests. We consider 20 simulation runs for each of the data generating models. Table 2 reports the average SHD between the estimated and true PAG skeleton by the nonFCI and FCI-stable algorithms.

Table 2. Comparison of the average structural Hamming distances (SHD) of nonFCI and FCI-stable algorithms across simulation studies.

		Normal		Copula		Mixture		Nonlinear SEM	
p	$\mathbb{E}(m)$	nonFCI	FCI-Stable	nonFCI	FCI-Stable	nonFCI	FCI-Stable	nonFCI	FCI-Stable
10	2.0	7.15	7.60	1.3	1.8	5.65	6.80	7.15	8.20
20	2.0	14.55	17.60	4.55	6.85	13.65	18.55	19.0	20.8
30	2.0	27.65	33.95	5.25	10.15	19.3	27.8	33.40	37.85
100	3.0	109.30	150.35	26.95	60.05	62.25	111.10	115.2	149.0
200	3.0	287.75	371.40	76.733	157.267	136.05	255.10	289.6	354.1

The results in Table 2 demonstrate that, in both the Gaussian and non-Gaussian examples, the nonFCI algorithm outperforms the FCI-stable in estimating the true PAG skeleton.

4.3. Real Data Example

A major difficulty in assessing whether nonPC and nonFCI provide more reasonable estimates compared to the parametric versions of the algorithms in high-dimensional real data settings is that the true causal graph is not known in most of the cases. In absence of the truth, we may only be able to draw some conclusions about sensible causal mechanisms by examining known or logical relationships among pairs of variables. However, this becomes increasingly difficult for larger networks, where even visualization becomes challenging. This is why we first choose a relatively smaller dataset in Section 4.3.1, where we can draw upon background knowledge to glean insight into potential causal mechanisms in a setting where the data are clearly non-Gaussian. This example highlights the main focus of the paper that, with non-Gaussian data (categorical, as in this example), nonPC is expected to perform better than the PC-stable in learning the true causal structure of the underlying DAG. In Section 4.3.2, we consider a larger example and examine the performance of PC-stable and nonPC in learning the DAG from both seemingly Gaussian data as well as a categorized version of the same data. This example clearly illustrates the potential limitations of PC-stable: in contrast to nonPC, the output of PC-stable can be strikingly different when applied to a categorized version of the original data.

4.3.1. Montana Poll Dataset

To demonstrate the flexibility of our proposed framework, we first apply the nonPC algorithm to the Montana Economic Outlook Poll dataset. The poll was conducted in May 1992 where a random sample of 209 Montana residents were asked whether their personal financial status was worse, the same or better than a year ago, and whether they thought the state economic outlook was better than the year before. Accompanying demographic information on the respondents' age, income, political orientation, and area of residence in the state were also recorded. We obtained the dataset from the Data and Story Library (DASL), available at https://math.tntech.edu/e-stat/DASL/page4.html (accessed on 25 March 2021). The study is comprised of the following seven categorical variables: AGE = 1 for under 35, 2 for 35–54, 3 for 55 and over; SEX = 0 for male, 1 for female; INC = yearly income: 1 for under \$20 K, 2 for \$20–35 K, 3 for over \$35 K; POL = 1 for Democrat, 2 for Independent, 3 for Republican; AREA = 1 for Western, 2 for Northeastern, 3 for Southeastern Montana; FIN (=Financial status): 1 for worse, 2 for same, 3 for better than a year ago; and STAT (=State economic outlook): 1 for better, 0 for not better than a year ago.

After removing the cases with missing values, we are left with $n = 163$ samples. Since all the variables are categorical, the Gaussianity assumption is outrightly violated. Thus, we would expect the nonPC to perform better than the PC-stable in learning the true causal structure among the variables in this case. Figure 1 below presents the CPDAGs estimated by the nonPC and PC-stable algorithms at a significance level $\alpha = 0.1$. We consider 199 bootstrap replicates for the CdCov-based conditional independence tests in the implementation of the nonPC algorithm.

It is quite intuitive that age and sex are likely to affect the income; one's financial status and the area of residence might also influence their political inclination; and improvements or downturns in the state economic outlook might impact an individual's financial status. The CPDAG estimated by the nonPC algorithm in Figure 1a affirms such common-sense understanding of these causal influences. However, in the CPDAG estimated by the PC-stable in Figure 1b, the edge between age and income is missing. In addition, the directed edges POL $\rightarrow$ AREA and POL $\rightarrow$ FIN seem to make little sense in this case.

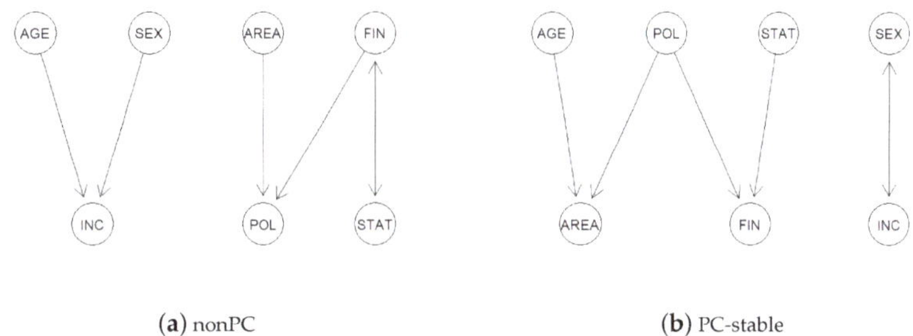

(**a**) nonPC   (**b**) PC-stable

Figure 1. CPDAGs estimated by the nonPC and PC-stable algorithms for the Montana poll dataset.

4.3.2. Protein Expression Data

We next consider a protein expression dataset of 410 patients with breast cancer from The Cancer Genome Atlas (TCGA). The dataset consists of $p = 118$ genes, and we randomly select a subset of $n = 100$ patients with PR-negative status. Since the true causal structure of the genes in the cancer cells may be different than that of normal cells [26], we apply both the nonPC and PC-stable algorithms to learn the causal structure. To put the performances of the nonPC and PC-stable under scrutiny as the data depart farther away from Gaussianity, we categorize the protein expression data for each of the p genes, denoted by $\{X_a^k\}_{k=1}^n$, $1 \leq a \leq p$, as follows. We compute the three quartiles $Q_{1;a}$, $Q_{2;a}$ and $Q_{3;a}$ of the protein expression values for every $1 \leq a \leq p$. Consequently, we obtain categorized protein expressions $\{X_{C;a}^k\}_{k=1}^n$ for $1 \leq a \leq p$, where

$$X_{C;a}^k := \begin{cases} 0 & \text{if } X_a^k \leq Q_{1;a} \\ 1 & \text{if } Q_{1;a} < X_a^k \leq Q_{2;a} \\ 2 & \text{if } Q_{2;a} < X_a^k \leq Q_{3;a} \\ 3 & \text{if } X_a^k > Q_{3;a}. \end{cases}$$

We apply the nonPC and PC-stable algorithms to both the original and the categorized protein expression data at a significance level $\alpha = 0.01$. We consider 199 bootstrap replicates for the CdCov-based conditional independence tests in the implementation of the nonPC algorithm. Table 3 below shows the SHD between the skeletons estimated from the original and the categorized data by the nonPC and PC-stable algorithms. It can be seen that the SHD between the skeletons estimated from the original and categorized data by the PC-stable algorithm is much larger than that for nonPC. This example highlights the potential limitation of parametric implementations of the PC algorithm: when the data deviate farther away from Gaussianity (in this case, being categorical), the estimates produced by the PC-stable may deviate considerably more from the estimates from the original data. In contrast, the nonparametric test in nonPC delivers more stable estimates regardless of the data distribution.

Table 3. Comparison of the SHD between the skeletons estimated from the original and the categorized protein expression data by the nonPC and PC-stable algorithms.

nonPC	PC-Stable
22	79

5. Discussion

We proposed nonparametric variants of the widely popular PC-stable and FCI-stable algorithms, which employ conditional distance covariance (CdCov) to test for conditional independence relationships in their sample versions. Our proposed algorithms broaden the applicability of the PC/PC-stable and FCI/FCI-stable algorithms to general distributions over DAGs, and enable taking into account nonlinear and non-monotone conditional dependence among the random variables, which partial correlations fail to capture. We show that the high-dimensional consistency of the PC-stable and FCI-stable algorithms carry over to more general distributions over DAGs when we implement CdCov-based nonparametric tests for conditional independence. These results are obtained without imposing any strict distributional assumptions and only require moment and tail conditions on the variables.

There are several intriguing potential directions for future research. First, it is generally difficult to select the tuning parameter (i.e., the significance threshold for the CdCov test) in causal structure learning. One possible strategy is to use ideas based on *stability selection* [27,28]. By assessing the stability of the estimated graphs in multiple subsamples, this strategy allows us to choose the tuning parameter in order to control the false positive error. However, the repeated subsampling increases the computational burden. Second, the computational and sample complexities of the PC and FCI algorithms (and hence those of the nonPC and nonFCI) scale with the maximum degree of the DAG, which is assumed to be small relative to the sample size. However, in many applications, one encounters sparse graphs containing a small number of highly connected 'hub' nodes. In such cases, ref. [29] proposed a low-complexity variant of the PC algorithm, namely the *reduced PC* (rPC) algorithm that exploits the local separation property of large random networks [30]. The rPC is shown to consistently estimate the skeleton of a high-dimensional DAG by conditioning only on sets of small cardinality. More recently, ref. [31] have generalized this approach to account for unobserved confounders. In this light, it would be intriguing to develop computationally faster variants of the nonPC and nonFCI in the future by exploiting the idea of local separation.

Author Contributions: Conceptualization, S.C. and A.S.; methodology, S.C. and A.S.; formal analysis, S.C.; investigation, S.C.; writing–original draft preparation, S.C.; writing–review and editing, S.C. and A.S.; supervision, A.S.; funding acquisition, A.S. All authors have read and agreed to the published version of the manuscript.

Funding: The authors gratefully acknowledge the funding from grants R01GM114029 and R01GM133848 from the US National Institutes of Health and grant DMS-1915855 from the US National Science Foundation.

Data Availability Statement: The Montana Poll dataset has been accessed from the Data and Story Library (DASL) at https://math.tntech.edu/e-stat/DASL/page4.html (accessed on 25 March 2021).

Conflicts of Interest: The authors declare no conflict of interest.

Appendix A. Preliminaries and Background

For the sake of completeness, we illustrate in this section the pseudocodes of the oracle versions of the PC-stable and FCI-stable algorithms. We also outline a local bootstrap procedure that can be used to approximate the threshold ζ_α mentioned in Section 3.1 and is used throughout the numerical studies in the paper.

Algorithm A1 presents the pseudocode of the oracle version of Step 1 of the PC-stable algorithm (Algorithm 4.1 of [7]), which estimates the skeleton of the underlying DAG. Algorithm A2 presents the pseudocode of Step 2 of the PC-stable algorithm (Algorithm 2 of [8]) that extends the skeleton estimated in Step 1 to the CPDAG. Algorithm A3 presents the pseudocode of the FCI-stable algorithm (Section 4.4 in [7]). It implements Algorithm A4 to obtain an initial skeleton of the underlying PAG, Algorithm A5 to orient the v-structures, and finally Algorithm A6 to obtain the final skeleton that the FCI-stable returns.

To approximate the threshold ζ_α to test for $H_0 : X \perp\!\!\!\perp Y|Z$ vs. $H_A : X \not\perp\!\!\!\perp Y|Z$ at level $\alpha \in (0,1)$ (see Section 3.1), we consider the following local bootstrap procedure in the light of Section 4.3 in [15]. Given the i.i.d. sample $\{W_i = (X_i, Y_i, Z_i)\}_{i=1}^{n}$ from the joint distribution of $W = (X, Y, Z)$, draw a local bootstrap sample $\{W_i^\dagger = (X_i^\dagger, Y_i, Z_i)\}_{i=1}^n$ and compute the bootstrap statistic. The detailed steps are as follows :

Algorithm A1 Step 1 of the PC-stable algorithm (oracle version).

Require : Conditional independence information among all variables in V, and an ordering $\mathrm{order}(V)$ on the variables.
Form the complete undirected graph $\mathcal{C}$ on the vertex set V.
Let $l = -1$;
repeat
    $l = l + 1$;
    for all vertices X_a in $\mathcal{C}$ **do**
        let $u(X_a) = adj(\mathcal{C}, X_a)$
    end for
    repeat
        Select a (new) ordered pair of vertices (X_a, X_b) that are adjacent in $\mathcal{C}$ such that $|u(X_a) \setminus \{X_b\}| \geq l$, using order (V);
        repeat
            Choose a (new) set $S \subseteq u(X_a) \setminus \{X_b\}$ with $|S| = l$, using $\mathrm{order}(V)$;
            if $X_a \perp\!\!\!\perp X_b \mid S$ **then**
                Delete the edge $X_a - X_b$ from $\mathcal{C}$;
                Let $sepset\,(X_a, X_b) = sepset\,(X_b, X_a) = S$;
            end if
        until X_a and X_b are no longer adjacent in $\mathcal{C}$ or all $S \subseteq u(X_a) \setminus \{X_b\}$ with $|S| = l$
have
        been considered
    until all ordered pairs of adjacent vertices (X_a, X_b) in $\mathcal{C}$ with $|u(X_a) \setminus \{X_b\}| \geq l$ have
been
    considered
until all pairs of adjacent vertices (X_a, X_b) in $\mathcal{C}$ satisfy $|u(X_a) \setminus \{X_b\}| \leq l$
Output : The estimated skeleton $\mathcal{C}$, separation sets $sepset$.

Algorithm A2 Step 2 of the PC-stable algorithm.

Require : Skeleton $\mathcal{C}$, separation sets $sepset$.
for all all pair of nonadjacent vertices X_a, X_c with common neighbor X_b in $\mathcal{C}$ **do**
    if $X_b \notin sepset(X_a, X_c)$ **then**
        Replace $X_a - X_b - X_c$ in $\mathcal{C}$ by $X_a \rightarrow X_b \leftarrow X_c$;
    end if
end for
In the resulting PDAG, try to orient as many undirected edges as possible by repeated applications of the following rules :
(R1) Orient $X_b - X_c$ into $X_b \rightarrow X_c$ whenever there is an arrow $X_a \rightarrow X_b$ such that X_a and X_c are nonadjacent (otherwise, a new v-structure is created).
(R2) Orient $X_a - X_c$ into $X_a \rightarrow X_c$ whenever there is a chain $X_a \rightarrow X_b \rightarrow X_c$ (otherwise, a directed cycle is created).
(R3) Orient $X_a - X_c$ into $X_a \rightarrow X_c$ whenever there are two chains $X_a - X_b \rightarrow X_c$ and $X_a - X_d \rightarrow X_c$ such that X_b and X_d are nonadjacent (otherwise, a new v-structure or a directed cycle is created).

Algorithm A3 The FCI-stable algorithm (oracle version).

Require : Conditional independence information among all variables in V_X given V_T.
Use Algorithm A4 to find an initial skeleton ($\mathcal{C}$), separation sets (sepset) and unshielded
triple list ($\mathcal{M}$);
Use Algorithm A5 to orient v-structures (update $\mathcal{C}$);
Use Algorithm A6 to find the final skeleton (update $\mathcal{C}$ and sepset);
Use Algorithm A5 to orient v-structures (update $\mathcal{C}$);
Use rules (R1)-(R10) of [6] to orient as many edge marks as possible (update $\mathcal{C}$);
Output : $\mathcal{C}$, sepset.

Algorithm A4 Obtaining an initial skeleton in the FCI-stable algorithm (Algorithm 4.1 in
the supplement of [4]).

Require : Conditional independence information among all variables in V_X given V_T,
and an ordering order(V_X) on the variables.
Form the complete undirected graph $\mathcal{C}$ on the vertex set V_X with edges $\circ\!-\!\circ$.
Let $l = -1$;
repeat
    $l = l + 1$;
    for all vertices X_a in $\mathcal{C}$ **do**
       let $u(X_a) = adj(\mathcal{C}, X_a)$
    end for
    repeat
       Select a (new) ordered pair of vertices (X_a, X_b) that are adjacent in $\mathcal{C}$ such that
       $|u(X_a) \setminus \{X_b\}| \geq l$, using order (V_X);
       repeat
          Choose a (new) set $Y \subseteq u(X_a) \setminus \{X_b\}$ with $|Y| = l$, using order(V_X);
          if $X_a \perp\!\!\!\perp X_b \mid Y \cup V_T$ **then**
             Delete the edge $X_a \circ\!-\!\circ X_b$ from $\mathcal{C}$;
             Let sepset(X_a, X_b) = sepset(X_b, X_a) = Y;
          end if
       until X_a and X_b are no longer adjacent in $\mathcal{C}$ or all $Y \subseteq u(X_a) \setminus \{X_b\}$ with $|Y| = l$
have
       been considered
    until all ordered pairs of adjacent vertices (X_a, X_b) in $\mathcal{C}$ with $|u(X_a) \setminus \{X_b\}| \geq l$ have
been
    considered
until all pairs of adjacent vertices (X_a, X_b) in $\mathcal{C}$ satisfy $|u(X_a) \setminus \{X_b\}| \leq l$
Form a list $\mathcal{M}$ of all unshielded triples $\langle X_c \cdot X_d \rangle$ (i.e., the middle vertex is left unspecified)
in $\mathcal{C}$ with $c < d$.
Output : $\mathcal{C}$, sepset, $\mathcal{M}$.

Algorithm A5 Orienting v-structures in the FCI-stable algorithm (Algorithm 4.2 in the
supplement of [4]).

Require : Initial skeleton ($\mathcal{C}$), separation sets (sepset) and unshielded triple list ($\mathcal{M}$).
for all elements $\langle X_a, X_b, X_c \rangle$ of $\mathcal{M}$ **do**
    if $X_b \notin$ sepset(X_a, X_c) **then** Orient $X_a \,\star\!\!-\!\circ X_b \circ\!-\!\star\, X_c$ as $X_a \star\!\!\rightarrow X_b \leftarrow\!\star X_c$
    end if
end for
Output : $\mathcal{C}$, sepset.

Algorithm A6 Obtaining the final skeleton in the FCI-stable algorithm (Algorithm 4.3 in the supplement of [4]).

Require: Partially oriented graph ($\mathcal{C}$) and separation sets (sepset).
for all vertices X_a in $\mathcal{C}$ **do**
    let $v(X_a) = \text{pds}(\mathcal{C}, X_a, \cdot)$;
    for all vertices $X_b \in adj(\mathcal{C}, X_a)$ **do**
        Let $l = -1$;
        repeat
            $l = l + 1$;
            repeat
                Choose a (new) set $Y \subseteq v(X_a) \setminus \{X_b\}$ with $|Y| = l$;
                if $X_a \perp\!\!\!\perp X_b \mid Y \cup V_T$ **then**
                    Delete the edge $X_a \star\!\!-\!\!\star X_b$ from $\mathcal{C}$;
                    Let $\text{sepset}(X_a, X_b) = \text{sepset}(X_b, X_a) = Y$;
                end if
                until X_a and X_b are no longer adjacent in $\mathcal{C}$ or all $Y \subseteq v(X_a) \setminus \{X_b\}$ with $|Y| = l$ have
                been considered
            until X_a and X_b are no longer adjacent in $\mathcal{C}$ or $|v(X_a) \setminus \{X_b\}| < l$
    end for
end for
Reorient all edges in $\mathcal{C}$ as $\circ\!\!-\!\!\circ$.
Form a list $\mathcal{M}$ of all unshielded triples $\langle X_c \cdot X_d \rangle$ in $\mathcal{C}$ with $c < d$.
Output : $\mathcal{C}$, sepset, $\mathcal{M}$.

A.   For $i = 1, \ldots, n$, draw $X_i^\dagger$ from

$$\widehat{F}_{X|Z=Z_i} = \frac{\sum_{j=1}^{n} K_{ij}\, \mathbf{1}(-\infty, X_j](x)}{\sum_{j=1}^{n} K_{ij}}.$$

Compute $\widehat{\rho}^{*\dagger}$ based on the local bootstrap sample $\{W_i^\dagger = (X_i^\dagger, Y_i, Z_i)\}_{i=1}^{n}$.

B.   Repeat Step A B times to obtain $\{\widehat{\rho}_b^{*\dagger}\}_{b=1}^{B}$. Obtain $\xi_{n,\alpha}^*$ as the $100(1-\alpha)^{th}$ percentile of $\{nh^{r/2}\,\widehat{\rho}_b^{*\dagger}\}_{b=1}^{B}$. Then, $\frac{1}{nh^{r/2}}\xi_{n,\alpha}^*$ can be considered as an approximation for ξ_α.

Appendix B. Proofs of the Theoretical Results

In this section, we provide detailed technical proofs of the theoretical results presented in the paper. We first state a concentration inequality in Lemma A1. The result in Lemma A1 is not new and can be seen as a corollary of Theorem A in Section 5.6.1 of [32]; however, it is a key technical ingredient in the proof of Theorem 1, which is the main theoretical innovation of our paper. For completeness, we include a short proof for Lemma A1.

Lemma A1. *Consider a U-statistic* $U_n = U(X_1, \ldots, X_n) = \binom{n}{m}^{-1} \sum_{i_1 < \cdots < i_m} h(X_{i_1}, \ldots, X_{i_m})$ *with a symmetric kernel h such that* $\mathbb{E}\, U_n = \mathbb{E}\, h(X_1, \ldots, X_m) = \theta$. *Further suppose* $|h(X_1, \ldots, X_m)| \leq M$ *for some $M > 0$. Then, for any $\epsilon > 0$, we have*

$$\mathbb{P}(|U_n - \theta| > \epsilon) \leq 2 \exp\left(-\frac{\epsilon^2 k}{2M^2}\right)$$

where $k := \lfloor \frac{n}{m} \rfloor$.

Proof of Lemma A1. Define

$$W(X_1, \ldots, X_n) := \frac{1}{k}\left[h(X_1, \ldots, X_m) + h(X_{m+1}, \ldots, X_{2m}) + \cdots + h(X_{km-m+1}, \ldots, X_{km}) \right].$$

Then, following Section 5.1.6 in [32], we can write

$$U_n \;=\; \frac{1}{n!} \sum_{\pi} W(X_{i_1}, \ldots, X_{i_n}) \tag{A1}$$

where $\sum_{\pi}$ denotes summation over all $n!$ permutations $(i_1, \ldots, i_n)$ of $(1, 2, \ldots, n)$. Thus, U_n can be expressed as an average of $n!$ terms, each of which is an average of k i.i.d. random variables. Using Markov's inequality, convexity of the exponential function and Jensen's inequality, we have, for any $t > 0$,

$$
\begin{aligned}
\mathbb{P}(U_n - \theta > \epsilon) &= \mathbb{P}\big(\exp\left(t\left(U_n - \theta\right)\right) > \exp(t\epsilon)\big) \\
&\leq \exp(-t\epsilon)\exp(-t\theta)\,\mathbb{E}\left[\exp\left(t\,U_n\right)\right] \\
&= \exp(-t\epsilon)\exp(-t\theta)\,\mathbb{E}\left[\exp\left(t\frac{1}{n!}\sum_{p} W(X_{i_1}, \ldots, X_{i_n})\right)\right] \\
&\leq \exp(-t\epsilon)\exp(-t\theta)\,\frac{1}{n!}\sum_{\pi}\mathbb{E}\left[\exp\left(t\,W(X_{i_1}, \ldots, X_{i_n})\right)\right] \\
&= \exp(-t\epsilon)\exp(-t\theta)\left[\mathbb{E}\left(\exp\left(\frac{t}{k}h\right)\right)\right]^k \\
&= \exp(-t\epsilon)\,\mathbb{E}^k\left[\exp\left(\frac{t}{k}(h - \theta)\right)\right]
\end{aligned}
\tag{A2}
$$

where, for notational simplicity, we use h to denote $h(X_1, \ldots, X_m)$. Using Hoeffding's Lemma, we have from (A2)

$$\mathbb{P}(U_n - \theta > \epsilon) \;\leq\; \exp\left(-t\epsilon + k\frac{1}{8}\frac{t^2}{k^2}(2M)^2\right) \;=\; \exp\left(-t\epsilon + \frac{t^2 M^2}{2k}\right).$$

Symmetrically, we obtain

$$\mathbb{P}(|U_n - \theta| > \epsilon) \;\leq\; 2\exp\left(-t\epsilon + \frac{t^2 M^2}{2k}\right). \tag{A3}$$

The right-hand side of (A3) is minimized at $t = \epsilon k / M^2$. Therefore, choosing $t = \epsilon k / M^2$, we obtain

$$\mathbb{P}(|U_n - \theta| > \epsilon) \;\leq\; 2\exp\left(-\frac{\epsilon^2 k}{2M^2}\right).$$

$\square$

Proof of Theorem 1. When $|S| = 0$, it can be shown in similar lines of Theorem 1 in Li et al. (2012) [33] that, for any $\epsilon > 0$, there exist positive constants A, B and $\gamma \in (0, 1/4)$ such that

$$\mathbb{P}(|\widehat{\rho}^*(X_a, X_b | X_S) - \rho_0^*(X_a, X_b | X_S)| > \epsilon) \;\leq\; O\left(2\exp\left(-A\,n^{1-2\gamma}\epsilon^2\right) + n\exp\left(-B\,n^{\gamma}\right)\right).$$

Now, consider the case $0 < |S| \leq m_p$.

For notational convenience, we treat X_a, X_b and X_S as X, Y and Z, respectively.

Denote $\delta_Z := \mathrm{CdCov}^2(X, Y | Z)$. Then, $\rho_0^* = \mathbb{E}[\delta_Z]$. Recall that

$$\hat{\rho}^{\,*}(X,Y|Z) \;:=\; \frac{1}{n}\sum_{u=1}^{n} \mathrm{CdCov}_n^2(X,Y|Z_u) \;:=\; \frac{1}{n}\sum_{u=1}^{n}\Delta_{i,j,k,l;u}$$

$$\text{where}\qquad \Delta_{i,j,k,l;u} \;:=\; \sum_{i,j,k,l} \frac{K_{iu}\,K_{ju}\,K_{ku}\,K_{lu}}{12\left(\sum_{i=1}^{n} K_{iu}\right)^4}\, d^S_{ijkl}\,. \tag{A4}$$

From (A4), we have

$$\mathbb{E}\left[\mathrm{CdCov}_n^2(X,Y|Z_u)\,|\,Z\right]$$

$$= \frac{1}{12}\,\mathbb{E}\left[\,d^S_{1234}\,|\,Z_1=Z_u,\ldots,Z_4=Z_u\right]\sum_{i,j,k,l} K_{iu}\,K_{ju}\,K_{ku}\,K_{lu}\,\Big/\,\left(\sum_{i=1}^{n} K_{iu}\right)^4 \tag{A5}$$

$$= \frac{1}{12}\,\mathbb{E}\left[\,d^S_{1234}\,|\,Z_1=Z_u,\ldots,Z_4=Z_u\right] \;=\; \delta_{Z_u}$$

where the last equality follows from Lemma 1 in [15]. Together, (A4) and (A5)

imply $\mathbb{E}\left[\hat{\rho}^{\,*}\right] = \rho_0^*$.

Now, consider the truncation

$$\rho_0^* = \rho_{01}^* + \rho_{02}^*$$
$$:= \mathbb{E}\left[\frac{1}{12}\,d^S_{i,j,k,l}\,\mathbf{1}\!\left(\left|\frac{1}{12}\,d^S_{i,j,k,l}\right|\le M\right)\right] + \mathbb{E}\left[\frac{1}{12}\,d^S_{i,j,k,l}\,\mathbf{1}\!\left(\left|\frac{1}{12}\,d^S_{i,j,k,l}\right|> M\right)\right] \tag{A6}$$

where $M>0$ will be specified later. Then, using triangle inequality,

$$\mathbb{P}(|\hat{\rho}^{\,*}-\rho_0^*|>\epsilon) \;=\; \mathbb{P}\!\left(\left|\frac{1}{n}\sum_{u=1}^{n}\Big(\sum_{i,j,k,l}\Delta_{i,j,k,l;u}-\rho_0^*\Big)\right|>\epsilon\right)$$

$$\le \mathbb{P}\!\left(\left|\frac{1}{n}\sum_{u=1}^{n}\Big(\sum_{i,j,k,l}\Delta_{i,j,k,l;u}\,\mathbf{1}\!\left(\left|\frac{1}{12}\,d^S_{i,j,k,l}\right|\le M\right)-\rho_{01}^*\Big)\right|>\epsilon/2\right) \tag{A7}$$

$$+\; \mathbb{P}\!\left(\left|\frac{1}{n}\sum_{u=1}^{n}\Big(\sum_{i,j,k,l}\Delta_{i,j,k,l;u}\,\mathbf{1}\!\left(\left|\frac{1}{12}\,d^S_{i,j,k,l}\right|> M\right)-\rho_{02}^*\Big)\right|>\epsilon/2\right)$$

$$=:\; \mathrm{I} + \mathrm{II}\,.$$

Clearly, from (A4), we have $|\Delta_{i,j,k,l;u}|\le M$ when $\left|\frac{1}{12}\,d^S_{i,j,k,l}\right|\le M$. With this observation, we have

$$\mathrm{I} \;\le\; 2\exp\!\left(-\frac{n\,\epsilon^2}{8\,M^2}\right) \tag{A8}$$

which follows from Lemma A1 by setting $m=1$, $k=\lfloor n\rfloor$ and $\epsilon=\epsilon/2$. Choosing $M=c\,n^\gamma$ for $\gamma\in(0,1/4)$ and some positive constant c, it follows from (A8) that

$$\mathrm{I} \;\le\; 2\exp\!\left(-C_1\,n^{1-2\gamma}\,\epsilon^2\right) \tag{A9}$$

for some $C_1>0$.

Now, to find a suitable upper bound for II, note that a simple application of triangle inequality yields

$$\frac{\epsilon}{2} < \left| \frac{1}{n} \sum_{u=1}^{n} \sum_{i,j,k,l} \Delta_{i,j,k,l;u} \, \mathbf{1}\left(\left| \frac{1}{12} d^{S}_{i,j,k,l} \right| > M \right) - \rho^{*}_{02} \right|$$

$$\leq \left| \frac{1}{n} \sum_{u=1}^{n} \sum_{i,j,k,l} \Delta_{i,j,k,l;u} \, \mathbf{1}\left(\left| \frac{1}{12} d^{S}_{i,j,k,l} \right| > M \right) \right| + \left| \rho^{*}_{02} \right| . \tag{A10}$$

For the choice of $M = c\, n^{\gamma}$, we have

$$\rho^{*}_{02} = \mathbb{E}\left[\frac{1}{12} d^{S}_{i,j,k,l} \, \mathbf{1}\left(\left| \frac{1}{12} d^{S}_{i,j,k,l} \right| > M \right) \right] < \frac{\epsilon}{4} \tag{A11}$$

for sufficiently large n (see, for example, Exercise 6 in Chapter 5, [34]). Combining (A10) and (A11), we obtain

$$\left\{ \left| \frac{1}{n} \sum_{u=1}^{n} \sum_{i,j,k,l} \Delta_{i,j,k,l;u} \, \mathbf{1}\left(\left| \frac{1}{12} d^{S}_{i,j,k,l} \right| > M \right) - \rho^{*}_{02} \right| > \epsilon/2 \right\}$$

$$\subseteq \left\{ \left| \frac{1}{n} \sum_{u=1}^{n} \sum_{i,j,k,l} \Delta_{i,j,k,l;u} \, \mathbf{1}\left(\left| \frac{1}{12} d^{S}_{i,j,k,l} \right| > M \right) \right| > \epsilon/4 \right\}$$

$$\subseteq \left\{ \left[\left| \frac{1}{12} d^{S}_{i,j,k,l} \right| > M \right] \text{ for some } 1 \leq i,j,k,l \leq n \right\},$$

which implies

$$\mathbb{P}\left(\left| \frac{1}{n} \sum_{u=1}^{n} \sum_{i,j,k,l} \Delta_{i,j,k,l;u} \, \mathbf{1}\left(\left| \frac{1}{12} d^{S}_{i,j,k,l} \right| > M \right) - \rho^{*}_{02} \right| > \epsilon/2 \right)$$

$$\leq \mathbb{P}\left(\left| \frac{1}{n} \sum_{u=1}^{n} \sum_{i,j,k,l} \Delta_{i,j,k,l;u} \, \mathbf{1}\left(\left| \frac{1}{12} d^{S}_{i,j,k,l} \right| > M \right) \right| > \epsilon/4 \right) \tag{A12}$$

$$\leq n^{4} \, \mathbb{P}\left(\left| \frac{1}{12} d^{S}_{i,j,k,l} \right| > M \right).$$

This is because, if $\left| \frac{1}{12} d^{S}_{i,j,k,l} \right| \leq M$ for all $1 \leq i,j,k,l \leq n$, then

$$n^{-1} \sum_{u=1}^{n} \sum_{i,j,k,l} \Delta_{i,j,k,l;u} \, \mathbf{1}\left(\left| \frac{1}{12} d^{S}_{i,j,k,l} \right| > M \right) = 0.$$

Under Condition (A1), Lemma 2 in the supplementary materials of [35] proves that there exists $s > 0$ for which $\mathbb{E}\left[\exp\left(s \left| d^{S}_{1234} \right| \right) \right]$ is finite. Using Markov's inequality, we have

$$\mathbb{P}\left(\left| \frac{1}{12} d^{S}_{i,j,k,l} \right| > M \right) \leq \mathbb{P}\left(\exp\left(s \left| \frac{1}{12} d^{S}_{i,j,k,l} \right| \right) > \exp(sM) \right)$$

$$\leq \exp(-sM)\, \mathbb{E}\left[\exp\left(s \left| \frac{1}{12} d^{S}_{i,j,k,l} \right| \right) \right] \tag{A13}$$

$$\leq C_{2} \exp(-sM) \leq C_{2} \exp(-s_{1}\, n^{\gamma})$$

for some positive constants C_{2} and s_{1}, where the last line uses the fact that $M = c\, n^{\gamma}$. Combining (A12) and (A13), we have

$$\text{II} \leq C_{2}\, n^{4} \exp(-s_{1}\, n^{\gamma}). \tag{A14}$$

Finally, combining (A7), (A9) and (A14), we obtain

$$\mathbb{P}(|\widehat{\rho}^* - \rho_0^*| > \epsilon/2) \;\leq\; 2\exp\!\left(-C_1 n^{1-2\gamma}\epsilon^2\right) + C_2\, n^4 \exp(-s_1 n^\gamma)$$

for some positive constants γ, C_1, C_2 and s_1. This completes the proof of the theorem. $\square$

Proof of Theorem 2. The first inequality in Theorem 2 simply follows by observing the fact that, for any generic random sequence $\{X_n\}_{n=1}^\infty$ and any $\epsilon > 0$,

$$P(|X_n| > \epsilon) \;\leq\; P\!\left(\sup_n |X_n| > \epsilon\right)$$

for all $n \geq 1$, which, in turn, implies

$$\sup_n P(|X_n| > \epsilon) \;\leq\; P\!\left(\sup_n |X_n| > \epsilon\right).$$

The second inequality follows from union bound and Theorem 1. $\square$

Proof of Theorem 3. Denote by $E_{ab|S}$ the event that "an error occurs while testing for $X_a \perp\!\!\!\perp X_b \mid X_S$" for $a, b \in V$ and $S \in J_{a,b}^{m_{p_n}}$. Then,

$$\mathbb{P}(\text{ an error occurs in the nonPC algorithm }) \;\leq\; \mathbb{P}\!\left(\bigcup_{\substack{a,b\in V \\ S\in J_{a,b}^{m_{p_n}}}} E_{ab|S}\right) \;\lesssim\; p_n^{m_{p_n}+2}\,\mathbb{P}(E_{ab|S}) \quad \text{(A15)}$$

which is essentially due to the union bound. Now, we can write $E_{ab|S} = E_{ab|S}^{\mathrm{I}} \cup E_{ab|S}^{\mathrm{II}}$, where

$$\text{(Type I error)} \qquad E_{ab|S}^{\mathrm{I}} : |\hat{\rho}_{ab|S}^*| > \xi_\alpha \qquad \text{when } \rho_{0;ab|S}^* = 0$$

$$\text{and} \qquad \text{(Type II error)} \qquad E_{ab|S}^{\mathrm{II}} : |\hat{\rho}_{ab|S}^*| \leq \xi_\alpha \qquad \text{when } \rho_{0;ab|S}^* > 0.$$

Then, by using triangle inequality,

$$\begin{aligned}
\mathbb{P}(E_{ab|S}^{\mathrm{I}}) &= \mathbb{P}(|\hat{\rho}_{ab|S}^*| > \xi_\alpha) = \mathbb{P}(|\hat{\rho}_{ab|S}^* - \rho_{0;ab|S}^* + \rho_{0;ab|S}^*| > \xi_\alpha) \\
&\leq \mathbb{P}(|\hat{\rho}_{ab|S}^* - \rho_{0;ab|S}^*| > \xi_\alpha - C_{max}) \\
&\lesssim 2\exp\left(-A\,n^{1-2\gamma}(\xi_\alpha - C_{max})^2\right) + n^4\exp\left(-Bn^\gamma\right)
\end{aligned} \quad \text{(A16)}$$

for positive constants A, B and $\gamma \in (0, 1/4)$, where the last inequality follows from Theorem 2. Similarly, using the definition of C_{min} and the identity $|a| - |b| \leq |a - b|$ for $a, b \in \mathbb{R}$, we have

$$\begin{aligned}
\mathbb{P}\!\left(E_{ab|S}^{\mathrm{II}}\right) &= \mathbb{P}(|\hat{\rho}_{ab|S}^*| \leq \xi_\alpha) = \mathbb{P}(-|\hat{\rho}_{ab|S}^*| \geq -\xi_\alpha) \\
&= \mathbb{P}(|\rho_{0;ab|S}^*| - |\hat{\rho}_{ab|S}^*| \geq |\rho_{0;ab|S}^*| - \xi_\alpha) \\
&\leq \mathbb{P}(|\rho_{0;ab|S}^* - \hat{\rho}_{ab|S}^*| \geq C_{min} - \xi_\alpha) \\
&\lesssim 2\exp\left(-A\,n^{1-2\gamma}(\xi_\alpha - C_{min})^2\right) + n^4\exp\left(-Bn^\gamma\right).
\end{aligned} \quad \text{(A17)}$$

Again, the last inequality follows from Theorem 2. Combining Equations (A15)–(A17), we have

$$\mathbb{P}\left(\text{ an error occurs in the nonPC algorithm }\right)$$
$$= O\left(p_n^{m_{p_n}+2} \left[2 \exp\left(-A\, n^{1-2\gamma}(\xi_\alpha - C_{max})^2 \right) + 2 \exp\left(-A\, n^{1-2\gamma}(\xi_\alpha - C_{min})^2 \right] \right.$$
$$\left. + n^4 \exp\left(-B\, n^\gamma \right) \right]\right)$$
$$= o(1)$$

where the last step follows from the fact that $\gamma \in (0, 1/4)$ and Assumption (A5). This implies that, as $n \to \infty$,

$$\mathbb{P}\left(\hat{G}_{\text{skel},n} = G_{\text{skel},n}\right) = 1 - \mathbb{P}\left(\text{ an error occurs in the nonPC algorithm }\right)$$
$$\to 1.$$

$\square$

Proof of Theorem 4. The proof follows similar lines of the proof of Theorem 4.2 in [4], replacing Lemma 1.4 in their supplement by Theorem 2 in our paper.
$\square$

References

1. Lauritzen, S.L. *Graphical Models*; Oxford University Press: Oxford, UK, 1996.
2. Maathuis, M.; Drton, M.; Lauritzen, S.; Wainwright, M. *Handbook of Graphical Models*; CRC Press: Boca Raton, FL, USA, 2019.
3. Spirtes, P.; Glymour, C.; Scheines, R. *Causation, Prediction, and Search*, 2nd ed; The MIT Press: Cambridge, MA, USA, 2000.
4. Colombo, D.; Maathuis, M.H.; Kalisch, M.; Richardson, T.S. Learning high-dimensional directed acyclic graphs with latent and selection variables. *Ann. Stat.* **2012**, *40*, 294–321. [CrossRef]
5. Spirtes, P. An anytime algorithm for causal inference. In Proceedings of the 8th International Workshop on Artificial Intelligence and Statistics, Key West, FL, USA, 3–6 January 2001; pp. 213–221.
6. Zhang, J. On the completeness of orientation rules for causal discovery in the presence of latent confounders and selection bias. *Artif. Intell.* **2008**, *172*, 1873–1896. [CrossRef]
7. Colombo, D.; Maathuis, M.H. Order-independent constraint-based causal structure learning. *J. Mach. Learn. Res.* **2014**, *15*, 3921–3962.
8. Kalisch, M.; Bühlmann, P. Estimating High-Dimensional Directed Acyclic Graphs with the PC-Algorithm. *J. Mach. Learn. Res.* **2007**, *8*, 613–636.
9. Loh, P-L.; Bühlmann, P. High-Dimensional Learning of Linear Causal Networks via Inverse Covariance Estimation. *J. Mach. Learn. Res.* **2014**, *15*, 3065–3105.
10. Voorman, A.; Shojaie, A.; Witten, D. Graph estimation with joint additive models. *Biometrika* **2014**, *99*, 1–25. [CrossRef]
11. Harris, N.; Drton, M. PC Algorithm for Nonparanormal Graphical Models. *J. Mach. Learn. Res.* **2013**, *14*, 3365–3383.
12. Sun, X.; Janzing, D.; Schölkopf, B.; Fukumizu, K. A kernel-based causal learning algorithm. In Proceedings of the 24th International Conference on Machine Learning, Corvalis, OR, USA, 20–24 June 2007; pp. 855–862.
13. Zhang, K.; Peters, J.; Janzing, D.; Schölkopf, B. Kernel-based conditional independence test and application in causal discovery. *arXiv* **2012**, arXiv:1202.3775.
14. Székely, G.J.; Rizzo, M.L.; Bakirov, N.K. Measuring and testing independence by correlation of distances. *Ann. Stat.* **2007**, *35*, 2769–2794. [CrossRef]
15. Wang, X.; Wenliang, P.; Hu, W.; Tian, Y.; Zhang, H. Conditional distance correlation. *J. Am. Stat. Assoc.* **2015**, *110*, 1726–1734. [CrossRef]
16. Pearl, J. *Causality*; Cambridge University Press: Cambridge, UK, 2000.
17. Verma, T.; Pearl, J. Equivalence and synthesis of causal models. In Proceedings of the Sixth Annual Conference on Uncertainty in Artificial Intelligence, Cambridge, MA, USA, 27–29 July 1990; pp. 255–270.
18. Richardson, T.S.; Spirtes, P. Ancestral graph markov models. *Ann. Stat.* **2002**, *30*, 962–1030. [CrossRef]
19. Ali, R.A.; Richardson, T.S.; Spirtes, P. Markov equivalence for ancestral graphs. *Ann. Stat.* **2009**, *37*, 2808–2837. [CrossRef]
20. Székely, G.J.; Rizzo, M.L. Partial distance correlation with methods for dissimilarities. *Ann. Stat.* **2014**, *42*, 2382–2412. [CrossRef]
21. Sheng, T.; Sriperumbudur, B.K. On distance and kernel measures of conditional independence. *arXiv* **2019**, arXiv:1912.01103.
22. Chakraborty, S.; Zhang, X. Distance Metrics for Measuring Joint Dependence with Application to Causal Inference. *J. Am. Stat. Assoc.* **2019**, *114*, 1638–1650. [CrossRef]
23. Liu, J.; Li, R.; Wu, R. Feature selection for varying coefficient models with ultrahigh-dimensional covariates. *J. Am. Stat. Assoc.* **2014**, *109*, 266–274. [CrossRef]

24. Uhler, C.; Raskutti, G.; Bühlmann, P.; Yu, B. Geometry of the faithfulness assumption in causal inference. *Ann. Stat.* **2013**, *41*, 436–463. [CrossRef]
25. Tsamardinos, I.; Brown, L.E.; Aliferis, C.F. The max-min hill-climbing Bayesian network structure learning algorithm. *Mach. Learn.* **2006**, *65*, 31–78. [CrossRef]
26. Shojaie, A. Differential network analysis: A statistical perspective. In *Wiley Interdisciplinary Reviews: Computational Statistics*; Wiley: New York, NY, USA, 2021; p. e1508.
27. Meinshausen, N.; Bühlmann, P. Stability selection. *J. R. Stat. Soc.* **2010**, *72*, 417–473. [CrossRef]
28. Shah, R.D.; Samworth, R.J. Variable selection with error control: Another look at stability selection. *J. R. Stat. Soc.* **2013**, *75*, 55–80. [CrossRef]
29. Sondhi, A.; Shojaie, A. The Reduced PC-Algorithm: Improved Causal Structure Learning in Large Random Networks. *J. Mach. Learn. Res.* **2019**, *20*, 1–31.
30. Anandkumar, A.; Tan, V.Y.F.; Huang, F.; Willsky, A.S. High-Dimensional Gaussian Graphical Model Selection: Walk Summability and Local Separation Criterion. *J. Mach. Learn. Res.* **2012**, *13*, 2293–2337.
31. Chen, W.; Drton, M.; Shojaie, A. Causal structural learning via local graphs. *arXiv* **2021**, arXiv:2107.03597.
32. Serfling, R. J. *Approximation Theorems of Mathematical Statistics*; Wiley: New York, NY, USA, 1980.
33. Li, R.; Zhong, W.; Zhu, L. Feature selection via distance correlation learning. *J. Am. Stat. Assoc.* **2012**, *107*, 1129–1139. [CrossRef] [PubMed]
34. Resnick, S. I. *A Probability Path*; Springer: Berlin/Heidelberg, Germany, 1999.
35. Wen, C.; Wenliang, P.; Huang, M.; Wang, X. Sure Independence Screening Adjusted for Confounding Covariates with Ultrahigh Dimensional Data. *Stat. Sin.* **2018**, *28*, 293–317.

Article

Transfer-Learning-Based Approach for the Diagnosis of Lung Diseases from Chest X-ray Images

Rong Fan [1] and Shengrong Bu [2],*

[1] School of Engineering and Applied Science, University of Pennsylvania, Philadelphia, PA 19104, USA; rongfan@seas.upenn.edu
[2] Department of Engineering, Brock University, St. Catharines, ON L2S 3A1, Canada
* Correspondence: sbu@brocku.ca

Abstract: Using chest X-ray images is one of the least expensive and easiest ways to diagnose patients who suffer from lung diseases such as pneumonia and bronchitis. Inspired by existing work, a deep learning model is proposed to classify chest X-ray images into 14 lung-related pathological conditions. However, small datasets are not sufficient to train the deep learning model. Two methods were used to tackle this: (1) transfer learning based on two pretrained neural networks, DenseNet and ResNet, was employed; (2) data were preprocessed, including checking data leakage, handling class imbalance, and performing data augmentation, before feeding the neural network. The proposed model was evaluated according to the classification accuracy and receiver operating characteristic (ROC) curves, as well as visualized by class activation maps. DenseNet121 and ResNet50 were used in the simulations, and the results showed that the model trained by DenseNet121 had better accuracy than that trained by ResNet50.

Keywords: transfer learning; deep learning; pretrained neural networks; chest X-ray images; lung diseases

Citation: Fan, R.; Bu, S. Transfer-Learning-Based Approach for the Diagnosis of Lung Diseases from Chest X-ray Images. *Entropy* **2022**, *24*, 313. https://doi.org/10.3390/e24030313

Academic Editors: S. Ejaz Ahmed and Farouk Nathoo

Received: 12 January 2022
Accepted: 15 February 2022
Published: 22 February 2022

Publisher's Note: MDPI stays neutral with regard to jurisdictional claims in published maps and institutional affiliations.

1. Introduction

Many people suffer from lung diseases such as pneumonia and emphysema every year. Chest X-ray images are one of the most widely used and low-cost diagnose tools for lung diseases [1]. However, since there might be more than one pathology to be detected from chest X-rays for a disease [2], diagnosing by doctors could be challenging sometimes. Computer-aided diagnosis for various diseases has been researched to improve the efficiency and accuracy of the diagnosis [3]. Various deep learning methods [4] for medical image classification have the potential of predicting and diagnosing diseases even more accurately than the average radiologist [5].

Since the global corona virus pandemic, researchers have developed methods to analyze radiographic chest images more efficiently to make the diagnosis of COVID-19 easier. Heidari et al. developed a novel deep learning model to detect non-pneumonia, non-COVID-19-infected pneumonia and COVID-19-infected pneumonia [6]. In [7], the authors presented a deep learning approach to realize the diagnosis of pulmonary hypertension by analyzing chest radiographs and compared the performance of ResNet50, Xception, and Inception V3. Yu et al. built a multi-task deep learning network consisting of an extraction architecture and three different routes for various functions by using chest X-rays from peripherally inserted central catheters [8]. Jaiswal et al. realized the localization and identification of pneumonia in chest X-ray images using a deep learning model derived from mask-RCNN [9]. In [5], a modified AlexNet with many handcrafted features was proposed to detect whether the chest X-ray images were in the normal or in the pneumonia class.

However, the medical image dataset could be too small to be used to train a neural network since the images have to be labeled by professionals. Transfer learning originated from terms such as knowledge transfer or inductive transfer in 1995 [10], and later, in 2005, it was defined as the technique of applying knowledge and skills learned in previous tasks to novel tasks [11]. Since then, many studies have employed transfer learning on small medical datasets and trained neural networks to realize image recognition and classification. Minaee et al. applied transfer learning to process chest X-ray images for the detection of COVID-19, and DenseNet121, ResNet18, ResNet50, and SqueezeNet were utilized as the pre-trained networks [12]. In [13], the advantages and challenges of deep transfer learning were studied. Ravishankar et al. realized ultrasound kidney images' detection using transfer learning [14]. A deep convolutional neural network (DCNN) was proposed to study the advantages of transfer learning in medicine [15]. Subspace-based techniques, such as in [16], can be used together with transfer learning to increase the accuracy when the dataset is small.

Class imbalance is a common challenging related to medical image diagnosis [17], since the amount of positive data and negative ones in each class might not be equivalent. In this kind of application, the rare or minor occurrences are much more important than the majority classes [18]. As a result, the contributions of the loss for these two kinds of data are not the same, and the small data size of some class will affect the overall training performance. Various methods could be used to handle imbalanced datasets, including setting appropriate class weights for the model and random under-sampling and over-sampling.

In this paper, a transfer learning method is proposed to classify 14 lung-related pathologies using frontal-view chest X-ray images. The contributions of this paper are as follows:

- We built image classification models using pretrained networks;
- We preprocessed the data including data augmentation of the ChestX-ray8 dataset and dealt with the class imbalance problem;
- We trained, validated, and tested the model using pretrained networks and compared the performance of each model using the ROC curves. We visualized the classification decision using Grad-CAM.

The structure of this paper is as follows. The methods and principles with respect to transfer learning, data augmentation, evaluation, and visualization are presented in Section 2. Section 3 then presents the experimental process and results. Finally, the conclusion of this paper is drawn in Section 4.

2. Proposed Transfer Learning Method

In our work, transfer learning was used for the chest X-ray image classification task. Transfer learning is an effective method in the image processing domain that can take advantage of well-developed models to solve new tasks [19]. There are two main ways to utilize pretrained networks in transfer learning: First, a pretrained model can be used as the feature extractor for the new dataset. Once the features are extracted, added layers such as a linear classifier can be trained for the new task. Second, the whole or some part of the pretrained network will be fine-tuned for the new classification task. Thus, the weights of the pretrained model are considered as the initial values and will be updated during the training process. In our work, the first method was used since the dataset was small and the computing power was limited. Two networks, i.e., DenseNet121 and ResNet50, were used as the base models for transfer learning. In the following, the principle of transfer learning, the framework of the networks, and the measures for the evaluation are discussed.

2.1. Transfer Learning with a Data Augmentation Approach

Two pretrained networks were employed as the training models in this project. The first one is called ResNet50, which won the first prize in the 2015 ImageNet competition. This model uses a shortcut connection, which is the basis of a residual network, and the connection ensures that the feature of one preceding layer is the input of the later layers, skipping some of the layers. Therefore, any layer in this framework has information from the preceding layers. The design overcomes the problem of learning rate reduction and invariant classification accuracy as a result of a deeper network. The second one is DenseNet121, which was the winner of the 2017 ImageNet competition and has been widely applied in deep learning. DenseNet consists of DenseBlock layers, each of which receives additional inputs from all preceding layers and transition layers. Additional inputs from all preceding layers together with the feature maps of the current layer are all passed on to other subsequent layers, and thus, the shortcuts of all the former layers and the latter layer are built densely. For comparison, the traditional CNN with l layers has l connections between adjacent layers, whereas DenseNet has $l(l+1)/2$ layers in total because of its shortcut feature [6]. Thus, the learned features could be reused and the network has less channels as a result of the collective knowledge feature of each layer. Besides, this also leads to better performance under the conditions of fewer parameters and little computing cost. It also has some other advantages such as vanishing gradient problem mitigation and parameter reduction. In contrast, since ResNet only has shortcuts between the former layer and the latter layer, and DenseNet has demonstrated better performance. Due to the aforementioned reasons, DenseNet is much deeper than ResNet and has more than 100 layers, and the training process could be more effective and the accuracy improved.

One basic problem of deep learning is the opposition of optimization and generalization [20]. Optimization is the learning process that adjusts the model to obtain the best performance, while generalization is the performance of the model on the testing of new data. The goal of learning is to realize a satisfactory generation, but this cannot be controlled, so the models are always adjusted based on the training data. When the training process begins, the generalization can become worse after a number of iterations, which means the model is overfitting, and this is a common problem in training neural networks. Among various methods used to prevent the neural networks from overfitting, data augmentation is the most effective one and is widely used in computer vision, especially when the dataset is small. In *Keras*, data augmentation can be realized by using the *ImageDataGenerator* class and transforming the image parameters randomly. Some commonly adjusted parameters include the following: *rotation_range* is the rotation range of the image; *width_shift* and *height_shift* are the range of shifting in the horizontal and vertical direction, respectively; *horizontal_flip* is the flip ratio; *sheer_range* is the random sheer angle of the image.

2.2. Evaluation Methods

The performance of the network needs to be evaluated after testing. Accuracy and receiver operating characteristic (ROC) curves with the AUCROC were used as the metrics for the evaluation. Accuracy shows the general performance of all testing images, and the ROC curves with the AUCROC indicate the classification performance for each label.

The classification task in our project was a multi-task classification because one image might correspond to more than one pathological condition. Therefore, the *Accuracy* can be calculated as follows, since there are 14 pathological conditions:

$$Accuracy = \frac{sum of\ truly\ predicted\ labels}{14 * (\#\ of\ testing\ images)} \tag{1}$$

The accurately predicted labels for all *testing images* were considered together instead of calculating the accuracy of each image and then averaging them. The *sum of the truly predicted labels* was calculated by first finding the number of truly predicted images for each label and adding them together.

An ROC curve is a classification evaluation tool in deep learning. In real-world applications, some datasets have the problem of class imbalance. For example, a common case is that the number of negative images is larger than that of the positive images for medical datasets. A stable evaluation curve could be achieved by using the ROC curve. To summarize, the ROC curve has the following features: First, the curve can be used to check the impact of a specific threshold value on the generalization ability of a classifier. Second, the ROC can help determine the best threshold value, since the closer it is to the upper-left corner, the better the classifier is. Third, the ROC is a good tool to compare the performance of many different classifiers for each class intuitively.

In the figure of a typical ROC curve, the horizontal coordinate, i.e., false positive rate (*FPR*), and the vertical coordinate, i.e., true positive rate (*TPR*), are defined as follows:

$$TPR = \frac{TP}{P} = \frac{TP}{TP + FN}.$$ (2)

$$FPR = \frac{FP}{N} = \frac{FP}{TN + FP},$$ (3)

where P is the number of real positive samples and N is the number of real negative samples. TP means true positive, which is the positive samples that are predicted positively by the model. FP mans false positive, which is the negative samples that are predicted positively by the model; FN means false negative, which is the positive samples that are predicted negatively by the model; TN means true negative, which is the negative samples that are predicted negatively by the model. For a specified classifier, a pair of TPR and FPR points can be obtained according to the testing performance. As a result, this classifier can be mapped into a point on the ROC plain. The area under ROC curve (AUCROC) is used to quantify the classification ability, and a larger AUCROC indicates better classification performance.

There are three methods to calculate the AUC manually, the namely trapezoidal rule, the Mann–Whitney statistics [21], and the parameter rule. The first method uses the vertical line of each point on the x-axis and calculates the sum of small trapezoidal areas. The second method is proper for medical images, because it calculates the value of the possibility that positive samples are larger than the negative samples. The third method uses the mean and variance value when the samples obey a Gaussian distribution. In our work, these two functions *roc_auc_score*, *roc_curve* can be used by directly importing them from the *sklearn.metrics* library. After the AUC value is calculated, the performance of the classifier can be analyzed: (1) If $AUC = 1$, the classifier is perfect. (2) If $0.5 < AUC < 1$, the performance is better than guessing randomly. If a proper threshold value is set, the classifier can predict most of the cases correctly. (3) If $AUC = 0.5$, the process of prediction is the same as a random guess, and there is no prediction value. (4) If $AUC < 0.5$, it is worse than guessing. However, if predicting inversely, it is similar to the second case.

2.3. Visualization Using Class Activation Maps

Visualization of neural networks increases the interpretability of the networks in the field of computer vision. The complexity of medical images always makes the visualization harder. In our work, the class activation map (CAM) was used for visualization. The basic principle of the CAM is that it will produce a heat map of the input images, indicating the degree of similarity between the real class and the predicted class. Specifically, the technique used in this work was gradient-weighted class activation mapping (Grad-CAM) [22]. This method generates a localization map with the significant parts of the image highlighted by extracting the gradient of the classification target and letting the gradient flow into the last layer.

A convolutional neural network normally consists of a feature extractor, which is used to extract useful features, and a classifier, which classifies according to the extracted features. There are two kinds of classification models. One is feature extraction with flatten and softmax layers: A flatten layer is used to transform the three-dimensional images into one-dimensional vectors. A dense layer will then be added, and finally, there is a softmax function as the activation function for the output. The other is feature extraction with global average pooling (GAP) and softmax, where a global average pooling layer is used to substitute the flatten layer: this has the advantages of reducing the number of parameters, making the training process easy and preventing from overfitting. Based on the classification model, the CAM is generated.

For a traditional CNN model that has a flatten layer, if the last layer of the CNN has n feature maps, which means there are n weights for a neuron in the classifier layer and each neuron relates to a class, then the class activation map [22] for class c can be calculated as follows:

$$L^c_{CAM} = \sum_{i=1}^{n} w^c_i A^i,$$

(4)

where the weights for the ith neuron are: $w^i_1, w^i_2, w^i_3, \cdots, w^i_n$, and A^i indicates the feature maps in the last layer. If a GAP is used to substitute the flatten layer, the classification score of class c [22] can be calculated as follows:

$$S_c = \sum_{i=1}^{n} w^c_i GAP(A^i) = \frac{1}{Z} \sum_{i=1}^{n} \sum_{k=1}^{c1} \sum_{j=1}^{c2} A^i_{kj} w^c_i,$$

(5)

where w^c_i is the weight for the GAP and the size of a feature map is $Z = c1 * c2$. The value of S_c is determined by the pixel value A^i_{kj} and weights w^c_i. If the multiplication of the pixel value and weights is larger than 1, the sample will be classified into this current class c, and the model considers the original image as related to this class. This equation helps decide which part of the original image corresponds to a specific pixel.

CAMs are a very powerful tool for the visualization of the neural network's decision-making process. However, they have certain limitations: (1) We can apply CAMs only if the CNN contains a GAP layer; (2) heat maps can be generated only for the last convolutional layer. To address these issues, gradient-weighted class activation mapping (Grad-CAM) is proposed. The class activation mapping for class c [22] can be generated by:

$$L^c_{Grad-CAM} = \frac{1}{Z} \sum_{i=1}^{n} \sum_{k=1}^{c1} \sum_{j=1}^{c2} \frac{\partial S_c}{\partial A^i_{kj}} A^i.$$

(6)

Grad-CAM is the generalization of the CAM, and the gradient operator indicates the backpropagation. Grad-CAM was employed in our work due to its advantages. The code implementation included the following steps: (1) The output of the batch normalization (BN) layer [23] and the output of the whole network were extracted. (2) Backpropagation was computed from the output of the whole network to the output of the BN layer by using function *gradients* in *TensorFlow* to calculate the gradient automatically. (3) We used the gradients as the weights and multiplied them with the output of the BN layer. (4) Function *resize* in the *OpenCV* library was used to compound the feature maps to visualize.

3. Simulation Results

Our simulation process can be divided into three parts: (1) The raw data need to be preprocessed, including checking the data leakage, handling the class imbalance, performing the data augmentation, and generating new images. (2) The training process was conducted. (3) The testing and evaluation results showed the generalization ability of the model. Simulations were conducted on a GPU-equipped computer, using *TensorFlow* and *Keras*.

3.1. Data Preprocessing

The data used in our work were frontal-view chest X-ray images from patients. The whole dataset was obtained from https://nihcc.app.box.com/v/ChestXray-NIHCC (accessed on 10 Febuary 2021). Each image in the dataset includes 14 labels for 14 pathological conditions, such as consolidation, effusion, edema, atelectasis and so on. For each label, 1 means positive and 0 means negative. After classification, the pathological conditions can be utilized by physicians to detect eight different diseases. The original datasets were divided into three groups for training, validation, and testing, respectively.

Data leakage is a common problem for processing medical images, because one patient may have multiple images. Data leakage will lead to the overfitting problem, since it is difficult for the model to learn from similar features and to predict other new features. To ensure that there is no data leakage between any two datasets, the datasets should not contain the images from the same patient. The identification of unique patents of each set was collected by using the *set* function in Python, and then, the *intersection* function was used to check whether the two datasets contain information from the same patient.

Neural networks can only process the data in the format of float tensor. Therefore, formatting is important, since the original dataset contains images in PNG files. In *Keras*, there is a class named *ImageDataGenerator*, which can be used to finish the following tasks in sequence: read image files; encode the PNG files into RGB pixels; transform these pixels into a float tensor; scale the pixels in the range of [0,1]. Then, three generators are defined to load the images into the network. Several parameters can be set to proper values in *ImageDataGenerator*:

- Batch size. The batch size, the number of samples for one training, influences the optimization degree and speed. Since the network was trained on a GPU (2×Tesla V100)-equipped machine, $batch\_size = 16$ matches the GPU's performance;
- Resolution. The original images provided in [24] have a size of 1024 × 1024, which is relatively too big to be processed. With the help of a Python generator in Keras, the images were scaled to 400 × 400, the value of which was chosen to balance the accuracy and learning speed.

The data augmentation module was added to the generator, which means that the data were already augmented before feeding them into the neural network. In order to compare the image before and after data augmentation, the first image of the dataset is shown in Figure 1 by using the *plt.imshow* function. As shown in Figure 2, the image was shifted and zoomed after augmentation.

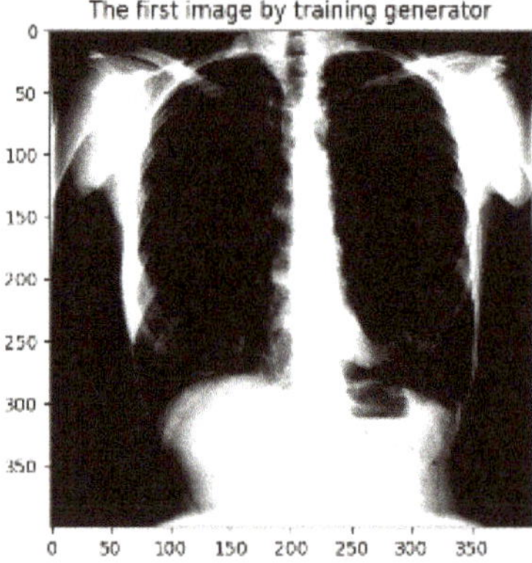

Figure 1. A chest X-ray image.

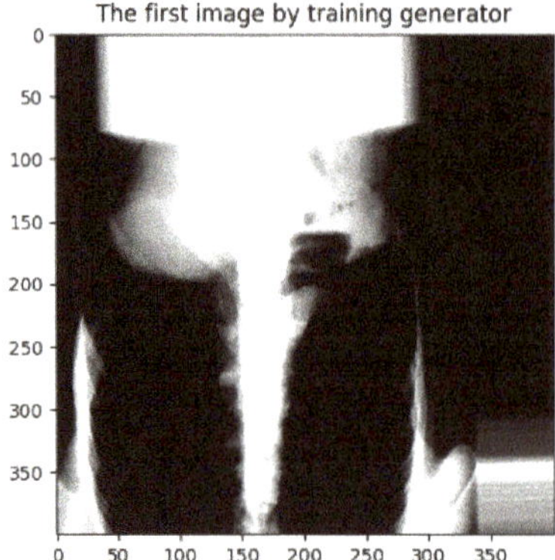

Figure 2. A chest X-ray image with data augmentation.

Class imbalance was handled by calculating the weight loss as the loss function. Specifically, for each label, the loss was weighted by the frequency of positive data (w_p) and that of negative data (w_n) as shown below:

$$
L(X,y) = \begin{cases} w_p * (-log(Y = 1|X)), & \text{if } y = 1, \\ w_n * (-log(Y = 1|X)), & \text{if } y = 0, \end{cases}
\tag{7}
$$

where Y stands for predication and X means input labels.

3.2. Training

The pretrained network was used as the base model. A global pooling layer was added using function *GlobalAveragePooling2D*, and a fully connected layer was placed as the output layer by employing the Dense function with *Softmax* activation. In our work, the aim was to realize the classification of 14 pathological conditions, which is a multi-task classification problem. In this scenario, the effective activation was *Softmax*. The final output of the model is called the prediction, which is a 14-length vector with each element indicating the probability of a certain pathological condition. In order to compile the whole model, function *compile* was used, and several related parameters were set. For example, compiling the model required the type of loss function and the optimizer. The weighted loss was considered as the loss function, since the class imbalance problem was handled by the weighted loss. *Adam* was used as the optimizer since it has better performance than the traditional optimizers, such as the *Momentum* and *RMSprop* optimizers. Since "accuracy" was used as the metric, the accuracy of each training step and each validation step was displayed while running the code.

After all the preparations were completed, the network was trained by using training labels and images. The goal of the training was optimization, which means the model itself builds the connection of the output and output and learns the features. By using the *fit_generator* function, the model first fits the data to realize training and then performs the validation. Some parameters are important for the training and/or validation process:

- Steps per epoch means the number of steps for each epoch. Data in its batch size were the input from the generator to the network for each step. The relationship between this parameter and the batch size was (# steps per epoch) × batch size = # total training samples. Since the batch size was set to 16 because of the GPU performance and the total samples for training were 402, the steps per epoch should be 25;
- The value of the validation steps needs to be assigned, after the steps per epoch are determined. The validation steps were the total number of steps in the validation dataset. The validation steps should be two, since there were forty images for validation, and (# validation steps) × batch size = # total validation samples;
- The value of the epoch decides the total number of training samples. In each epoch, the network learns the features from all of the input images. In this work, the epoch was set to 80. The reason was that the plots with 80 epochs could clearly show the variation tendency of the accuracy and loss, and also, overfitting might occur if the network is trained for too many epochs. Early stopping was also used by stopping training if the accuracy did not increase for 10 epochs, which can help mitigate the overfitting problem to some extent.

When each epoch of training was finished, the weights of the current trained network were saved in a weight file, by calling the *model.save* function. The later training was based on the formerly saved weights.

The next step was to plot the loss curve for training and validation, which is useful for observing network convergence and the overfitting problem. Function *Matplotlib* in *Keras* was used for plotting. After all training and validation epochs, the loss for each epoch can be retrieved by calling the *history* function.

The training loss and validation loss of DenseNet121 without DA, DenseNet121 with DA, and ResNet50 with DA as the base model are shown in Figure 3. The results without

DA and with DA were firstly compared. Figure 3a,b shows that both of the losses with or without DA for the training decreased from one to nearly zero with the increase of the epoch, while those for validation increased from one to almost five, which means that the model was overfit. Ideally, training loss and validation loss should have the same trends, if the model is well fit. The figures also demonstrate that the model with DA had better performance than that without DA. The figures show that the model with DA learned the model slower than that without DA, since more images needed to be fed into the the network after data augmentation. The curves for DenseNet121 without DA fluctuated more than those with DA. The loss curves by using ResNet50 as the base model with DA are also presented. Compared with DenseNet121, ResNet50 took more time to train because the training loss converged at around the 70th epoch.

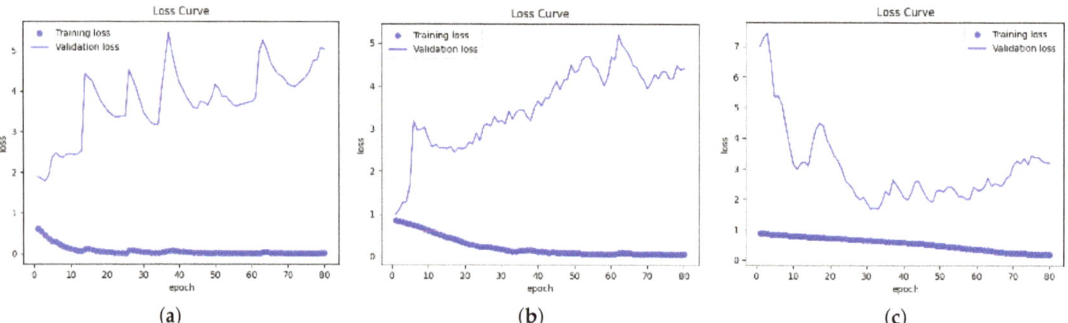

Figure 3. Loss curves for xx with/without DA. (**a**) DenseNet121 without DA. (**b**) DenseNet121 with DA. (**c**) ResNet50 with DA.

The training accuracy of using these three models is shown in Table 1. DenseNet121 with DA had the highest training accuracy, followed by DenseNet121 without DA and then ResNet50. The reason was that the dataset became larger and more diversified after DA, and thus, the network was trained to be optimal. ResNet50 had the lowest training accuracy, since there were fewer shortcut connections inside of the base model, and consequently, the learning ability was poorer.

Table 1. Training accuracy for different networks.

Networks	Type of Data Processing	Training Accuracy
DenseNet121	Without data augmentation	0.89
	With data augmentation	0.92
ResNet50	With data augmentation	0.84

3.3. Testing and Evaluation

All the testing images were fed into the model, and the prediction results could be obtained. To test the network, function *predict_generator* was used as the major function. The output of this function was a list, which included the probability of classification for each label. When this probability was larger than the threshold value of 0.5, the program considered the prediction as correct. After comparing the prediction results with the real label of each image, the generalization ability of the model could be known with the self-defined function to calculate the testing accuracy. The classification accuracy for testing the datasets using DenseNet121 without DA, DenseNet121 with DA, and ResNet50 with DA is shown in Table 2. This table shows that DenseNet121 had better performance than ResNet50, and DA was beneficial for improving the classification accuracy.

Table 2. Testing accuracy for different networks.

Networks	Type of Data Processing	Testing Accuracy
DenseNet121	Without data augmentation	0.82
	With data augmentation	0.84
ResNet50	With data augmentation	0.76

In order to evaluate the model, the receiver operating characteristic (ROC) curves were generated, and the area under the curve (AUC) was calculated. *Keras* has a library, *sklearn*, which can conduct some advanced computations in machine learning and computer vision. For the evaluation, functions *roc_auc_score* and *roc_curve* were imported from the library to calculate the AUCROC and to derive the ROC curve. Figure 4 illustrates the ROC curves and the AUCROC values of DenseNet121 without DA for the 14 pathological conditions. The horizontal axis indicates the false positive rate, while the vertical axis indicates the true positive rate. The AUCROC score for each class is listed at the lower-right corner of this figure, e.g., for cardiomegaly, the AUCROC was 0.51, which means that the area under curve for the label was 0.51. The figure shows that the ROC curves for several pathologies lie below the straight line that passes through points (0,0) and (1,1). For these pathologies, the classifier worked even worse than random guessing. The AUCROC values of five pathologies, i.e., emphysema, infiltration, pneumothorax, pleural thickening, and pneumonia, were all less than 0.5, which means that the classifier could not diagnose most of the images in these classes correctly. Therefore, this figure indicates that the classification ability of DenseNet121 without DA was relatively poor.

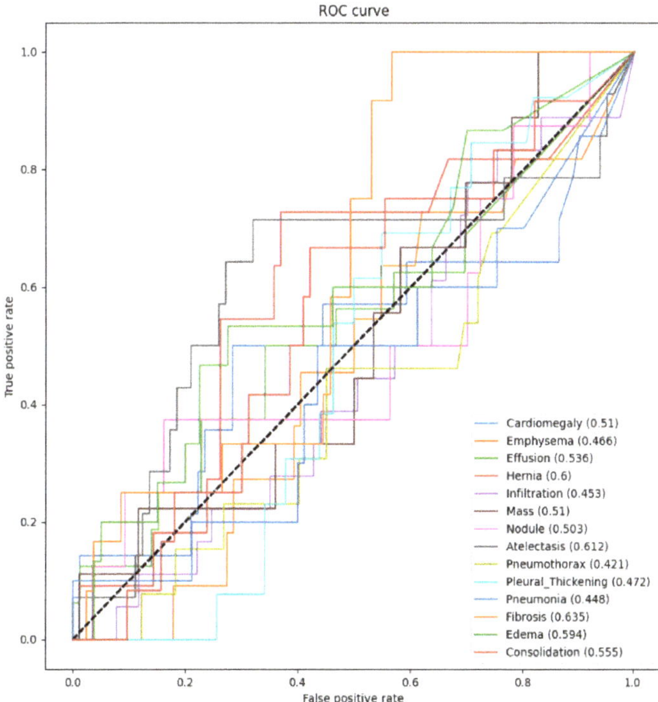

Figure 4. The ROC and AUCROC for DenseNet121 without DA.

Figure 5 illustrates the ROC curves and the AUCROC values of DenseNet121 with DA for the 14 pathological conditions. This figure shows that most of these ROC curves are located above the dotted line that passes through points (0,0) and (1,1), and all of the AUCROC values are larger than 0.5. The reason was that the images were preprocessed with DA, which led to a better-trained network. For fibrosis, the ROC curve lies significantly higher than the other curves and is mostly close to the upper-left corner, and its AUCROC was the largest with a value of 0.775, which means that its classifier had the best performance among all 14 classifiers. For nodule and infiltration, their AUCROC values were just slightly larger than 0.5, which means that these classifiers could help predict these pathological conditions, but the performance was relatively poor.

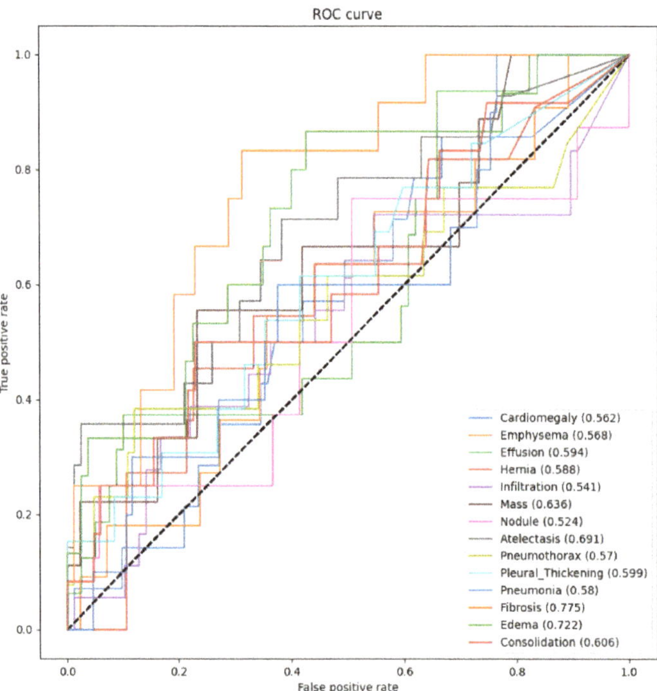

Figure 5. The ROC and AUCROC for DenseNet121 with DA.

Figure 6 illustrates the ROC curves and the AUCROC values of ResNet50 with DA for the 14 pathological conditions. Compared to Figure 5, more ROC curves using ResNet50 with DA lie below the straight dotted line that passes through points (0,0) and (1,1) than those using DenseNet121 with DA. The largest AUCROC value was for fibrosis, with the value of 0.68, which was smaller than that of using DenseNet121 with DA. The AUCROC values of three classes, i.e., emphysema, pneumothorax, and pneumonia, were smaller than 0.5, which means that these classifiers could not help predict these pathological conditions.

The comparison of the ROC curves and AUCROC values for different networks demonstrated that the classifiers trained by DenseNet121 had better performance than those trained by ResNet50. The results also indicated that DA improved the classification capability for all of the classes. Most of the ROC curves lie above the straight dotted line that passes through points (0,0) and (1,1), but they are not close to the upper-left corner enough, because the dataset used for testing was relatively small.

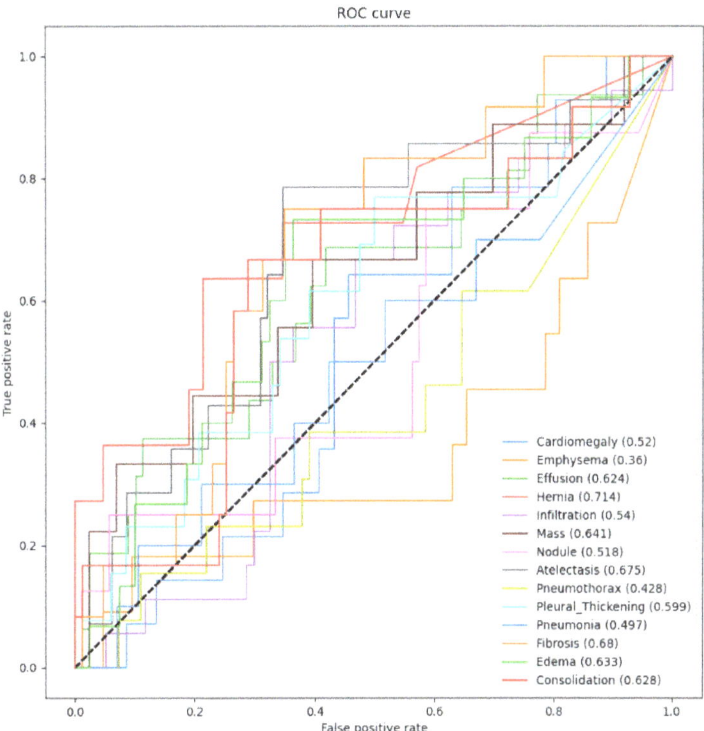

Figure 6. The ROC and AUCROC for ResNet50 with DA.

3.4. Visualization

The visual explanation of classification decision-making was produced by using Grad-CAM techniques. The heat maps of using DenseNet121 as the base model are shown in Figures 7 and 8. These chest X-rays were randomly selected from the datasets, and only the four most probable diagnosis heat maps are shown in the figure. The probability of diagnosing a certain pathological condition is demonstrated in each of the subfigures. For example, in Figure 7, the original chest X-ray image is shown in the first subfigure. The second and third subfigures indicate that it is impossible for the image to be classified as cardiomegaly or hernia. The fourth and fifth subfigures mean that the image has a probability of 0.763 and 0.593 to be diagnosed as nodule and edema, respectively. Figure 8 shows that the original image has the possibility of being diagnosed into four pathological conditions, and the most probable one is nodule with a probability of 0.822.

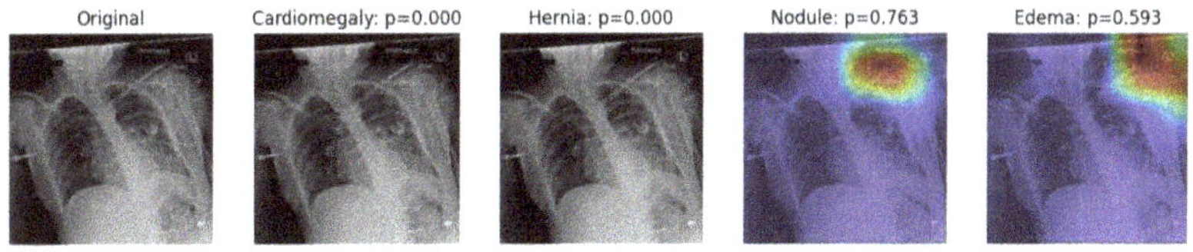

Figure 7. Visualization of the diagnosis heat maps of one image example by the use of Grad-CAM.

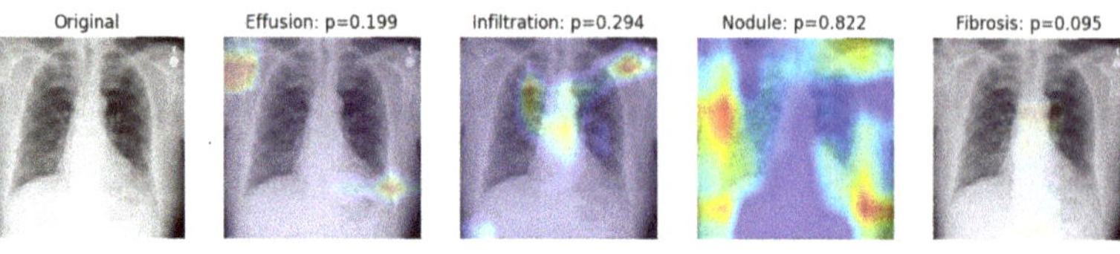

Figure 8. Visualization of the diagnosis heat maps of the second example by the use of Grad-CAM.

4. Conclusions

A deep learning approach was proposed to use transfer learning and pretrained networks to recognize and classify chest X-ray images into 14 pathological conditions, and therefore help with diagnosing diseases related to these pathological conditions. The performance of the two adopted pretrained networks DenseNet121 and ResNet50 was compared, and DA was also used to further improve the performance. Evaluation metrics, such as the accuracy, ROC curves, and AUCROC curves were utilized. The simulation results showed that the network using DenseNet121 as the base model with DA had a better generalization ability on the testing datasets. In the future, multiple transfer learning methods could be used together with ensemble classifiers to further improve the performance of the proposed work. The potential use of the other datasets, such as PadChest, ChexPpert, and MIMIC-CXR, will be explored in our future work.

Author Contributions: Formal analysis, R.F.; Funding acquisition, S.B.; Investigation, R.F.; Methodology, R.F.; Software, R.F.; Supervision, S.B.; Validation, R.F.; Writing—original draft, R.F.; Writing—review & editing, S.B. All authors have read and agreed to the published version of the manuscript.

Funding: This work was supported by start-up funds provided by Brock University.

Institutional Review Board Statement: Not applicable.

Informed Consent Statement: Not applicable.

Data Availability Statement: The data analyzed in this study are openly available at https://nihcc.app.box.com/v/ChestXray-NIHCC accessed on 10 Febuary 2021.

Conflicts of Interest: The authors declare no conflict of interest.

References

1. Fancourt, N.; Knoll, M.; Barger-Kamate, B.; de Campo, J.; Diallo, M.; Ebruke, B.E.; Feikin, D.; Gleeson, F.; Gong, W.; Hammitt, L.; et al. Standardized Interpretation of Chest Radiographs in Cases of Pediatric Pneumonia From the PERCH Study. *Clin. Infect. Dis.* **2017**, *64*, S253–S261. [CrossRef] [PubMed]
2. Popper, H. *Pathology of Lung Disease: Morphology—Pathogenesis—Etiology*; Springer International Publishing: Berlin/Heidelberg, Germany, 2021.
3. Liu, N.; Wan, L.; Zhang, Y.; Zhou, T.; Huo, H.; Fang, T. Exploiting Convolutional Neural Networks with Deeply Local Description for Remote Sensing Image Classification. *IEEE Access* **2018**, *6*, 11215–11228. [CrossRef]
4. Yin, X.; Han, J.; Yang, J.; Yu, P.S. Efficient classification across multiple database relations: A CrossMine approach. *IEEE Trans. Knowl. Data Eng.* **2006**, *18*, 770–783. [CrossRef]
5. Bhandary, A.; Prabhu, G.A.; Rajinikanth, V.; Thanaraj, K.P.; Satapathy, S.C.; Robbins, D.E.; Shasky, C.; Zhang, Y.; Tavares, J.M.; Raja, N.S.M. Deep-learning framework to detect lung abnormality—A study with chest X-Ray and lung CT scan images. *Pattern Recognit. Lett.* **2020**, *129*, 271–278. [CrossRef]
6. Heidari, M.; Mirniaharikandehei, S.; Khuzani, A.Z.; Danala, G.; Qiu, Y.; Zheng, B. Improving the performance of CNN to predict the likelihood of COVID-19 using chest X-ray images with preprocessing algorithms. *Int. J. Med. Inform.* **2020**, *144*, 104284. [CrossRef] [PubMed]
7. Zou, X.L.; Ren, Y.; Feng, D.; Guo, Y.; Yang, H.; Li, X.; Fang, J.; Li, Q.; Ye, J.; Han, L.; et al. A promising approach for screening pulmonary hypertension based on frontal chest radiographs using deep learning: A retrospective study. *PLoS ONE* **2020**, *15*, e0236378. [CrossRef] [PubMed]
8. Yu, D.; Zhang, K.; Huang, L.; Zhao, B.; Zhang, X.; Guo, X.; Li, M.; Gu, Z.; Fu, G.; Hu, M.; et al. Detection of peripherally inserted central catheter (PICC) in chest X-ray images: A multi-task deep learning model. *Comput. Methods Programs Biomed.* **2020**, *197*, 105674. [CrossRef] [PubMed]

9. Jaiswal, A.K.; Tiwari, P.; Kumar, S.; Gupta, D.; Khanna, A.; Rodrigues, J.J.P.C. Identifying pneumonia in chest X-rays: A deep learning approach. *Meas. J. Int. Meas. Confed.* **2019**, *145*, 511–518. [CrossRef]
10. Zhu, X.; Wu, X. Class Noise Handling for Effective Cost-Sensitive Learning by Cost-Guided Iterative Classification Filtering. *IEEE Trans. Knowl. Data Eng.* **2006**, *18*, 1435–1440.
11. Pan, S.J.; Tsang, I.W.; Kwok, J.T.; Yang, Q. Domain Adaptation via Transfer Component Analysis. *IEEE Trans. Neural Netw.* **2011**, *22*, 199–210. [CrossRef] [PubMed]
12. Minaee, S.; Kafieh, R.; Sonka, M.; Yazdani, S.; Soufi, G.J. Deep-COVID: Predicting COVID-19 from chest X-ray images using deep transfer learning. *Med. Image Anal.* **2020**, *65*, 101794. [CrossRef] [PubMed]
13. Sufian, A.; Ghosh, A.; Sadiq, A.S.; Smarandache, F. A Survey on Deep Transfer Learning to Edge Computing for Mitigating the COVID-19 Pandemic. *J. Syst. Archit.* **2020**, *108*, 101830. [CrossRef]
14. Ravishankar, H.; Sudhakar, P.; Venkataramani, R.; Thiruvenkadam, S.; Annangi, P.; Babu, N.; Vaidya, V. *Understanding the Mechanisms of Deep Transfer Learning for Medical Images*; Springer International Publishing: Berlin/Heidelberg, Germany, 2016; pp. 188–196.
15. Alzubaidi, L.; Fadhel, M.; Al-Shamma, O.; Zhang, J.; Santamaria, J.; Duan, Y.; Oleiwi, S. Towards a Better Understanding of Transfer Learning for Medical Imaging: A Case Study. *Appl. Sci.* **2020**, *10*, 4523. [CrossRef]
16. Ahmed, S.E.; Amiri, S.; Doksum, K. Ensemble linear subspace analysis of high-dimensional data. *Entropy* **2021**, *23*, 324. [CrossRef] [PubMed]
17. Sun, Y.; Wong, A.; Kamel, M. Classification of imbalanced data: A review. *Int. J. Pattern Recognit. Artif. Intell.* **2009**, *23*, 687–719. [CrossRef]
18. Somasundaram, A.; Reddy, S. Data Imbalance: Effects and Solutions for Classification of Large and Highly Imbalanced Data. In Proceedings of the 1st International Conference on Research in Engineering, Computers and Technology(ICRECT), Tiruchirappalli, India, 8–10 September 2016.
19. Pan, S.J.; Yang, Q. A Survey on Transfer Learning. *IEEE Trans. Knowl. Data Eng.* **2010**, *22*, 1345–1359. [CrossRef]
20. Chollet, F. *Deep Learning with Python*; Manning: Shelter Island, NY, USA, 2017.
21. Brumback, L.C.; Pepe, M.S.; Alonzo, T.A. Using the ROC curve for gauging treatment effect in clinical trials. *Stat. Med.* **2006**, *25*, 575–590. [CrossRef] [PubMed]
22. Selvaraju, R.R.; Cogswell, M.; Das, A.; Vedantam, R.; Parikh, D.; Batra, D. Grad-CAM: Visual Explanations from Deep Networks via Gradient-Based Localization. *Int. J. Comput. Vis.* **2020**, *128*, 336–359. [CrossRef]
23. Ioffe, S.; Szegedy, C. Batch normalization: Accelerating deep network training by reducing internal covariate shift. In Proceedings of the 32nd International Conference on International Conference on Machine Learning, Lille, France, 6–11 July 2015.
24. Wang, X.; Peng, Y.; Lu, L.; Lu, Z.; Bagheri, M.; Summers, R.M. ChestX-ray8: Hospital-scale Chest X-ray Database and Benchmarks on Weakly-Supervised Classification and Localization of Common Thorax Diseases. In Proceedings of the IEEE Conference on Computer Vision and Pattern Recognition (CVPR), Honolulu, HI, USA, 21–26 July 2017; pp. 3462–3471.

Article

Associations between Longitudinal Gestational Weight Gain and Scalar Infant Birth Weight: A Bayesian Joint Modeling Approach

Matthew Pietrosanu [1], Linglong Kong [1], Yan Yuan [2], Rhonda C. Bell [3], Nicole Letourneau [4] and Bei Jiang [1,*]

[1] Department of Mathematical and Statistical Sciences, University of Alberta, Edmonton, AB T6G 2G1, Canada; pietrosa@ualberta.ca (M.P.); lkong@ualberta.ca (L.K.)
[2] School of Public Health, University of Alberta, Edmonton, AB T6G 1C9, Canada; yyuan@ualberta.ca
[3] Department of Agricultural, Food & Nutritional Science, University of Alberta, Edmonton, AB T6G 2P5, Canada; bellr@ualberta.ca
[4] Faculty of Nursing and Cumming School of Medicine, University of Calgary, Calgary, AB T2N 1N4, Canada; nicole.letourneau@ucalgary.ca
* Correspondence: bei1@ualberta.ca

Abstract: Despite the importance of maternal gestational weight gain, it is not yet conclusively understood how weight gain during different stages of pregnancy influences health outcomes for either mother or child. We partially attribute this to differences in and the validity of statistical methods for the analysis of longitudinal and scalar outcome data. In this paper, we propose a Bayesian joint regression model that estimates and uses trajectory parameters as predictors of a scalar response. Our model remedies notable issues with traditional linear regression approaches found in the clinical literature. In particular, our methodology accommodates nonprospective designs by correcting for bias in self-reported prestudy measures; truly accommodates sparse longitudinal observations and short-term variation without data aggregation or precomputation; and is more robust to the choice of model changepoints. We demonstrate these advantages through a real-world application to the Alberta Pregnancy Outcomes and Nutrition (APrON) dataset and a comparison to a linear regression approach from the clinical literature. Our methods extend naturally to other maternal and infant outcomes as well as to areas of research that employ similarly structured data.

Keywords: Bayesian modeling; functional regression; gestational weight; infant birth weight; joint modeling; longitudinal data; maternal weight gain

Citation: Pietrosanu, M.; Kong, L.; Yuan, Y.; Bell, R.C.; Letourneau, N.; Jiang, B. Associations between Longitudinal Gestational Weight Gain and Scalar Infant Birth Weight: A Bayesian Joint Modeling Approach. *Entropy* **2022**, *24*, 232. https://doi.org/10.3390/e24020232

Academic Editors: S. Ejaz Ahmed and Farouk Nathoo

Received: 16 December 2021
Accepted: 29 January 2022
Published: 2 February 2022

Publisher's Note: MDPI stays neutral with regard to jurisdictional claims in published maps and institutional affiliations.

1. Introduction

Maternal weight gain supports fetal growth and holds important health implications for both mother and child during and after pregnancy [1–3]. Insufficient weight gain is associated with preterm birth and low infant birth weight, while excessive weight gain is linked to postpartum weight retention, gestational diabetes, hypertension, infant macrosomia, and other complications [3–5]. A growing amount of clinical literature further implicates maternal gestational weight gain outside of recommendations in adverse, long-term health outcomes for the child, including a heightened future risk of cardiovascular disease [6,7].

It is not yet conclusively understood how weight gain in different stages of pregnancy affects health outcomes for either mother or child. This is despite previous findings that gestational weight trajectories are similar across human populations with varying genetic, cultural, and lifestyle traits [8]. As an example central to this article, previous studies present conflicting conclusions on the effect of first- and second-trimester weight gain on infant birth weight [8–12]. We attribute this in part to differences in and the validity of the statistical methods currently used to jointly analyze scalar outcomes and longitudinal data. Thus, developments in methodology for analyzing how patterns in longitudinal data (e.g., gestational weight gain) influence scalar outcomes (e.g., infant birth weight) are both statistically and clinically relevant.

Retnakaran et al. [13] investigates the relationship between infant birth weight and gestational weight gain in different periods of pregnancy using traditional linear regression. The work's models include, as predictors, demographic covariates together with pregravid weight and interval-specific average weight gain. The authors opt for clinical data in order to avoid bias in self-reported pregravid measurements that they claim is prevalent in other studies [5,12]. The resulting preconception study design presents a few practical problems: this design is more difficult to implement, limits the use of secondary data, and can introduce other sampling biases and restrict model generalizability (e.g., through the exclusion of unplanned pregnancies). Despite the supposed benefit of bias reduction, the work's average weight gain measurements are precomputed (as differences in average weight between gestational intervals) and may be highly variable due to clinical measurement error and the small number of observations in each gestational interval. As Richardson notes, ignoring this measurement error can lead to unreliable effect estimates and misleading conclusions [14]. This linear regression approach furthermore does not account for gestational age at each weight measurement and, through its initial precomputing stage, reduces the amount of data used to fit the model. The consequent coarsening of information may contribute to unreliable effect estimates and conclusions.

To address these issues, we turn to other approaches for modeling longitudinal data. Joint models that simultaneously consider longitudinal responses and scalar health outcomes are well established in the statistical literature [15–22]. These models were originally motivated by HIV/AIDS and cancer research to predict patient outcomes using a time-dependent covariate trajectory. Relevant methodology has since evolved to incorporate techniques from functional data analysis, semiparametric inference, robust estimation, and Bayesian methods [23].

In this paper, we consider a joint model for infant birth weight and gestational weight gain trajectories that also incorporates clinical covariates. Our approach efficiently uses information from estimated mean weight trajectories—including estimated pregravid weight, interval-specific rates of weight gain, and individual residual variance—to predict infant birth weight. As a result, our model can correct for bias in self-reported weight measurements (when combined with clinical observations) and permits nonprospective study designs with unbalanced longitudinal observations.

We employ the Bayesian joint modeling approach of Jiang et al. [23]. Our model uses parameter estimates that describe individual gestational weight trajectories to model the association between infant birth weight and gestational weight gain. We model the mean [24,25] and measurement error [26,27] of these trajectories using a robust, semiparametric mixed effects model and a Bayesian linear spline approach [23].

Our joint model remedies the issues noted above for linear regression [13]. First, by using estimated mean trajectory parameters as predictors of infant birth weight, our approach obtains more-efficient estimates of the time-dependent effects of gestational weight gain. More generally, our joint modeling method, implemented in a Bayesian framework, borrows information from all observations and patients in a one-stage procedure. On the other hand, the predictors in the traditional linear model, such as interval-specific weight gain, are precomputed in an initial step independently for each patient using only a small proportion of the available data at a time. Second, our approach truly accommodates longitudinal data by explicitly accounting for gestational age at each weight measurement when estimating weight gain trajectories. Third, unlike other studies that treat within-patient residual variance as a nuisance parameter, our method models measurement error variance and uses it as a random effect to predict infant birth weight.

Our approach to mean trajectory modeling mitigates bias in self-reported prestudy measurements and accounts for variability inherent in observed data. These are notable advantages over traditional methods such as the linear regression approach above, where the amalgamation of data from different sources can negatively impact an analysis. Another advantage of the proposed model is its potential to be used for prediction and intervention: our model can be applied to predict infant birth weight well before term and can thus be conveniently deployed in clinical settings. More generally, while infant birth weight is the primary focus of the present paper, our approach and discussions apply to other maternal and infant outcomes and to other areas of research that employ similarly structured data.

In Section 2, we introduce the pregnancy outcomes dataset used in this article and the proposed model. This section also presents our chosen prior distributions and computational methods. We present estimates for the effect of time-specific maternal weight gain on infant birth weight obtained under the proposed model in Section 3, and compare these estimates to those obtained using the linear regression approach described above [13]. In Section 4, we discuss our results and provide some concluding remarks on the general significance of our approach and future directions.

2. Materials and Methods

2.1. Data

Throughout this paper, we use data from the 2009–2012 Alberta Pregnancy Outcomes and Nutrition (APrON) study [28]. The 2189 women in the APrON study, all of whom were at least 16 years of age and at most 27 weeks into gestation, are part of a longitudinal cohort [28,29]. As part of the APrON study, maternal weight and gestational age were measured at each trimester following registration. Participants recruited before 13 weeks gestation have measurements corresponding to all three trimesters, while those recruited between 14 and 27 weeks gestation have measurements only for the second and third trimesters. Pregravid weight, along with other demographic characteristics, were self-reported by each participant upon recruitment. Gestational age at delivery was assessed postpartum. In addition to the APrON data, clinical weight measurements were collected from all participants at regularly scheduled prenatal visits. The number of weight measurements for each participant varies due to missing appointments or data. The longitudinal weight data in this study may be considered sparse and has been previously examined in the functional data analysis literature [30].

We only include participants with a live, singleton birth in the following analyses. We exclude individuals without a reported pregravid weight; those with less than three weight measurements during pregnancy; and those with missing gestational age at delivery, infant birth weight, marital status, education level, income level, ethnic origin, parity, or age. We do not consider any postpartum weight measurements in our analyses.

The final analytic sample consists of $n = 1340$ participants with $N = 15{,}183$ weight observations. Demographic characteristics for this sample, stratified by infant birth weight class, are summarized in Table 1. We use <2.5 kg, $\geq$2.5 kg and <4 kg, and $\geq$4 kg as criteria defining low, normal, and high infant birth weight classes [31]. Clinical weight measurements (i.e., not including self-reported pregravid measurements) were taken at gestational ages ranging from 4.4 to 41.7 weeks, with a median of 30.3 weeks. Participants have a median of 12 recorded weight measurements each.

Table 1. Summary of demographic covariates for the analytic sample in the APrON dataset. For categorical variables, counts and relative percentages are reported. A * indicates the chosen reference category. For continuous variables, means (and standard deviations, in parentheses) are reported.

		Infant Birth Weight Class	
	Low (<2.5 kg)	Normal (≥2.5 and <4 kg)	High (≥4 kg)
Mother characteristics			
Participants	56 (4.18%)	1163 (86.79%)	121 (9.03%)
Age, years	32.66 (4.76)	31.33 (4.27)	31.91 (4.03)
Marital status			
Married *	55 (98.21%)	1122 (96.47%)	118 (97.52%)
Single	1 (1.79%)	41 (3.53%)	3 (2.48%)
Education			
Graduate degree	11 (19.64%)	273 (23.47%)	27 (22.31%)
Some post-secondary *	37 (66.07%)	775 (66.64%)	83 (68.60%)
High school	8 (14.29%)	115 (9.89%)	11 (9.09%)
Income level			
<70 k	11 (19.64%)	234 (20.12%)	19 (15.70%)
≥70 k *	45 (80.36%)	929 (79.88%)	102 (84.30%)
Ethnic origin			
Asian	8 (14.29%)	77 (6.62%)	0 (0.00%)
Black	4 (7.14%)	11 (0.95%)	0 (0.00%)
Caucasian *	37 (66.07%)	956 (82.20%)	114 (94.22%)
Latin American	1 (1.79%)	38 (3.27%)	3 (2.48%)
Southeast Asian	4 (7.14%)	53 (4.56%)	2 (1.65%)
Other	2 (3.57%)	28 (2.41%)	2 (1.65%)
Parity			
0 *	35 (62.50%)	667 (57.35%)	48 (39.67%)
1	18 (32.14%)	387 (33.28%)	52 (42.98%)
≥2	3 (5.46%)	109 (9.37%)	21 (17.36%)
Child characteristics			
Birth weight, kg	2.23 (0.34)	3.33 (0.35)	4.25 (0.21)
Gestational age at delivery, weeks	36.01 (2.57)	39.51 (1.27)	40.20 (1.02)

2.2. Joint Model

We now present our joint model for infant birth weight and longitudinal gestational weight gain. As a main feature, the model estimates the former using parameter estimates from patient-specific maternal weight trajectories:

$$Y_i \mid \boldsymbol{b}_i = (1, \boldsymbol{z}_i^\top, \boldsymbol{b}_i^\top, \ln \sigma_i^2)\boldsymbol{\theta} + \varepsilon_i$$

$$\varepsilon_i \stackrel{\text{i.i.d.}}{\sim} N(0, \sigma^2)$$

for $i = 1, \ldots, n$, where Y_i denotes an observed infant birth weight; z_i an observed demographic covariate vector; $\boldsymbol{b}_i$ a vector of random weight trajectory parameters; and σ_i^2 the trajectory's residual variance for the ith patient. The vector $\boldsymbol{\theta}$ contains the corresponding fixed and random effects.

Individual longitudinal weight trajectories influence Y_i through the random trajectory parameters $\boldsymbol{b}_i$ in the longitudinal submodel

$$X_{ij} = f(t_{ij}; \boldsymbol{b}_i) + \varepsilon_{ij}$$

$$\varepsilon_{ij} \stackrel{\text{i.i.d.}}{\sim} N(0, \sigma_i^2)$$

$$\boldsymbol{b}_i \stackrel{\text{i.i.d.}}{\sim} N(\boldsymbol{\beta}, \Sigma)$$

for $j = 1, \ldots, n_i$, where X_{ij} is the observed weight of the ith patient at gestational age t_{ij} and n_i is the total number of longitudinal observations for the ith patient. We consider a piecewise linear weight trajectory (as a function of gestational age $t \geq 0$) [32]

$$f(t; \boldsymbol{b} = (b_0, b_1, \ldots, b_K)^\top) = b_0 + \sum_{k=1}^{K} b_k (t - t_k^*)_+,$$

where $x_+ = \max\{0, x\}$ for $x \in \mathbb{R}$ and $(t_1^* = 0, \ldots, t_K^*, t_{K+1}^* = \infty)$ is a fixed, increasing sequence of changepoint locations. Consequently, b_0 is the mean pregravid weight and $\sum_{k=1}^{k_0} b_k$ is the mean rate of weight gain in the gestational age interval $[t_{k_0}^*, t_{k_0+1}^*)$, for $k_0 = 1, \ldots, K$. Following common trimester boundaries [13], we take $K = 8$ with $t_2^* = 13$, $t_3^* = 18$, $t_4^* = 23$, $t_5^* = 27$, $t_6^* = 32$, $t_7^* = 37$, and $t_8^* = 45$.

Under the proposed model, $\boldsymbol{\beta}$ describes an average, "prototype" trajectory, while the random $\boldsymbol{b}_i$s describe patient-specific trajectories and deviations from $\boldsymbol{\beta}$. Our longitudinal model accounts for short-term variation and measurement error in patient trajectories by using $\ln \sigma_i^2$ as a predictor of Y_i.

2.3. Bayesian Framework and Model Estimation

We take a Bayesian approach to parameter estimation in the proposed model.

In the longitudinal submodel, we model random trajectory parameters as $\boldsymbol{b}_i \overset{\text{i.i.d.}}{\sim} N(\boldsymbol{\beta}, \Sigma)$ under the diffuse prior $\boldsymbol{\beta} \sim N(0, 10I)$. Additional tests, not presented here, indicate no need to consider a Gaussian mixture [23] in the distribution of the $\boldsymbol{b}_i$s for our APrON dataset. To avoid issues with unbounded likelihood [33] when using an unstructured random effect covariance matrix Σ, we implement the empirical Bayes Wishart prior [34]

$$\Sigma \sim W\left(m = 2 + \frac{K+1}{2}, \; \Lambda = \sum_{i=1}^{n} \widehat{\mathrm{Cov}}(\hat{\boldsymbol{b}}_i^{(\mathrm{OLS})})^{-1}\right),$$

where $\widehat{\mathrm{Cov}}(\hat{\boldsymbol{b}}_i^{(\mathrm{OLS})})$ is an estimate of the covariance matrix of the ordinary least squares (OLS) estimator of $\boldsymbol{b}_i$. For the σ_i^2s, the trajectory residual variances, we assume a log-normal prior $\ln \sigma_i^2 \overset{\text{i.i.d.}}{\sim} N(\mu, \tau^2)$ under the diffuse hyperpriors $\mu \sim N(0, 10^3)$ and $\tau^2 \sim$ Inv-Gamma$(10^{-4}, 10^{-4})$. For the scalar response Y_i, we take $\boldsymbol{\theta} \sim N(\boldsymbol{0}, 10I)$ and $\sigma^2 \sim$ Inv-Gamma$(10^{-4}, 10^{-4})$.

For notational simplicity, let $\boldsymbol{\varphi} = \{\boldsymbol{\theta}, \sigma^2, \boldsymbol{\beta}, \Sigma, \mu, \tau^2\}$ be the collection of model parameters. We assume that all elements of $\boldsymbol{\varphi}$ have independent prior distributions and denote the joint prior of $\boldsymbol{\varphi}$ by π. Define $\eta_i^\mu = (1, \boldsymbol{z}_i^\top, \boldsymbol{b}_i^\top, \ln \sigma_i^2)\boldsymbol{\theta}$ as the linear predictor corresponding to Y_i.

The full likelihood of $\boldsymbol{\varphi}$ for our model is

$$L(\boldsymbol{\varphi}) = \pi(\boldsymbol{\varphi}) \prod_{i=1}^{n} \left[|\Sigma|^{-0.5} \exp\left\{ -0.5(\boldsymbol{b}_i - \boldsymbol{\beta})^\top \Sigma^{-1}(\boldsymbol{b}_i - \boldsymbol{\beta}) \right\} \right.$$

$$\times \prod_{j=1}^{n_i} \left[\sigma_i^{-1} \exp\left\{ -0.5\sigma_i^{-2}(x_{ij} - f(t_{ij}; \boldsymbol{b}_i))^2 \right\} \right]$$

$$\times \tau^{-1} \exp\left\{ -0.5\tau^{-2}(\ln \sigma_i^2 - \mu)^2 \right\}$$

$$\left. \times \sigma^{-1} \exp\left\{ -0.5\sigma^{-2}(y_i - \eta_i^\mu)^2 \right\} \right].$$

We implement a Gibbs sampler to perform posterior draws. For analytic derivations of the posterior distributions, see Jiang et al. [23]. As the full conditional posterior of σ_i^2 has no closed form, we obtain draws using the inverse cumulative distribution function method. In our Markov Chain Monte Carlo (MCMC) procedure, we run a chain of 150,000 iterations and use the first 50,000 iterations as a burn-in period; however, in this particular application, we observe that the model converges very quickly and that even 10,000 total iterations are sufficient. To reduce autocorrelation in subsequent draws, we thin posterior draws by saving only every 10th. We implement our model in C++ using the Scythe open-source statistical library [35] and R [36].

We consider two models, each accounting for a different set of demographic covariates. The first model (JM1) includes education level, income level, ethnic origin, parity, age at pregnancy, and gestational age at delivery. The second model (JM2) includes only demographic variables whose 95% credible interval in JM1 do not contain zero.

2.4. Comparison to Linear Regression

We compare our proposed method against the previously noted traditional linear regression (LR) approach. We focus specifically on differences in the effects of maternal weight gain rate in different gestational age periods on infant birth weight. To make this comparison easier, we use the rate of weight gain in each gestational period (rather than period-specific absolute weight gain) as a predictor of infant birth weight Y_i.

We use the same gestational age intervals in both models: $[0, 13)$, $[13, 18)$, $[18, 23)$, $[23, 27)$, $[27, 32)$, $[32, 37)$, and $[32, 45)$. To compute the average rate of weight gain $\tilde{b}_k$ in the kth interval, we first calculate the averages, μ_k and μ_{k-1}, of weight measurements taken in the kth and $(k-1)$th intervals, respectively. We then calculate the rate of weight gain as $\tilde{b}_k = (\mu_k - \mu_{k-1})/(m_k - m_{k-1})$, where m_k is the midpoint of the kth gestational age interval. For the sake of notation, we let $k = 0$ refer to pregravid measurements (i.e., at week zero).

As noted previously, our joint model addresses numerous shortcomings of the LR approach. First, the LR model does not fully take into account the timing of individual maternal weight measurements, while our JM approach estimates patient-specific weight trajectories as functions of time. Second, LR model estimates are subject to short-term measurement error and variability: this is because only a small number of measurements contribute to pregravid weight and the estimated rates of weight gain. Our hierarchical Bayesian framework borrows information from all observations to estimate these quantities via patient-specific trajectory parameters. As another feature that may be clinically relevant in some applications, our model also estimates and uses short-term variability in maternal weight as another predictor.

We similarly consider two linear regression models in the following analyses. The first (LR1) uses estimated rates of weight gain (i.e., the $\tilde{b}_k$s), average pregravid weight $\tilde{b}_0 = \mu_0$, and the same demographic variables as JM1. Similar to JM2, the second model (LR2) includes only the demographic covariates whose 95% confidence intervals in LR1 do not contain zero.

3. Results and Discussion

Table 2 presents parameter estimates for all four of the models described in the previous section. Model convergence for the joint models were assessed visually and numerically using five parallel chains. Trace plots for each of the coefficients in Table 2 suggest adequate convergence and mixing. Numerically, Rubin–Gelman statistics [37] for these coefficients range from 1.005 to 1.027 and also imply model convergence.

We observe major differences in the estimated effects of weight gain between the LR and JM approaches. Both LR models find rate of weight gain to be a useful predictor of infant birth weight only after 18 weeks gestation. On the other hand, the JM models find this to be true throughout gestation, including before 18 weeks.

Table 2. Parameter estimates obtained using the LR and the proposed JM models, with 95% confidence and credible intervals, respectively. For JM model interpretability, we present estimates for $\sum_{j=1}^{k} b_j$ (rather than for just b_k), which can be interpreted as the effect of weight gain rate in the kth gestational interval. Boldface indicates an estimate whose corresponding credible (or confidence) interval does not contain zero.

		Model			
		JM1	**JM2**	**LR1**	**LR2**
Demographic variables					
Marital status					
	Single	0.151 $(-0.161, 0.467)$		**0.277** **(0.001, 0.553)**	
Education					
	Graduate	0.016 $(-0.115, 0.146)$		0.059 $(-0.058, 0.175)$	
	High school	-0.034 $(-0.222, 0.152)$		-0.114 $(-0.340, 0.112)$	
Income level					
	<70 k	-0.039 $(-0.193, 0.114)$		-0.137 $(-0.291, 0.016)$	
Ethnic origin					
	Asian	-0.038 $(-0.264, 0.183)$		-0.025 $(-0.227, 0.177)$	
	Black	-0.322 $(-0.856, 0.208)$		0.056 $(-0.805, 0.917)$	
	Latin American	-0.038 $(-0.356, 0.272)$		-0.144 $(-0.419, 0.131)$	
	Southeast Asian	-0.098 $(-0.369, 0.178)$		-0.139 $(-0.396, 0.118)$	
	Other	-0.09 $(-0.445, 0.269)$		**0.348** **(0.027, 0.670)**	
Parity					
	1	**0.147** **(0.028, 0.269)**	**0.136** **(0.020, 0.254)**	**0.137** **(0.017, 0.258)**	**0.121** **(0.005, 0.238)**
	≥ 2	**0.246** **(0.052, 0.444)**	**0.215** **(0.032, 0.400)**	**0.384** **(0.206, 0.561)**	**0.331** **(0.159, 0.503)**
Age at pregnancy		-0.013 $(-0.073, 0.047)$		-0.025 $(-0.082, 0.032)$	
Gestational age at delivery		**0.162** **(0.127, 0.197)**	**0.166** **(0.130, 0.200)**	**0.092** **(0.039, 0.145)**	**0.103** **(0.051, 0.155)**
Pre-pregnancy weight					
Clinical measure				**0.006** **(0.002, 0.011)**	**0.006** **(0.002, 0.011)**
Trajectory estimate ($\hat{b}_0$)		**0.007** **(0.003, 0.012)**	**0.007** **(0.003, 0.012)**		
Trajectory estimator variance ($\ln \hat{\Sigma}_{11}$)		**0.162** **(0.127, 0.197)**	0.003 $(-0.063, 0.066)$		
Rate of weight gain (by GA interval)					
	$[0, 13)$	**0.701** **(0.264, 1.138)**	**0.718** **(0.300, 1.153)**	0.061 $(-0.186, 0.307)$	0.085 $(-0.157, 0.327)$
	$[13, 18)$	**1.256** **(0.527, 1.972)**	**1.291** **(0.581, 2.014)**	0.076 $(-0.186, 0.333)$	0.123 $(-0.129, 0.375)$
	$[28, 23)$	**1.703** **(0.697, 2.708)**	**1.758** **(0.780, 2.728)**	**0.201** **(0.032, 0.371)**	**0.200** **(0.034, 0.365)**
	$[23, 27)$	**1.929** **(0.665, 3.183)**	**1.997** **(0.754, 3.219)**	**0.191** **(0.026, 0.356)**	**0.193** **(0.031, 0.356)**
	$[27, 32)$	**2.010** **(0.490, 3.525)**	**2.082** **(0.613, 3.538)**	**0.270** **(0.102, 0.437)**	**0.223** **(0.06, 0.385)**
	$[32, 37)$	**2.009** **(0.390, 3.673)**	**2.086** **(0.455, 3.662)**	**-0.277** **$(-0.507, -0.048)$**	**-0.285** **$(-0.513, -0.056)$**
	$[37, 45)$	**2.027** **(0.330, 3.720)**	**2.108** **(0.439, 3.729)**	**0.277** **(0.002, 0.551)**	**0.304** **(0.033, 0.574)**

Further, we note a difference in the direction of the estimated effect of weight gain during weeks 32–37 between the JM and LR models. Our JM approach estimates this effect to be positive, while the LM model estimates a negative effect. Given the positive estimates for other gestational intervals and the positive estimate originally reported in Retnakaran et al. [13], we suspect that the LR model is inaccurate here. As discussed previously, this could be attributed to the loss of time information or the precomputation of average weight gain measurements. These results illustrate how the LR approach might not yield reliable conclusions, even with relatively large datasets. Towards the end of this section, we also discuss the sensitivity of the LR approach to the choice of gestational intervals.

Other differences in the effect of rate of weight gain are less drastic but important nonetheless. In general, effect estimates in the LR models (relative to those in the JM models) are shrunk towards zero. We attribute this shrinkage to attenuation bias in the LR models due to self-reporting bias (in pregravid measurements) and the LR models' inability to account for short-term variation in the weight trajectories. As discussed previously, this can be due to the small number of observations used to compute each patient's pregravid weight ($\tilde{b}_0$) and interval-specific rates of weight gain (the $\tilde{b}_k$s).

Figure 1 illustrates the importance of accounting for deviation in patient-level trajectories (described by the b_is) from the prototype trajectory (described by β) in our JM approach. While an overall trend in individual fitted trajectories is apparent, we see significant amounts of variation in gestational weight gain trajectories between patients. Figure 2 illustrates our proposed model's ability to accommodate individual longitudinal trajectories even in the presence of between-patient variability.

In a separate analysis not shown in Table 2, we consider a different set of gestational intervals (i.e., the sequence of t_k^*s): $[0, 15)$, $[15, 20)$, $[20, 25)$, $[25, 30)$, $[30, 35)$, and $[35, 45)$, this time chosen out of convenience. The JM models yield similar conclusions with these different intervals while the LR models find weight gain during only 20–30 weeks gestation to be associated with infant birth weight. This demonstrates that the LR model is not robust with respect to the precomputation of interval-specific weight gain measurements and, as above, calls into question the validity of this approach.

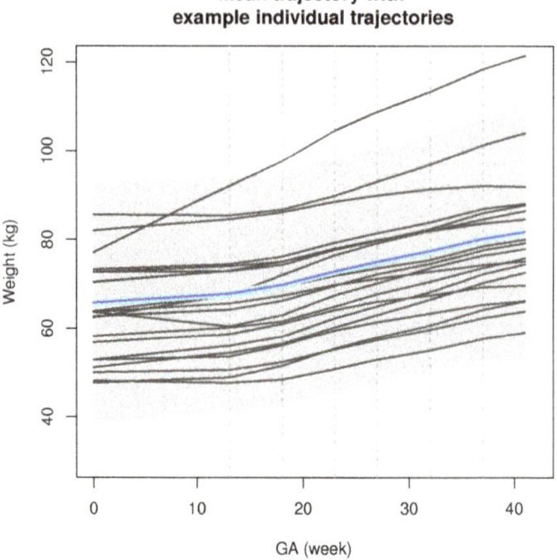

Figure 1. Posterior mean estimates from the proposed JM1 model for the mean weight gain trajectory β (solid blue) and twenty randomly selected individual trajectories b_i (solid grey), both as functions of gestational age (GA). The light blue and grey regions describe 95% credible bands for β and b_i, respectively. Dotted grey lines indicate model changepoints (i.e., at GA $= t_k^*$).

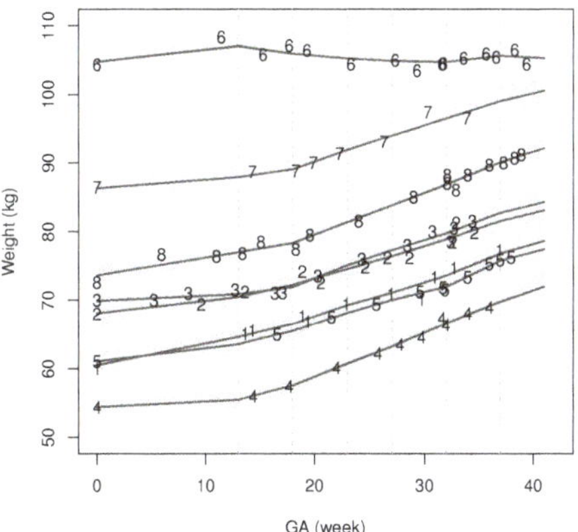

Figure 2. Eight randomly selected estimates of individual trajectories b_i from the JM1 model as functions of gestational age (GA) (solid grey) and corresponding observed weights X_{ij}. Observed weights from the eight patients are denoted by $1, 2, \ldots, 8$. Light grey regions denote 95% credible bands for X_{ij} (each for a fixed i). Dotted grey lines indicate model changepoints (i.e., at GA $= t_k^*$).

4. Conclusions

In this paper, we provided a hierarchical Bayesian model for the joint analysis of scalar and longitudinal data based on Jiang et al. [23]. Our work was motivated by a question in maternal health research on the relationship between (scalar) infant birth weight and (longitudinal) gestational weight gain during different periods of pregnancy. We contrasted our joint modeling approach with one using traditional linear regression that has appeared in the clinical literature [13] and is reminiscent of analyses commonly seen in applied research.

This comparative LR approach was originally proposed for a preconception cohort study to eliminate self-reporting bias in pregravid measurements [13]. However, in addition to the design's inconvenience, this approach does not fully account for gestational age or clinical measurement error and uses only a small number of observations to pre-estimate (i.e., in an initial stage separate from model estimation) weight gain in each gestational period. This results in high-variance model estimates that are not robust to the choice of gestational intervals. In contrast, through a one-stage, hierarchical Bayesian framework, our JM approach accounts for gestational age and short-term variability in longitudinal measurements, and borrows information from all observations to reduce bias and obtain more-reliable estimates.

The benefits of our model over the LR approach are apparent in our real-world study using the APrON pregnancy outcomes dataset. Beyond the LR model's questionable negative estimated association between infant birth weight and maternal weight gain for 32–37 weeks gestation, we observed relative shrinkage in LR effect estimates towards zero. This illustrates the unreliability of the LR methodology and the impact of attenuation bias on effect estimates. On the other hand, our JM approach produced estimates that were reasonable and stable, even when considering different gestational periods.

We have demonstrated the usefulness of our joint modeling approach in settings with continuous scalar and longitudinal responses. Our approach extends naturally to other submodels and data types such as ordinal health outcomes (e.g., through an appropriate (cumulative) probit or logit link function at the response level of the model) [23]. While our focus in this paper was on comparing the JM and LR approaches, the proposed model can be further optimized for predictive purposes. Our developments hold immediate implications for clinical interventions, such as the early identification of pregnant women at risk of birth complications (e.g., extreme infant birth weight or other outcomes, whether scalar or ordinal) using self-reported prepregnancy data or sparse clinical observations.

Author Contributions: Conceptualization, B.J. and Y.Y.; methodology, B.J. and M.P.; software, B.J. and M.P.; validation, M.P.; formal analysis, M.P.; investigation, M.P.; resources, B.J., R.C.B. and N.L.; data curation, Y.Y, R.C.B. and N.L.; writing—original draft preparation, M.P.; writing—review and editing, M.P., B.J., Y.Y., R.C.B. and N.L.; visualization, M.P.; supervision, B.J. and L.K.; project administration, B.J., L.K., R.C.B. and N.L.; funding acquisition, B.J., R.C.B. and N.L. All authors have read and agreed to the published version of the manuscript.

Funding: B.J.'s research is supported by a Discovery Grant from the Natural Sciences and Engineering Research Council of Canada. L.K.'s research is supported by a Discovery Grant from the Natural Sciences and Engineering Research Council of Canada and a Canada Research Chair in Statistical Learning. M.P.'s graduate studies are supported by a Canadian Graduate Scholarship (Master's and Doctoral) from the Natural Sciences and Engineering Research Council of Canada. Y.Y.'s research is supported by a Discovery Grant from the Natural Sciences and Engineering Research Council of Canada (RGPIN-2019-04-862) and the Women and Children's Health Research Institute. The APrON cohort was established by an interdisciplinary team grant from Alberta Innovates Health Solutions (formerly the Alberta Heritage Foundation for Medical Research). Additional funding from the Alberta Children's Hospital Foundation assisted with the collection and analysis of data presented in this manuscript.

Institutional Review Board Statement: The study was conducted according to the guidelines of the Declaration of Helsinki, and approved by the Human Ethics Review Board (Biomedical Panel) of the University of Alberta (Pro00002954; 4 March 2009) and the Research Ethics Board of the University of Calgary (14-702; 2008, renewed 4 November 2021).

Informed Consent Statement: Informed consent was obtained from all participants involved in the study.

Data Availability Statement: Data are available from the Secondary Analyses to Generate Evidence (SAGE) databases held within the Policy Wise for Children and Families (nongovernmental) organization in Alberta, Canada: https://policywise.com/, accessed on 16 December 2021. Data are available subject to appropriate review and approvals. Access to Alberta Pregnancy Outcomes and Nutrition (APrON) data is administered by SAGE: requests can be made to data@policywise.com.

Acknowledgments: The authors are grateful to the families who took part in the APrON study and the APrON team (http://APrONstudy.ca, accessed on 16 December 2021), investigators, research assistants, graduate and undergraduate students, volunteers, clerical staff, and managers.

Conflicts of Interest: The authors declare no conflict of interest. The APrON cohort was established by an interdisciplinary team grant from Alberta Innovates Health Solutions (formerly the Alberta Heritage Foundation for Medical Research). Additional funding from the Alberta Children's Hospital Foundation assisted with the collection and analysis of data presented in this manuscript.

References

1. Caughey, A.B. Gestational weight gain and outcomes for mothers and infants. *J. Am. Med. Assoc.* **2017**, *317*, 2175–2176. [CrossRef] [PubMed]
2. Sidebottom, A.C.; Brown, J.E.; Jacobs, D.R. Pregnancy-related changes in body fat. *Eur. J. Obstet. Gynecol. Reprod. Biol.* **2001**, *94*, 216–223. [CrossRef]
3. Hediger, M.L.; Scholl, T.O.; Schall, J.I.; Healey, M.F.; Fischer, R.L. Changes in maternal upper arm fat stores are predictors of variation in infant birth weight. *J. Nutr.* **1994**, *124*, 24–30. [CrossRef] [PubMed]
4. Nohr, E.A.; Vaeth, M.; Baker, J.L.; Sørensen, T.I.; Olsen, J.; Rasmussen, K.M. Combined associations of prepregnancy body mass index and gestational weight gain with the outcome of pregnancy. *Am. J. Clin. Nutr.* **2008**, *87*, 1750–1759. [CrossRef] [PubMed]
5. Ismail, L.C.; Bishop, D.C.; Pang, R.; Ohuma, E.O.; Kac, G.; Abrams, B.; Rasmussen, K.; Barros, F.C.; Hirst, J.E.; Lambert, A.; et al. Gestational weight gain standards based on women enrolled in the fetal growth longitudinal study of the INTERGROWTH-21st Project: A prospective longitudinal cohort study. *BMJ* **2016**, *352*, i555. [CrossRef] [PubMed]
6. Gluckman, P.D.; Hanson, M.A.; Cooper, C.; Thornburg, K.L. Effect of in utero and early-life conditions on adult health and disease. *N. Engl. J. Med.* **2008**, *359*, 61–73. [CrossRef] [PubMed]
7. Barker, D.J.P. Fetal origins of cardiovascular disease. *Ann. Med.* **1999**, *31*, 3–6. [CrossRef] [PubMed]

8.   Catov, J.M.; Abatemarco, D.; Althouse, A.; Davis, E.M.; Hubel, C. Patterns of gestational weight gain related to fetal growth among women with overweight and obesity. *Obesity* **2015**, *23*, 1071–1078. [CrossRef] [PubMed]

9.   Ruchat, S.M.; Allard, C.; Doyon, M.; Lacroix, M.; Guillemette, L.; Patenaude, J.; Battista, M.C.; Ardilouze, J.L.; Perron, P.; Bouchard, L.; et al. Timing of excessive weight gain during pregnancy modulates newborn anthropometry. *J. Obstet. Gynaecol. Can.* **2016**, *38*, 108–117. [CrossRef] [PubMed]

10. Karachaliou, M.; Georgiou, V.; Roumeliotaki, T.; Chalkiadaki, G.; Daraki, V.; Koinaki, S.; Dermitzaki, E.; Sarri, K.; Vassilaki, M.; Kogevinas, M.; et al. Association of trimester-specific gestational weight gain with fetal growth, offspring obesity, and cardiometabolic traits in early childhood. *Am. J. Obstet. Gynecol.* **2015**, *212*, 502.e1–502.e14. [CrossRef] [PubMed]

11. Strauss, R.S.; Dietz, W.H. Low maternal weight gain in the second or third trimester increases the risk for intrauterine growth retardation. *J. Nutr.* **1999**, *129*, 988–993. [CrossRef] [PubMed]

12. Sridhar, S.B.; Xu, F.; Hedderson, M.M. Trimester-specific gestational weight gain and infant size for gestational age. *PLoS ONE* **2016**, *11*, e0159500. [CrossRef]

13. Retnakaran, R.; Wen, S.W.; Tan, H.; Zhou, S.; Ye, C.; Shen, M.; Smith, G.N.; Walker, M.C. Association of timing of weight gain in pregnancy with infant birth weight. *JAMA Pediatr.* **2018**, *172*, 136. [CrossRef] [PubMed]

14. Richardson, S.; Gilks, W.R. A Bayesian approach to measurement error problems in epidemiology using conditional independence models. *Am. J. Epidemiol.* **1993**, *138*, 430–442. [CrossRef]

15. Tsiatis, A.A.; Degruttola, V.; Wulfsohn, M.S. Modeling the relationship of survival to longitudinal data measured with error. Applications to survival and CD4 counts in patients with AIDS. *J. Am. Stat. Assoc.* **1995**, *90*, 27. [CrossRef]

16. Muthén, B.; Shedden, K. Finite mixture modeling with mixture outcomes using the EM algorithm. *Biometrics* **1999**, *55*, 463–469. [CrossRef] [PubMed]

17. Wang, Y.; Taylor, J.M.G. Jointly modeling longitudinal and event time data with application to acquired immunodeficiency syndrome. *J. Am. Stat. Assoc.* **2001**, *96*, 895–905. [CrossRef]

18. Law, N.J. The joint modeling of a longitudinal disease progression marker and the failure time process in the presence of cure. *Biostatistics* **2002**, *3*, 547–563. [CrossRef] [PubMed]

19. Song, X.; Davidian, M.; Tsiatis, A.A. A semiparametric likelihood approach to joint modeling of longitudinal and time-to-event data. *Biometrics* **2002**, *58*, 742–753. [CrossRef]

20. Brown, E.R.; Ibrahim, J.G. A Bayesian semiparametric joint hierarchical model for longitudinal and survival data. *Biometrics* **2003**, *59*, 221–228. [CrossRef] [PubMed]

21. Brown, E.R.; Ibrahim, J.G. Bayesian approaches to joint cure-rate and longitudinal models with applications to cancer vaccine trials. *Biometrics* **2003**, *59*, 686–693. [CrossRef]

22. Yu, M.; Taylor, J.M.G.; Sandler, H.M. Individual prediction in prostate cancer studies using a joint longitudinal survival–cure model. *J. Am. Stat. Assoc.* **2008**, *103*, 178–187. [CrossRef]

23. Jiang, B.; Wang, N.; Sammel, M.D.; Elliott, M.R. Modelling short- and long-term characteristics of follicle stimulating hormone as predictors of severe hot flashes in the Penn ovarian aging study. *J. R. Stat. Soc. Ser. C (Appl. Stat.)* **2015**, *64*, 731–753. [CrossRef]

24. Lang, S.; Brezger, A. Bayesian P-splines. *J. Comput. Graph. Stat.* **2004**, *13*, 183–212. [CrossRef]

25. Eilers, P.H.C.; Marx, B.D. Flexible smoothing with B-splines and penalties. *Stat. Sci.* **1996**, *11*, 89–121. [CrossRef]

26. Jiang, B.; Elliott, M.R.; Sammel, M.D.; Wang, N. Joint modeling of cross-sectional health outcomes and longitudinal predictors via mixtures of means and variances. *Biometrics* **2015**, *71*, 487–497. [CrossRef] [PubMed]

27. Elliott, M.R. Identifying latent clusters of variability in longitudinal data. *Biostatistics* **2007**, *8*, 756–771. [CrossRef]

28. Kaplan, B.J.; Giesbrecht, G.F.; Leung, B.M.Y.; Field, C.J.; Dewey, D.; Bell, R.C.; Manca, D.P.; Obeirne, M.; Johnston, D.W.; Pop, V.J.; et al. The Alberta pregnancy outcomes and nutrition (APrON) cohort study: Rationale and methods. *Matern. Child Nutr.* **2012**, *10*, 44–60. [CrossRef]

29. Begum, F.; Colman, I.; Mccargar, L.J.; Bell, R.C. Gestational weight gain and early postpartum weight retention in a prospective cohort of Alberta women. *J. Obstet. Gynaecol. Can.* **2012**, *34*, 637–647. [CrossRef]

30. Che, M.; Kong, L.; Bell, R.C.; Yuan, Y. Trajectory modeling of gestational weight: A functional principal component analysis approach. *PLoS ONE* **2017**, *12*, e0186761. [CrossRef] [PubMed]

31. Martin, J.A.; Hamilton, B.E.; Ventura, S.J.; Osterman, M.J.; Kirmeyer, S.; Mathews, T.J.; Wilson, E.C. Births: Final data for 2009. *Natl. Vital Stat. Rep.* **2011**, *60*, 1–70. [PubMed]

32. Ruppert, D.; Wand, M.P.; Carroll, R.J. *Semiparametric Regression*; Cambridge Series in Statistical and Probabilistic Mathematics; Cambridge University Press: Cambridge, UK, 2003. [CrossRef]

33. Day, N.E. Estimating the components of a mixture of normal distributions. *Biometrika* **1969**, *56*, 463–474. [CrossRef]

34. Kass, R.E.; Natarajan, R. A default conjugate prior for variance components in generalized linear mixed models (Comment on article by Browne and Draper). *Bayesian Anal.* **2006**, *1*, 535–542. [CrossRef]

35. Pemstein, D.; Quinn, K.M.; Martin, A.D. The Scythe statistical library: An open source C library for statistical computation. *J. Stat. Softw.* **2011**, *42*, 1–26. [CrossRef]

36. R Core Team. *R: A Language and Environment for Statistical Computing*; R Foundation for Statistical Computing: Vienna, Austria, 2021.

37. Gelman, A.; Carlin, J.B.; Stern, H.S.; Rubin, D.B. Inference and assessing convergence In *Bayesian Data Analysis*, 3rd ed.; Chapman and Hall/CRC: Boca Raton, FL, USA, 2014; Chapter 11.4, pp. 281–285.

Article

Multivariate Functional Kernel Machine Regression and Sparse Functional Feature Selection

Joseph Naiman and Peter Xuekun Song *

Department of Biostatistics, University of Michigan, Ann Arbor, MI 48109, USA; jnaiman@umich.edu
* Correspondence: pxsong@umich.edu

Abstract: Motivated by mobile devices that record data at a high frequency, we propose a new methodological framework for analyzing a semi-parametric regression model that allow us to study a nonlinear relationship between a scalar response and multiple functional predictors in the presence of scalar covariates. Utilizing functional principal component analysis (FPCA) and the least-squares kernel machine method (LSKM), we are able to substantially extend the framework of semi-parametric regression models of scalar responses on scalar predictors by allowing multiple functional predictors to enter the nonlinear model. Regularization is established for feature selection in the setting of reproducing kernel Hilbert spaces. Our method performs simultaneously model fitting and variable selection on functional features. For the implementation, we propose an effective algorithm to solve related optimization problems in that iterations take place between both linear mixed-effects models and a variable selection method (e.g., sparse group lasso). We show algorithmic convergence results and theoretical guarantees for the proposed methodology. We illustrate its performance through simulation experiments and an analysis of accelerometer data.

Keywords: functional principal component analysis; functional predictor; linear mixed-effects model; mobile device; sparse group regularization; wearable device data

Citation: Naiman, J.; Song, P.X. Multivariate Functional Kernel Machine Regression and Sparse Functional Feature Selection. *Entropy* **2022**, *24*, 203. https://doi.org/10.3390/e24020203

Academic Editors: S. Ejaz Ahmed and Farouk Nathoo

Received: 4 January 2022
Accepted: 26 January 2022
Published: 28 January 2022

Publisher's Note: MDPI stays neutral with regard to jurisdictional claims in published maps and institutional affiliations.

1. Introduction

Data captured by mobile devices have lately received much attention in the data science community. Such data are typically recorded at a high frequency, giving rise to an ample volume of information at a very fine scale, and thus present many methodological challenges in statistical modeling and data analyses. In this paper, we plan to utilize the strength of the classical kernel machine method that enjoys fast computing speed via the linear mixed-effects model to deal with such high-frequency data using a functional data analysis approach. The motivation for our proposed framework come from data collected from a tri-axis accelerometer. Accelerometers, worn on the hip or wrist as a way of monitoring physical activity, are becoming more and more common [1–4]. There are several different accelerometers available such as ActiGraph GT3X+ (ActiGraph, Pensacola, FL, USA) and Actical (Phillips Respironics, Bend, OR). Raw accelerometer data are often collected in high-resolution signals with a sampling frequency ranging from 30–100 Hz. The commercial software on these devices provides activity counts (ACs) [2,4], which are calculated from the raw accelerometer data using proprietary algorithms. As an example from our motivating dataset, Figure 1 displays a three-dimensional time series of ACs per minute, each on one axis, from one subject wearing the GT3X+ over a period of 7 days (d).

Oftentimes, different types of summaries of the tri-axis ACs are suggested in the literature as opposed to the utility of all three raw functionals [5–8]. These summary-data-based approaches may be regarded as a quick and dirty dimension reduction strategy that comes up with summarized data with computationally manageable volumes, which would be then analyzed by existing methods and software. One concern with the use of summarized data would be the loss of potential fine features that can only be captured in data of high resolution. Recently, some researchers have attempted to use the entire functional AC curve through functional data analysis techniques [6,9,10]. Further details on current methods being used to retrieve and interpret accelerometer data can be found in [11]. Our contribution in this paper pertains to a new framework in that tri-axis accelerometer data are used as three-dimensional correlated functional predictors in an association analysis with a potential health outcome such as the Body Mass Index (BMI). The relationship between physical activities and childhood obesity has long been a central interest of public health sciences, and our new scalar-on-functional regression model can provide some new insights into this important scientific problem.

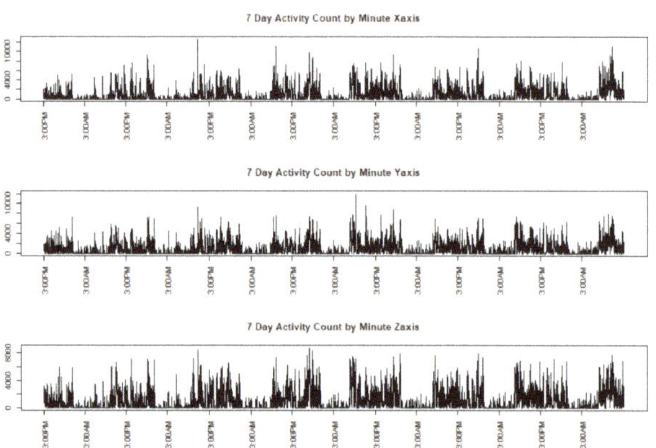

Figure 1. Activity counts over 7 d from a tri-axis (*X*-, *Y*- and *Z*-axis) accelerometer of a subject.

We begin with a brief review of existing functional data models, the least-squares kernel machine model, and different variable selection techniques, which prelude the framework for this paper.

1.1. Functional Regression

There has been much attention in recent years given to functional data analysis (FDA) where either covariates, or response, or both are functional as opposed to scalar in nature [12–17]. In this paper, we focused on the methodology that allows us to relate multiple functional covariates to a scalar outcome in a nonlinear way in the presence of other scalar covariates. To proceed, let us introduce some notation. Let $L^2(\mathcal{T})$ be the class of square-integrable functions on a compact set $\mathcal{T}$. This is a separable Hilbert space with inner product $< f, g >:= \int_{\mathcal{T}} fg$ for $f, g \in L^2(\mathcal{T})$. Consider a probability space $(\Omega, \mathcal{F}, P)$, where Z denotes a functional random variable that maps into $L^2(\mathcal{T})$, namely $Z : \Omega \mapsto L^2(\mathcal{T})$. Define $L^2(\Omega) := \{Z : (\int_{\Omega} \|Z\|^2 dP)^{\frac{1}{2}} < \infty\}$, where P is a certain probability measure, $\|Z\|^2 = < Z, Z >$, and assume $Z \in L^2(\Omega)$ in the rest of this paper. For convenience, we also assume that Z is mean centered, namely $E(Z) = 0$.

The class of functional linear models (FLM) (e.g., [13–15]) is proposed to relate a functional covariate Z with a mean-centered scalar outcome y, which is also known as scalar-on-functional regression: $y = < b, Z > + \epsilon$, where the error term ϵ is a mean zero random variable uncorrelated with Z. An optimal solution of the unknown functional parameter $b \in L^2(\mathcal{T})$ is typically obtained by minimizing the mean-squared error: $\inf_{b \in L^2(\mathcal{T})} E(y - < b, Z >)^2$. Moreover, the mean model for the mean-centered scalar y takes the form $E(y|Z) = \int_{\mathcal{T}} Z(t)b(t)dt$.

As suggested in the literature, we may obtain an optimal estimator of b by expanding functional predictor Z under certain basis functions. In this paper, we focus on the utility of functional principal component analysis (FPCA) to perform the decomposition of the functional Z. By the Karhunen–Loève expansion (e.g., [18–20]), we may write

$Z(t) = \sum_{k=1}^{\infty} \sqrt{\varsigma_k}\tilde{\xi}_k\phi_k(t)$, where $\varsigma_k > 0$ are the eigenvalues, and the loadings are given by $\tilde{\xi}_k := \frac{1}{\sqrt{\varsigma_k}} < Z, \phi_k >$. These coefficients satisfy (i) mean zero, $E(\tilde{\xi}_k) = 0$; (ii) variance one, $E(\tilde{\xi}_k^2) = 1$; (iii) uncorrelated, $E(\tilde{\xi}_k\tilde{\xi}_j) = 0$ for $k \neq j$. Then, the mean model may be rewritten as follows,

$$E(y|Z) = \sum_{k=1}^{\infty} \beta_k\tilde{\xi}_k, \tag{1}$$

where coefficients $\beta_k = < b, \sqrt{\varsigma_k}\phi_k >, k = 1, \cdots$, which are unknown due to the unknown b. Equation (1) presents a linear projection of scalar outcome y on the space spanned by the standardized principal components (PCs) $\tilde{\xi}_k$'s of functional predictor Z. On these lines of research, Müller and Yao (2008) proposed a class of functional additive models (FAMs) that extends Equation (1) by allowing a nonparametric form of the projection:

$$E(y|Z) = \sum_{k=1}^{\infty} f_k(\tilde{\xi}_k), \tag{2}$$

where f_k is a fully unspecified nonlinear smooth function to be estimated. It is obvious that Müller and Yao's extension given in (2) takes an additive model on individual coefficient (or feature) components $\tilde{\xi}_k$'s. Regularization is often needed for both (1) and (2) in order to deal with these infinite-dimensional unknowns. One of the challenges concerning regularization for (2) lies in the technical treatment in the functional space. Müller and Yao (2008) [21] proposed truncation (or a hard threshold) of the eigenspace to retain only the leading components that explain the majority of the total variation in Z. Zhu, Yao, and Zhang (2014) [15] proposed another regularization for the functions f_k using the powerful COSSO method [22]. One advantage for this kind of regularization method is that sums of higher-order functional principal components are allowed to be potentially included in the fit model, if they make stronger contributions to the functional relationship than the leading functional principal components. This regularization method [15] begins with an additive model $E(y|Z) = \sum_{k=1}^{s} f_k(\tilde{\xi}_k)$, where s represents some initial degrees of truncation to specify the total number of additive components to be considered. Then, COSSO helps simultaneously regularize and select important functional components among the s functions f_k. Although the above discussion is based on a single functional predictor Z in mind, it is appealing to extend such a framework with multiple functional predictors for a broad range of problems.

When multiple functional predictors, *say* $Z^1, \ldots, Z^p$, are considered, it is not clear if the above additive model specification remains suitable to handle the complexity, especially a non-additive relationship (e.g., interactions) may be of interest to understand the association between a scalar outcome and multiple functional predictors. In effect, from both the perspectives of theoretical advances and application needs, relaxing the additive relationship is an important task in functional data analysis. Alternatively, there are some methods (e.g., [16,17]) in the literature that do not use the strategy of decomposing Z into its functional components. In this paper, we adopt the framework of kernel machine regression models to extend the methodologies with non-additive relationships between multiple functional predictors and the scalar outcome.

1.2. Least-Squares Kernel Machine

Liu, Lin, and Ghosh (2007) [23] proposed a semi-parametric regression model $y_i = \mathbf{x}_i^{\top}\boldsymbol{\beta} + h(\mathbf{z}_i) + \epsilon_i$ for subject $i = 1, \ldots, n$, where they used the least-squares kernel machine (LSKM) to analyze multidimensional genetic pathways denoted by a vector $\mathbf{z}_i$. The key feature of this model is the nonlinear relationship between the outcome y_i and a vector of gene expressions $\mathbf{z}_i$, which is characterized by a nonparametric smooth function h. Under the theory of smoothing splines, function h is assumed to lie in a reproducing kernel Hilbert space (RKHS), $\mathcal{H}_{\mathcal{K}}$, generated by a positive-definite kernel function $\mathcal{K}(\cdot, \cdot)$. For the ease of exposition, we suppress the bandwidth for the kernel $\mathcal{K}$ in the following discussion.

Then, both parameter β and function h are estimated by maximizing the scaled penalized likelihood function:

$$J(h, \beta) = -\frac{1}{2} \sum_{i=1}^{n} \{ y_i - \mathbf{x}_i^{\top} \beta - h(\mathbf{z}_i) \}^2 - \frac{1}{2} \lambda_1 \|h\|_{\mathcal{H}_{\mathcal{K}}}^2, \tag{3}$$

where $\lambda_1 > 0$ is the tuning parameter and $\|\cdot\|_{\mathcal{H}_{\mathcal{K}}}$ is the norm of the RKHS. For a function $h \in L^2(\mathcal{H}_{\mathcal{K}})$, we have $h(\cdot) = \sum_{i=1}^{n} \alpha_i \mathcal{K}(\cdot, \mathbf{z}_i)$. Then, $\|h\|_{\mathcal{H}_{\mathcal{K}}}^2 = \alpha^{\top} \mathbf{K} \alpha$, where $\mathbf{K}$ is an $n \times n$ matrix whose (i, j) entry is $\mathcal{K}(\mathbf{z}_i, \mathbf{z}_j)$ and $\alpha = (\alpha_1, \ldots, \alpha_n)^{\top}$.

It is known in the literature (e.g., [23,24]) that maximizing $J(h, \beta)$ in (3) turns out to be equivalent to solving the normal equations from the following linear mixed-effects model (LMM): $\mathbf{Y} = \mathbf{X}\beta + \mathbf{h} + \epsilon$, where $\mathbf{h}$ is an $n \times 1$ vector of random effects with distribution $N(\mathbf{0}, \tau\mathbf{K})$ and an n-dimensional vector error term $\epsilon \sim N(\mathbf{0}, \sigma^2 \mathbf{I})$, with $\tau = \lambda_1^{-1}\sigma^2 > 0$. One remarkable advantage of solving (3) through the existing numerical procedure of the LMM is most advocated in the literature [25], where we can determine the smoothing parameter λ_1 as part of the estimation of the variance components of the LMM. Therefore, instead of using cross-validation or other information-based tuning methods on λ_1, we can solve simultaneously for all the model parameters in (3), as shown in [23]. Utilizing this numerical strength of the kernel machine regression model, we propose a semi-parametric regression model by incorporating functional principal components of functional predictors (i.e., the $\mathbf{z}_i$) to evaluate a nonlinear relationship of a scalar outcome with multiple functional covariates in a non-additive way. Assuming that function h belongs to an RKHS, we can use existing software packages for solving LMMs to obtain estimates of all model parameters and the smoothing parameter.

1.3. Feature Selection

To deal with high-dimensional functional principal components from functional covariates, we invoked the sparse regularization approach in the kernel machine regression model. Note that for both mean models (1) and (2), one needs to truncate the series from the Karhunen–Loève expansion. Regularization helps reduce from an infinite number of terms to a sum of finite terms. To introduce some notations, here we present a brief review on the group lasso (GL) [26], sparse group lasso (SGL) [27], and non-negative garrote [28]. See also the series of work originated by COSSO [22]. Yuan and Lin (2007) [26] proposed the group lasso, which solves the convex optimization problem: $\min_{\beta \in \mathcal{R}^p} \left\| \mathbf{Y} - \sum_{\ell=1}^{L} \mathbf{X}^{\ell} \beta^{\ell} \right\|_2^2 + \lambda \sum_{\ell=1}^{L} \left\| \beta^{\ell} \right\|_2$, where L is the total number of groups of covariates and $\mathbf{X}^{\ell}$ refers to a subset of covariates associated with group ℓ. Friedman, Hastie, and Tibshirani [27] extended the group lasso to allow within-group sparsity, namely SGL, given as $\min_{\beta \in R^p} \left\| \mathbf{Y} - \sum_{\ell=1}^{L} \mathbf{X}^{\ell} \beta^{\ell} \right\|_2^2 + \lambda(1 - \delta) \sum_{\ell=1}^{L} \left\| \beta^{\ell} \right\|_2 + \lambda\delta\|\beta\|_1$, where $\delta \in [0, 1]$. The additional ℓ_1-norm penalty term on β encourages individual sparsity, while the first penalty targets sparsity at the group level. It is easy to see that group lasso is a special case of the SGL when $\delta = 0$.

The non-negative garrote proposed by Breiman (1995) [28] is another useful means of variable selection. It invokes a scaled version of least-squares estimation given by: $\arg\min_{\mathbf{d}} \frac{1}{2} \left\| \mathbf{Y} - \tilde{\mathbf{X}}\mathbf{d} \right\|_2^2 + \lambda \sum_{j=1}^{p} d_j$, subject to $d_j \geq 0, j = 1, \ldots, p$. Here, $\tilde{\mathbf{X}} = (\tilde{\mathbf{x}}_1, \ldots, \tilde{\mathbf{x}}_p)$ is an $n \times p$ matrix with columns $\tilde{\mathbf{x}}_j = \mathbf{x}_j \hat{\beta}_j^{OLS}$, with $\hat{\beta}_j^{OLS}$ being the least-squares estimates from $\arg\min_{\beta} \frac{1}{2} \|\mathbf{Y} - \mathbf{X}\beta\|_2^2$ with no constraints. Obviously, estimate $\hat{d}_j = 0$ implies that covariate x_j would be excluded from the fit model. Breiman's formulation that turns a variable selection problem into a parameter estimation problem will be applied for the development of feature selection on functional principal components in this paper.

This paper is organized as follows. Section 2 introduces our proposed high-dimensional kernel machine regression. Section 3 outlines a simple step-by-step algorithm that is used to implement the sparse estimation method. Section 4 concerns asymptotic properties for our proposed sparse kernel machine regression. Section 5 provides simulation results to examine the performance of our method, with comparisons with existing methods. Section 6 illustrates the proposed method by an association analysis of the relationship between the BMI and functional accelerometer data. Section 7 includes our conclusions. The Appendix A contains some key technical details, including the proofs of the theoretical results, while Appendix B presents a discussion on the model identifiability issue.

2. Model and Estimation

Consider a regression analysis of a scalar outcome y on p functional covariates, Z^ℓ, $\ell = 1, \ldots, p$. Let $\mathbf{z}_i^\ell = (\xi_1^\ell, \ldots, \xi_{s_\ell}^\ell)_i^\top$ be the s_ℓ-element vector of functional principal component (FPC) features from the ith observation of the ℓth functional covariate Z^ℓ, and let $\bar{\mathbf{z}}_i = [(\mathbf{z}_i^1)^\top, \ldots, (\mathbf{z}_i^p)^\top]^\top$ be the grand vector of all FPC features from all p functional covariates for subject i, $i = 1, \ldots, n$. Clearly, the set of FPC features from each functional covariate forms a group, and in total, there are p groups with $s = \sum_{\ell=1}^p s_\ell$ many FPC features and $\bar{\mathbf{z}}_i \in \mathcal{R}^s$. The high dimensionality of FPC features presents the key methodological challenge in the analysis. We consider the following functional kernel machine regression (FKMR) model:

$$y_i = \mathbf{x}_i^\top \beta + h(\bar{\mathbf{z}}_i) + \epsilon_i, \ i = 1, \cdots, n, \tag{4}$$

where $\beta \in \mathcal{R}^q$ is a set of parameters for the effects of q scalar covariates $\mathbf{x} = (x_1, \ldots, x_q)^\top$, $h \in \mathcal{H}_\mathcal{K}$ is an s-variate smooth nonparametric function with $\mathcal{H}_\mathcal{K}$ being the functional space generated by a *Mercer kernel* $\mathcal{K}$ and error terms $\epsilon_i \overset{iid}{\sim} N(0, \sigma^2)$. The FKMR model (4) allows for not only nonlinear, but also non-additive relationships with multiple functional covariates Z^ℓ via their FPC features, $\ell = 1, \ldots, p$, and a scalar outcome, y. The statistical task is to estimate and select important functional covariates that are related to the outcome of interest through regularizing the FPC features within each functional covariate. To proceed, following Beiman's [28] non-negative garrote method, we here introduce a new s-dimensional scaling vector $\gamma \in \mathcal{R}^s$, $\gamma = (\gamma_1, \ldots, \gamma_{s_1}, \ldots, \gamma_s)^\top$, by which we can set $\gamma \circ \bar{\mathbf{z}}_i = (\gamma_1 \xi_1^1, \ldots, \gamma_{s_1} \xi_{s_1}^1, \ldots, \gamma_s \xi_{s_p}^p)_i^\top$ a new vector of weighted FPC features by γ via the Hadamard product (i.e., elementwise product). Note that γ is grouped and denoted by $\gamma = ((\gamma^1)^\top, \ldots, (\gamma^p)^\top)^\top$ where γ^ℓ is an s_ℓ-element vector of FPC features $\mathbf{z}^\ell$ of the ℓth functional covariate Z^ℓ. When the element, say γ_j, is equal to zero, the corresponding FPC feature ξ_j will not be selected in the set of important FPCs, and moreover, functional covariate Z^ℓ is excluded from the FKMR model when the entire vector $(\gamma^\ell)^\top = 0$.

We estimate the unknowns in the FKMR model (4), as well as the scaling parameters γ by minimizing the penalized objective function $J_1(h, \beta, \gamma)$, whose expression is given on the right-hand side of the following Equation (5):

$$\min_{h,\beta,\gamma} J_1(h, \beta, \gamma) = \min_{h,\beta,\gamma} \frac{1}{2n} \sum_{i=1}^n \{y_i - \mathbf{x}_i^\top \beta - h(\gamma \circ \mathbf{z}_i)\}^2 + \frac{1}{2}\lambda_1 \|h\|_{\mathcal{H}_K}^2 + \lambda_2 \rho(\gamma; \delta), \tag{5}$$

where $\lambda_1 > 0$ and $\lambda_2 > 0$ are two tuning parameters, and penalty $\rho(\gamma; \delta)$ may be specified according to a certain regularization method. For the case of sparse group lasso (SGL), we take $p(\gamma; \delta) = (1 - \delta) \sum_{\ell=1}^p \left\| \gamma^\ell \right\|_2 + \delta \|\gamma\|_1, \delta \in [0, 1]$. Typically, δ is predetermined and set to 0.95 or 0.05 depending on the trade-off between group and within-group sparsity, while the factor $(1 - \delta)$ controls the relative group sparsity to individual sparsity of each functional predictor Z^ℓ. Meanwhile, a large tuning parameter for λ_2 would remove a certain group of FPC features from the FKMR model when all elements in the vector γ^ℓ are zero. Given $h \in \mathcal{H}_\mathcal{K}$, an equivalent optimization to the above (5) can be formulated as follows:

$$\min_{\alpha,\beta,\gamma} J_2(\alpha, \beta, \gamma) = \min_{\alpha,\beta,\gamma} \frac{1}{2n} \sum_{i=1}^n \left\{ y_i - \mathbf{x}_i^\top \beta - \sum_{k=1}^n \alpha_k \mathcal{K}(\gamma \circ \bar{\mathbf{z}}_i, \gamma \circ \bar{\mathbf{z}}_k) \right\}^2$$
$$+ \frac{1}{2}\lambda_1 \alpha^\top \mathbf{K}(\gamma; Z)\alpha + \lambda_2 \rho(\gamma; \delta), \tag{6}$$

where $\mathbf{K}(\gamma; Z)$ is an $n \times n$ matrix whose (i,k)th element is $[\mathbf{K}(\gamma; \mathbf{Z})]_{ik} = \mathcal{K}(\gamma \circ \vec{z}_i, \gamma \circ \vec{z}_k)$. Lemma 1 below establishes the equivalency of optimization solutions between (5) and (6), which is crucial in our estimation procedure.

Lemma 1. *A solution $(\hat{h}, \hat{\beta}, \hat{\gamma})$ is a minimizer of (5) if and only if $(\hat{\alpha}, \hat{\beta}, \hat{\gamma})$ is a minimizer of (6), where $\hat{h}(\hat{\gamma} \circ \vec{z}) = \sum_{k=1}^{n} \hat{\alpha}_k \mathcal{K}(\hat{\gamma} \circ \vec{z}, \hat{\gamma} \circ \vec{z}_k)$.*

The proof of Lemma 1 is given in Appendix A.1.

Theorem 1 (Existence of optimizers). *If the kernel $\mathcal{K}(\cdot, \gamma \circ \vec{z})$ is continuous with respect to $\gamma \in \mathcal{R}^s$, then there exists a global minimizer $(\hat{h}, \hat{\beta}, \hat{\gamma})$ for the optimization problem (5).*

The proof of Theorem 1 is given in Appendix A.3. Note that there may exist multiple optimal minimizers for (5); Theorem 1 ensures only the existence of optimal solutions, but provides no guarantees for uniqueness due to the fact that (5) or (6) is a nonlinear and non-convex optimization problem. It is worth noting that in both (5) and (6), we set the bandwidth for the kernel at a fixed value due to the identifiability issue with respect to the scaling parameters γ. Refer to Appendix B for more detailed discussions on the issue of parameter identifiability.

3. Implementation and Algorithm

We propose an iterative algorithm to implement our proposed estimation procedure in which we require the differentiability of the kernel with respect to the scaling factor γ and some additional assumptions presented below in order to ensure algorithmic convergence. One part of the algorithm solving (5) is carried out under fixed γ, where the resulting minimization problem reduces to the equivalent maximization problem in the least-squares kernel machine (3) with the FPC features, $\vec{z}_i$, being replaced by $\gamma \circ \vec{z}_i$. As pointed out in Section 1.2, the step of numerical calculation can be easily executed in the same fashion as the solution from the linear mixed model, including the REML estimation of the smoothing parameter λ_1. The other part of the algorithm is performed under fixed α, β and λ_1, where we solve the nonlinear and non-convex optimization problem to update estimates of γ. Lemma 2 below helps us solve for the scaling parameter γ.

Lemma 2. *For fixed $(\alpha, \beta, \lambda_1)$, minimizing (6) over γ is equivalent to minimizing over γ the following objective function:*

$$\frac{1}{2n}\left\|\mathbf{F}(\gamma) - \tilde{\mathbf{Y}}\right\|_2^2 + \lambda_2 \rho(\gamma; \delta), \text{ for } \lambda_2 > 0, \tag{7}$$

where $\mathbf{F}(\gamma) = \mathbf{K}(\gamma; Z)\alpha$ and $\tilde{\mathbf{Y}} = \mathbf{Y} - \mathbf{X}\beta - \frac{n}{2}\lambda_1\alpha$.

The proof of Lemma 2 is given in Appendix A.2. Linearizing the function $\mathbf{F}(\gamma)$ in (7) leads to an equivalent form:

$$\min_{\gamma} \frac{1}{2n}\left\|\tilde{\mathbf{Y}} - \sum_{\ell=1}^{p} \nabla_{\gamma}\mathbf{F}^{(\ell)}(\tilde{\gamma})\gamma^{\ell}\right\|_2^2 + \lambda_2 \rho(\gamma; \delta), \tag{8}$$

where $\tilde{\mathbf{Y}} = \left(\mathbf{Y} - \mathbf{X}\beta - \frac{n}{2}\lambda_1\alpha\right) - \mathbf{F}(\tilde{\gamma}) + \nabla_{\gamma}\mathbf{F}(\tilde{\gamma})\tilde{\gamma}$, with $\nabla_{\gamma}\mathbf{F}(\tilde{\gamma})$ being the gradient of the function $\mathbf{F}$ with respect to γ evaluated at $\tilde{\gamma}$ for some $\tilde{\gamma}$, and $\nabla_{\gamma}\mathbf{F}^{(\ell)}(\tilde{\gamma})$ being the columns of $\nabla_{\gamma}\mathbf{F}(\tilde{\gamma})$ associated with the ℓth group of γ^{ℓ}. This is precisely the form of the standard sparse group regularization problem: $\min_{\beta \in \mathcal{R}^p} \frac{1}{2n}\left\|\mathbf{Y} - \sum_{\ell=1}^{p}\mathbf{X}^{\ell}\beta^{\ell}\right\|_2^2 + \lambda_2 \rho(\gamma; \delta)$. This implies that (8) presents a standard sparse group regularization problem with a specific choice of penalty function $\rho(\gamma; \delta)$.

The convergence of the above iterative search algorithm for updating $\tilde{\gamma}$ for fixed $(\alpha, \beta, \lambda_1)$ can be justified by the proximal Gauss–Newton method [29]. Readers are referred to [30] for details on the proximal Gauss–Newton method. One of the key assumptions of the proximal Gauss–Newton method is the existence of a local minimizer. This condition is satisfied in the above (8). This is because according to Theorem 1, there exists a global minimizer.

Algorithm 1 summarizes these iterative steps, which is showed to satisfy a descent property: $J_2(\alpha^{(r+1)}, \beta^{(r+1)}, \gamma^{(r+1)}) \leq J_2(\alpha^{(r)}, \beta^{(r)}, \gamma^{(r)})$ under the convergence of the proximal Gauss–Newton algorithm for Step 2.2.

Algorithm 1 An iterative algorithm for optimization in FKMR.

1.1   Perform FPCA (e.g., the R package `fdapace`) to extract the functional component features for the p functional predictors, and store them in a grand vector for each individual subject $\vec{z}_i = [(\mathbf{z}_i^1)^\top, \ldots, (\mathbf{z}_i^p)^\top)]^\top, i = 1, \cdots, n$;

1.2   Initialize γ to be a vector of ones. which translates to mapping the original component scores to itself. Set up a grid of possible tuning parameters for λ_1 and λ_2, respectively. Set the kernel bandwidth parameter, which may depend on λ_1. For each pair of (λ_1, λ_2) from our grid, perform Steps 2.1-2.3 and 3.1 below.

2.1   At the $(r+1)$-th step in the algorithm, first solve the LSKM problem with fixed $(\gamma^{(r)}, \lambda_1)$ (based on a closed-form solution) to update $\beta^{(r+1)}$ and $\alpha^{(r+1)}$.

2.2   Solve the group regularity problem (8) with fixed $\tilde{\gamma} = \gamma^{(r)}$ and fixed $(\alpha^{(r+1)}, \beta^{(r+1)}, \lambda_1, \lambda_2)$ using the $r+1$ updates from the previous iteration. At this step, the proximal Gauss–Newton algorithm produces an update $\gamma^{(r+1)}$ at convergence.

2.3   Repeat Steps 2.1–2.2 until convergence.

3.1   Perform cross-validation over all pairs of (λ_1, λ_2) to determine the final (α, β, γ).

To speed up Algorithm 1, we propose the following operational schemes that avoid setting up the pairs of (λ_1, λ_2) and performing Step 3.1. Here are a few remarks on the two algorithms. (i) Algorithm 2 depends on good starting values in order to enjoy a fast search. (ii) The main difference between Algorithms 1 and 2 is that λ_2 is fixed in Algorithm 1, while it is changing in Algorithm 2. Some similar algorithms with changing tuning parameters have been proposed in the literature, such as the single index model [31]. (iii) There is no guarantee that both algorithms converge to a global minimizer, and the proximal Gauss–Newton method used in the implementation can only find stationary points. Numerical solvers for the optimization problem in (5) or in (6) indeed remain an open problem in the field of nonlinear and nonconvex optimization.

Algorithm 2 A fast operational scheme of Algorithm 1.

---

1.   Step 2.1 of Algorithm 1 is performed by running the linear mixed model with our initial fixed γ from Step 1.2 of Algorithm 1 to obtained updated values of λ_1, β, and α.

2.   Step 2.2 is performed with solving the group regularity problem (8) through the Gauss–Newton algorithm using cross-validation-based tuning (e.g., R package `oem`).

3.   Rerun Step 2.1 using the updated γ from Step 2.2 to obtain the estimates for β and α.

---

4. Theoretical Guarantees

Our theoretical analysis focuses on the finite-sample L_2 error bounds for the estimators $(\hat{h}, \hat{\gamma})$ obtained by (5) or (6). Consequently, we are able to establish the estimation consistency. For simplicity, we set $\beta = 0$ and consider a general setting of random vectors $\mathbf{z}_1, \ldots, \mathbf{z}_n$ so that the FPC features $\vec{z}_1, \ldots, \vec{z}_n$ correspond to a special case. Along similar lines as those of [15,32], the estimation consistency is proven in the case of the SGL penalty function. We define a map Γ with an s-element vector $\gamma \in \mathcal{R}^s$, which gives rise to a collection of all scaling map functions. $\mathcal{A} = \{\Gamma . \mathcal{R}^s \mapsto \mathcal{R}^s \mid \Gamma(\mathbf{z}) = \gamma \cup \mathbf{z}, \mathbf{z} \in \mathcal{R}^s \text{ and } \gamma \in \mathcal{R}^s\}$. Since Γ is a linear (and bounded) operator, $\mathcal{A}$ is a real vector space where $(c_1\Gamma_1 + c_2\Gamma_2)(\mathbf{z}) = c_1\Gamma_1(\mathbf{z}) + c_2\Gamma_2(\mathbf{z})$ with any $c_1, c_2 \in \mathcal{R}$ and $\Gamma_1, \Gamma_2 \in \mathcal{A}$. To perform a group regularization estimation, we define an SGL penalty by a norm on $\mathcal{A}$ for a fixed $\delta \in [0,1]$ as follows:

$$\|\Gamma\|_{SGL} = \delta \sum_{\ell=1}^{p} \left\|\gamma^{\ell}\right\|_2 + (1-\delta)\|\gamma\|_1. \tag{9}$$

Consequently, the SGL regularization estimation requires the following constrained optimization:

$$\min_{\Gamma \in \mathcal{A},\, h \in \mathcal{H}_{\mathcal{K}}} J_3(\Gamma, h) = \min_{\Gamma \in \mathcal{A},\, h \in \mathcal{H}_{\mathcal{K}}} \|\mathbf{Y} - h \circ \Gamma\|_n^2 + \lambda_1 \|h\|_{H_K}^2 + \lambda_2 \|\Gamma\|_{SGL}, \tag{10}$$

where $\|\mathbf{Y} - h \circ \Gamma\|_n^2 = \frac{1}{n} \sum_{i=1}^{n} \{y_i - (h \circ \Gamma)(\mathbf{z}_i)\}^2$. Lemma 3 below provides the essential finite-sample inequalities that lead to the estimation consistency.

Lemma 3 (Basic inequality). *Let $\hat{h} \circ \hat{\Gamma}$ be the minimizer of* (10). *Let $h_0 \circ \Gamma_0$ be the true function. Then, we have:*

$$J_3(\hat{\Gamma}, \hat{h}) \leq 2(\boldsymbol{\epsilon}, \hat{h} \circ \hat{\Gamma} - h_0 \circ \Gamma_0)_n + \lambda_1 \|h_0\|_{\mathcal{H}_K}^2 + \lambda_2 \|\Gamma_0\|_{SGL}, \tag{11}$$

where $2(\boldsymbol{\epsilon}, \hat{h} \circ \hat{\Gamma} - h_0 \circ \Gamma_0)_n = \frac{2}{n} \sum_{i=1}^{n} \epsilon_i \left\{ (\hat{h} \circ \hat{\Gamma})(\mathbf{z}_i) - (h_0 \circ \Gamma_0)(\mathbf{z}_i) \right\}.$

We need the following notation before presenting our theoretical guarantees. Let $\mathcal{N}(\delta, M, P_n)$ denote the minimal δ covering number of the function set M under the empirical metric P_n based on the random vectors $\mathbf{z}_1, \cdots, \mathbf{z}_n$. Let $N = \mathcal{N}(\delta, M, P_n)$ be a shorthand notation. This means that there exist functions $m_1, \cdots, m_N$ (not necessarily in the set M) such that for every function $m \in M$, there exists a $j \in \{1, \cdots, N\}$ such that $\|m - m_j\|_{P_n} \leq \delta$, with $\|m - m_j\|_{P_n} := \sqrt{\frac{1}{n} \sum_{i=1}^{n} \{m(\mathbf{z}_i) - m_j(\mathbf{z}_i)\}^2}$. Define the δ-entropy of M for the empirical metric, P_n, as $H(\delta, M, P_n) := log(\mathcal{N}(\delta, M, P_n))$. Consider a functional space of the form:

$$\mathcal{B} = \left\{ b := b(h, \Gamma) = \frac{h \circ \Gamma - h_0 \circ \Gamma_0}{\|h\|_{\mathcal{H}_K}^2 + \|h_0\|_{\mathcal{H}_K}^2 + \|\Gamma\|_{SGL}^2 + \|\Gamma_0\|_{SGL}^2} \middle| h \in \mathcal{H}_K, \Gamma \in \mathcal{A} \right\}.$$

We postulate the following assumptions.

Assumption 1. *The error term* $\boldsymbol{\epsilon} = (\epsilon_1, \ldots, \epsilon_n)^{\top}$ *is uniformly sub-Gaussian; that is, for constants* C_1 *and* C_2,

$$\max_{n \geq 1} \max_{i=1,\cdots,n} C_1^2 \left[E\left\{ exp\left(\frac{\epsilon_i^2}{C_1^2} \right) \right\} - 1 \right] \leq C_2.$$

Clearly, the moment condition is bounded below from zero.

Assumption 2. $\|\Gamma_0\|_{SGL}^2 + \|h_0\|_{\mathcal{H}_K}^2 > 0$, *and the entropy of space* $\mathcal{B}$ *with respect to the empirical metric* P_n *is bounded as follows:*
$$H(\delta, \mathcal{B}, P_n) \leq C_3 \delta^{-2\psi},$$

where C_3 *is some constant and* $\psi \in (0,1)$.

Assumption 3. $\sup_{b \in \mathcal{B}} \|b\|_{P_n} \leq C_4$ *for some constant* C_4.

Theorem 2. *(Consistency) Under Assumptions 1-3 above, if tuning parameters* λ_1 *and* λ_2 *satisfy*

$$\lambda_2^{-1} = n^{\frac{1}{1+\psi}} \left(\|h_0\|_{\mathcal{H}_K}^2 + \|\Gamma_0\|_{SGL} \right)^{\frac{1-\psi}{1+\psi}}, \text{ and } \lambda_1 = O_p(1)\lambda_2,$$

then we have

$$\left\| \hat{h} \circ \hat{\Gamma} - h_0 \circ \Gamma_0 \right\|_n = O_p(n^{-\frac{1}{2+2\psi}})\left(\|h\|^2_{\mathcal{H}_K} + \|\Gamma\|_{SGL} \right)^{\frac{\psi}{1+\psi}}, \text{ and} \tag{12}$$

$$\left\| \hat{h} \right\|^2_{\mathcal{H}_K} + \|\hat{\Gamma}\|_{SGL} = O_p(1)\left(\|h_0\|^2_{\mathcal{H}_K} + \|\Gamma_0\|_{SGL} \right). \tag{13}$$

Theorem 2 implies estimation consistency under the right rates for the two tuning parameters λ_1 and λ_2. Due to the potential identifiability issues explained in detail in Appendix B, although the estimator $(\hat{h}, \hat{\Gamma})$ may not be unique, the sum of $\hat{h}$ and $\hat{\Gamma}$ is not too far away from the sum of the true h_0 and Γ_0.

Corollary 1. *If the RKHS, $\mathcal{H}_K$, contains differentiable functions $\nabla h(\mathbf{z})$ whose norm $\|\nabla h(\mathbf{z})\|_{\mathcal{H}_K}$ is uniformly bounded for all functions $h \in \mathcal{H}_K$ and $\mathbf{z} \in R^s$, then Assumption 2 holds when Theorem 2 is replaced by $H(\delta, \mathcal{H}_K, P_n) \leq C_1\delta^{-2\psi}$, for all $\delta \geq 0$.*

The proofs of Theorem 2 and Corollary 1 are given in Appendices A.4 and A.5, respectively. Often, when we are only interested in a subset of functions in the RKHS (e.g., functions with norm less than one), we can substitute the full space $\mathcal{H}_K$ in Corollary 1 with the subspace of interest. Refer to [15] or [32], where both considered an RKHS (i.e., Sobolev space) with functions of norm less than or equal to one.

5. Simulation Experiments

We performed extensive simulation to investigate the performance of our proposed procedure, including the performance of SGL variable selection and its overall accuracy. Due to the limitations of space, we include results from two simulation experiments in this section, and more results may be found in the first author's Ph.D. dissertation [30].

5.1. Setup

In the evaluation of the performance accuracy, following [15], we used both quasi-R^2 and adjusted quasi-R^2 defined as follows:

$$R^2_Q := 1 - \frac{\sum_{i=1}^n (y_i - \hat{y}_i)^2}{\sum_{i=1}^n (y_i - \bar{y}_i)^2}, \text{ and } R^2_{AQ} := 1 - \left(1 - R^2_Q\right)\left(\frac{n-1}{n-(k+1)}\right).$$

The latter is known to be appealing for the comparison of the estimation sparsity. There is another performance metric of interest in addition to model accuracy. Performance in variable selection is summarized in terms of the stability measured by sensitivity and specificity for both functional and variable selections under these simulation experiments. Our algorithm uses existing R packages, including `emmreml`, `kspm`, and `oem`.

Specifically, we designed the following two simulation settings.

Scenario 1: A single functional predictor with sparsity in the FPC features.

Scenario 2: Multiple functional predictors with sparsity in the functional predictors and with sparsity in the FPC features of important functional predictors.

Each of these two scenarios would be handled using certain suitable penalty functions to address the designed sparsity; for example, in Scenario 2 we used a two-level variable selection penalty (e.g., SGL) to deal with two types of sparsity in the true model. In all analyses, we used the Gaussian kernel $\mathcal{K}(u, v) = \exp(-\frac{1}{p}\|u - v\|^2)$ in our estimation, where p was set as the number of features, which is equivalent to dividing the γ vector by $\sqrt{p}$. This scaling parameter may be either estimated or set to the number of features to overcome the identifiability issue according to [33], where theoretical justification was given for the use of the number of features for the bandwidth parameter in the case of the Gaussian kernel.

According to [23], due to the difficulty of the graphical display for the estimated s-dimensional function $h(\cdot)$ of $\mathbf{z}$, we summarized the goodness-of-fit by regressing the true h on the estimated $\hat{h}$, with both being evaluated at the design points. From this concordance regression analysis, we may measure the goodness-of-fit on $\hat{h}$ through the average intercepts, slopes, and R-squared (also known as the coefficient of determination) obtained over the number of replications. Clearly, a high-quality fit is reflected by (i) the intercept being close to zero, (ii) the slope being close to one, and (iii) the R-squared being close to one. Moreover, we graphically display the estimated function $\hat{h}$ by setting all variables equal to 0.5 except the one of interest over a grid of 100 equally spaced points on the interval $[0, 1]$. Such visualization of the functional estimation at each margin further facilitates the evaluation of the proposed algorithm in addition to the results obtained from the concordance regression analyses.

In all scenarios, we generated 1000 IID functional paths, of which 750 paths were assigned to the training set and 250 paths were assigned to the test set for an external performance evaluation. It is the test set that we used to display the performance accuracy. We used a one-dimensional covariate x_i to show the flexibility of our model in a semi-parametric setting, with independent copies of $x_i \sim N(0, 1)$. We chose the true coefficients in the kernel machine model similar to those given in [23].

5.2. Simulation in Scenario 1

In this simple scenario with a single functional predictor, we simulated data from a model with sparsity in its FPC features. To do so, we generated a single functional predictor based on the first 15 eigenbasis of the Fourier basis functions over the interval $[0, 1]$: $Z(t) = \sum_{j=1}^{15} \sqrt{\varsigma_j} \xi_j \phi_j(t)$. That is, a functional predictor was created as a linear combination of the 15 basis functions, where $\phi_j(\cdot)$ is the jth Fourier basis function, ς_j is the jth eigenvalue of Z, and ξ_j is the jth FPC feature that is simulated from a normal distribution detailed as follows.

There were 100 sampled points that were first equally spaced in the interval $[0, 1]$ and then varied with certain small deviations drawn from $\nu \sim N(0, 0.001)$. Set $\varsigma_j = 45 \times 0.64^j$ and $\xi_j \sim N(0, 1)$ independently over $j = 1, \ldots, 15$. As was done in [17], instead of directly using ξ_j, we used $\zeta_j = \Phi(\xi_j)$, where Φ is the CDF of the standard normal. This resulted in $\bar{\mathbf{z}} = (\zeta_1, \ldots, \zeta_{15})^\top$. We chose the second, ζ_2, and ninth, ζ_9, features as important features in the following true nonlinear non-additive model:

$$y_i = 2x_i + 20 \cos(2\pi\zeta_{i2}) - 10 \sin(2\pi\zeta_{i9}) + \zeta_{i2}\zeta_{i9} + \epsilon_i,$$

with $\epsilon_i \overset{iid}{\sim} N(0, 1)$. FPCA was performed by the R package PACE [34], producing the estimated FPC scores, $\hat{\xi}_j$, as well as the estimated eigenvalues, $\hat{\varsigma}_j$, which in turn enabled us to compute $\hat{\zeta}_j$, $j = 1, \ldots, 15$.

We applied both LASSO and MCP penalty functions in our implementation, termed as $FKMR_{Lasso}$ and $FKMR_{MCP}$, respectively. We compared the results of our method with the standard linear approach with both LASSO and MCP under the assumption of linear functional relationships, as well as the COSSO method for functional additive regression [15] using the R package COSSO [15,34]. Since the COSSO package is built for nonparametric regression (and not partial linear models), we adopted the backfitting strategy and regressed the residuals with our estimated effect of x_i removed.

In addition, we compared our method with an oracle FKMR estimator, called $FKMR^{oracle}$, that assumed the full knowledge of the true ζ_j containing two true nonzero signals, ζ_2 and ζ_9. We also considered two oracle versions of our proposed algorithm, $FKMR^{oracle}_{Lasso}$ and $FKMR^{oracle}_{MCP}$, both of which used the knowledge of true ζ_j in order to evaluate the performance of the FPCA procedure. This evaluation is important as our proposed procedure can be in principle used in simpler cases that do not involve functional covariates. Note that once we used FPCA to obtain $\hat{\zeta}_j$ features, our algorithm essentially works in a standard regression setting with the sparsity of covariates. Thus, our proposed procedure can be in principle used in simpler cases with scalar covariates. In Scenario 1, due to the highly nonlinear relationships between the FPC features and the outcome, as expected, the naive linear model performed poorly in terms of both model selection and model consistency. The detailed simulation results for Scenario 1 can be found in the first author's Ph.D. dissertation [30]. In brief, our proposed method worked well in all aspects. In this setting, COSSO also worked well in terms of model fit, but it tended to select noisy features more frequently than our proposed method, leading to more false positives.

5.3. Simulation in Scenario 2

Now, we generated four functional predictors of the form: $Z^\ell(t) = \sum_{j=1}^{9} \sqrt{\varsigma_j^\ell} \xi_j^\ell \phi_j^\ell(t)$, $\ell = 1, \ldots, 4$, where ϕ_j^ℓ, ς_j^ℓ, and ξ_j^ℓ were set in the same way as those given in Scenario 1. It follows that $\vec{z} = (\zeta_1^1, \ldots, \zeta_9^1, \ldots, \zeta_1^4, \ldots, \zeta_9^4)^\top$, where ζ_j^ℓ is the jth Φ-transformed feature for the ℓth functional covariate. Sparsity was specified as follows: the first and second functional covariates, Z^1 and Z^2, were chosen as important signals in which these transformed FPC features, $\{\zeta_1^1, \zeta_3^1, \zeta_4^1, \zeta_2^2, \zeta_7^2\}$, are five important features (three features from the Z^1 and two features from Z^2) that are related to the outcome:

$$
\begin{aligned}
y_i = {} & 2x_i + \zeta_{i1}^1 + \zeta_{i3}^1 + \zeta_{i4}^1 + \zeta_{i2}^2 + \zeta_{i7}^2 + 10\cos(2\pi\zeta_{i1}^1) - 10\left(\zeta_{i2}^2\right)^2 + 10\left(\zeta_{i7}^2\right)^2 - 10\left(\zeta_{i3}^1\right)^2 \\
& + 10\exp(-\zeta_{i3}^1)\zeta_{i4}^1 - 8\sin(2\pi\zeta_{i7}^2)\cos(2\pi\zeta_{i3}^1) + 20\zeta_{i1}^1\zeta_{i7}^2 + \epsilon_i, \quad i = 1, \ldots, n,
\end{aligned}
$$

where $\epsilon_i \overset{iid}{\sim} N(0,1)$. This model specifies both group sparsity (two of the four functional predictors) and within-group sparsity (three of the nine FPC features in Z^1 and two of the nine FPC features in Z^2). In addition, we specified non-additive relationships in the true model across multiple functional covariates.

We fit the data using the proposed methods, including $FKMR_{GMCP}^{oracle}$, $FKMR_{Lasso}$, $FKMR_{GLasso}$, $FKMR_{SGL}$, $FKMR_{MCP}$, and $FKMR_{GMCP}$, and the results based on 100 replicates are summarized in Table 1. For comparison, we also fit the simulated data by existing methods, including the linear model (denoted by LM + penalty), COSSO functional additive regression, and the oracle method using the knowledge of true important features in the analysis, as done in the above simulation of Scenario 1. From Table 1 regarding the goodness-of-fit, we see that all of our FKMR estimators outperformed the standard linear estimators in terms of R_{AQ}^2 among all of our penalty functions, and they outperformed COSSO for penalties that accounted for group sparsity. In the concordance regression analysis, we see that all intercepts were close to zero, all slopes close to one, and all R^2 close to one, indicating a high goodness-of-fit for functional estimation. COSSO tended to perform on par for penalties that did not account for group sparsity (LASSO and MCP). It is evident that using a group sparsity penalty function (SGL, GLasso, and GMCP) clearly outperformed the methods that did not regularize the grouping of covariates (Lasso and MCP). In addition, our FKMR estimators (except $FKMR_{Lasso}$) performed as well as the oracle estimator $FKMR_{GMCP}^{oracle}$ both in terms of R_{AQ}^2 and in terms of our estimate of functional h. The results also indicated that there were little differences between using a concave (MCP or GMCP) penalty function or using a convex (GLasso or SGL) penalty function.

As regards the group sparsity, Table 2 indicates that the all methods had a high sensitivity of detecting functional signals, while the proposed FKMR methods had better specificity than both sparse linear models and COSSO. Concerning the within-group sparsity, it is interesting to note that a bigger difference was seen in terms of what type of penalty function was being used in feature selection. As shown in Tables 3 and 4, using a general penalty (e.g., Lasso and MCP) that does not take the grouping structure into account tended to under-select important features within a group. COSSO tended to perform well within group sparsity. Moreover, Figure 2 shows that the FKMR method estimated the five signal functions (Z^1 and Z^2) well.

Table 1. Goodness-of-fit and the concordance regression for Scenario 2.

Model	R^2_{AQ}	β	Reg of h on $\hat{h}$		
			Intercept	Slope	R^2
$FKMR_{Lasso}$	0.830	2.00	-0.062	1.01	0.848
$FKMR_{GLasso}$	0.937	1.99	-0.055	1.01	0.972
$FKMR_{SGL}$	0.928	2.00	-0.051	1.01	0.955
$FKMR_{MCP}$	0.835	2.01	-0.062	1.01	0.856
$FKMR_{GMCP}$	0.935	1.99	-0.056	1.01	0.970
$FKMR_{GMCP}^{oracle}$	0.911	1.99	-0.049	1.01	0.937
COSSO	0.832	–	–	–	–
LM + Lasso	0.453	–	–	–	–
LM + GLasso	0.324	–	–	–	–
LM + SGL	0.450	–	–	–	–
LM + MCP	0.513	–	–	–	–
LM + GMCP	0.307	–	–	–	–

Table 2. Sensitivity and specificity of functional selection for Scenario 2.

Model	Selection Frequency			
	$\hat{Z}^1$	$\hat{Z}^2$	$\hat{Z}^3$	$\hat{Z}^4$
$FKMR_{Lasso}$	100	100	0	0
$FKMR_{GLasso}$	100	100	4	4
$FKMR_{SGL}$	100	100	0	0
$FKMR_{MCP}$	100	100	0	0
$FKMR_{GMCP}$	100	100	3	4
COSSO	100	100	5	6
LM + Lasso	100	100	19	21
LM + GLasso	94	99	7	8
LM + SGL	100	100	19	18
LM + MCP	100	100	20	19
LM + GMCP	93	99	7	8

Table 3. FPC feature selection for signal functional Z^1 in Scenario 2.

Model	Selection Frequency								
	$\hat{\zeta}^1_1$	$\hat{\zeta}^1_2$	$\hat{\zeta}^1_3$	$\hat{\zeta}^1_4$	$\hat{\zeta}^1_5$	$\hat{\zeta}^1_6$	$\hat{\zeta}^1_7$	$\hat{\zeta}^1_8$	$\hat{\zeta}^1_9$
$FKMR_{Lasso}$	100	1	97	0	0	0	0	0	0
$FKMR_{GLasso}$	100	100	100	100	100	100	100	100	100
$FKMR_{SGL}$	100	21	100	71	26	20	17	16	15
$FKMR_{MCP}$	100	1	99	1	0	0	0	0	0
$FKMR_{GMCP}$	100	100	100	100	100	100	100	100	100
COSSO	100	2	100	93	1	0	0	1	0
LM + Lasso	100	10	100	100	10	8	7	10	5
LM + GLasso	94	94	94	94	94	94	94	94	94
LM + SGL	100	12	100	100	10	8	8	11	5
LM + MCP	100	10	100	100	9	8	9	7	5
LM + GMCP	93	93	93	93	93	93	93	93	93

Table 4. FPC feature selection for signal functional Z^2 in Scenario 2.

Model	Selection Frequency								
	$\hat{\zeta}^2_1$	$\hat{\zeta}^2_2$	$\hat{\zeta}^2_3$	$\hat{\zeta}^2_4$	$\hat{\zeta}^2_5$	$\hat{\zeta}^2_6$	$\hat{\zeta}^2_7$	$\hat{\zeta}^2_8$	$\hat{\zeta}^2_9$
$FKMR_{Lasso}$	0	3	0	0	0	0	100	0	0
$FKMR_{GLasso}$	100	100	100	100	100	100	100	100	100
$FKMR_{SGL}$	16	100	14	7	16	23	100	15	7
$FKMR_{MCP}$	0	11	0	0	0	1	100	0	0
$FKMR_{GMCP}$	100	100	100	100	100	100	100	100	100
COSSO	8	97	5	5	5	15	100	3	3
LM + Lasso	17	100	14	7	16	23	100	15	6
LM + GLasso	99	99	99	99	99	99	99	99	99
LM + SGL	17	100	14	7	16	23	100	15	7
LM + MCP	17	100	13	6	16	23	100	15	8
LM + GMCP	99	99	99	99	99	99	99	99	99

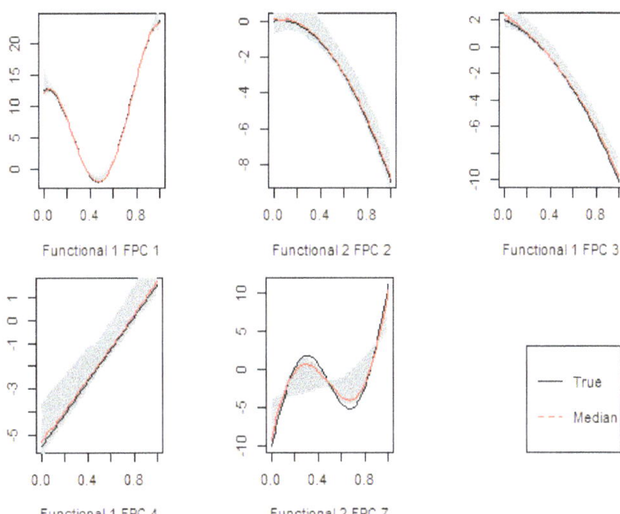

Figure 2. Five marginal estimates of important feature functions with 95% shaded confidence bands evaluated at 100 grid points while holding all other components equal to 0.5 in Scenario 2.

6. Data Example

To show the usefulness of our proposed methodology, we analyzed data of 550 children recruited by the ELEMENTS study [35], who had consent to wear an actigraph (ActiGraph GT3X+; ActiGraph LLC. Pensacola, FL, USA). This wearable was to be placed on their non-dominant wrist for five to seven days with no interruption. The actigraph measured tri-axis accelerometer data sampled at 30 Hz, which captured three different directions of a person's movement. The BMI was the outcome of interest as it is biomarker of obesity. Sex and age were confounding factors used in the analysis. Due to some missing data, our analysis only included children who wore the device properly for 85% or more over the study period, which resulted in 395 participants, consisting of 189 males and 206 females. Other studies such as [36] have excluded days of accelerometer data with more than five percent missing. The mean ± SD BMI of the study cohort was 21.5 ± 4.1. The mean age of the study participants was 14.3 ± 2.1 y. A more detailed description of the dataset used for this paper can be found in [37]. Our primary interest was to see if the BMI is associated with physical activity in the presence of other covariates, specifically sex and age. We preprocessed the activity counts over the 7 d of wear by taking the median in the 1 min epoch over the entire 7 d of wear. For example, since all the participants started wearing the device at 3 p.m., the first data point for each individual was a median of 7 ACs (each for one day) for the 1 min epoch of 3:00–3:01 p.m. This procedure that takes the medians across the minutes from different days has been considered in other applications such as [36]. See Figure 3 as an example of the resulting time series of medians derived from the AC data displayed in Figure 1.

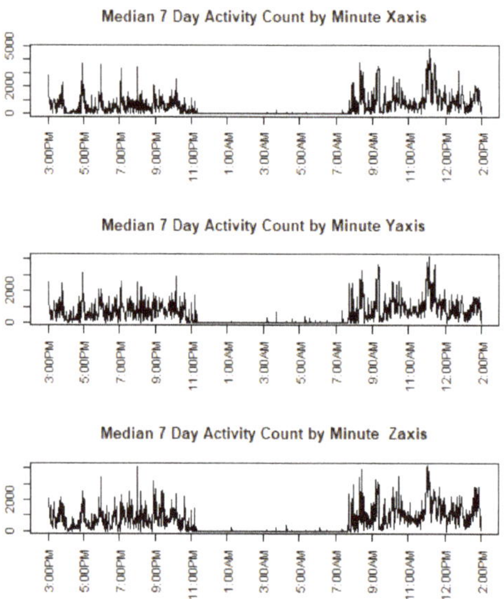

Figure 3. The 24 h minute-by-minute medians of 7 d ACs for one subject.

We applied the following five models, labeled as M0–M4 for convenience, to analyze the data with the 24 h median ACs as functional predictors. Let ξ_{ij}^k be the ith person's kth FPC score for functional predictor j.

M0: Linear model (LM) with only the fixed features: $BMI_i \sim \beta_0 + \beta_1 Age_i + \beta_2 Sex_i$;

M1: Linear model with SGL penalty (LM+SGL) using the FPCA features: $BMI_i \sim \beta_0 + \beta_1 Age_i + \beta_2 Sex_i + \sum_{j=1}^{3} \sum_{k=1}^{s_k} \beta_j^k \xi_{ij}^k$;

M2: LSKM using the FPCA features: $BMI_i \sim \beta_0 + \beta_1 Age_i + \beta_2 Sex_i + h(\mathbf{z}_i)$;

M3: FKMR model with SGL penalty ($FKMR_{SGL}$) using the FPCA features: $BMI_i \sim \beta_0 + \beta_1 Age_i + \beta_2 Sex + h(\gamma \circ \mathbf{z}_i)$;

M4: COSSO using the FPCA features: $res(BMI_i)|\mathbf{z}_i \sim \sum_{j=1}^{3} \sum_{k=1}^{s_k} f_{ij}(\xi_{ij}^k)$. In order for a direct application of the COSSO R package, we used residuals $res(BMI_i) = BMI_i - \hat{\beta}_0 + \hat{\beta}_1 Age_i + \hat{\beta}_2 Sex_i$ in the COSSO model fit, with $\hat{\beta}_0, \hat{\beta}_1$ and $\hat{\beta}_2$ being the estimates of the coefficients from Model M0.

The BMI and age were mean centered and scaled to be a standard deviation of one, so β_0 was absent in the models. Here are some key findings from the data analyses. First, in terms of the goodness-of-fit, Table 5 suggests that M3, i.e., our proposed model FKMR with the SGL penalty, gave the best performance, where the adjusted R^2 of M3 was nearly twice as big as all the other four models. Second, it is interesting to note that both the COSSO and the $FKMR_{SGL}$ did not select the FPC scores associated with the Z-axis. Third, as shown in Table 6, all of the FPC components chosen by COSSO were also chosen by the $FKMR_{SGL}$. It is worth noting that the linear model together with the SGL penalty selected the highest number of FPC components, yet performed the worst in terms of the model fit.

Table 5. Goodness-of-fit for the five models used in the data analysis.

Model	Adjusted R^2
M0: LM	0.07
M1: LM + SGL	0.13
M2 : LSKM	0.18
M3: $FKMR_{SGL}$	0.30
M4: COSSO	0.14

Table 6. Axis-specific FPC feature selection.

Model	X-Axis						Y-Axis					Z-Axis			
	$\hat{\zeta}_1^1$	$\hat{\zeta}_2^1$	$\hat{\zeta}_3^1$	$\hat{\zeta}_4^1$	$\hat{\zeta}_5^1$	$\hat{\zeta}_6^1$	$\hat{\zeta}_1^2$	$\hat{\zeta}_2^2$	$\hat{\zeta}_3^2$	$\hat{\zeta}_4^2$	$\hat{\zeta}_5^2$	$\hat{\zeta}_1^3$	$\hat{\zeta}_2^3$	$\hat{\zeta}_3^3$	$\hat{\zeta}_4^3$
$FKMR_{SGL}$		✓	✓	✓		✓	✓		✓		✓				
COSSO				✓			✓		✓						
LM + SGL	✓			✓	✓	✓	✓	✓	✓						✓

7. Conclusions

In this paper, we proposed a method to model the nonlinear relationship between multiple functional predictors and a scalar outcome in the presence of other scalar confounders. We used the FPCA to decompose the functional predictors for feature extraction and used the LSKM framework to model the functional relationship between the outcome and principal components. We developed a simultaneous procedure to select important functional predictors and important features within selected functionals. We proposed a computationally efficient algorithm to implement our regularization method, which was easily programmed in R with the utility of multiple existing R packages. It should be noted that although we focused on functional regression in this paper, the method proposed can be applied to non-functional predictors. In effect, by using functional principal components, we essentially bypassed the infinite-dimensional problem and worked effectively in a non-functional framework with the FPC features. Through simulation and using data from the ELEMENT dataset, we demonstrated how the FKMR estimator outperformed existing methods in terms of both variable selection and model fit. It should be noted that the existing COSSO method did perform well in terms of variable selection, as shown in Section 5.

A technical issue pertains to identifiability limitations with regard to the bandwidth parameter and to the RKHS estimator. To overcome this, we suggested fixing the bandwidth parameter; see the detailed discussion in Section 3. We established key theoretical guarantees for our proposed estimator. In the case where there are multiple proposed estimators (and thus the identifiability issues arise), the established theoretical properties in Section 4 apply to any of those estimators.

Variable section on functional predictors presents many technical challenges, and there are many methodological problems that remain unsolved. This paper demonstrated a possible framework to regularize estimation with a bi-level sparsity of functional group sparsity and within-group sparsity. In the LSKM paper [23], it was briefly mentioned that if the relationship between the scalar outcome and p genetic pathways is additive, we can tweak the model as $y_i = x_i^\top \beta + h_1(z_i^1) + \cdots + h_p(z_i^p) + \epsilon_i$ where each h_j belongs to its own RKHS. It is easy to extend our method and algorithms to handle this case. For future research, an extension on longitudinal outcomes may be considered via a mixed-effects model $y_{ij} = x_i^\top \beta + h(z_{ij}) + u_{ij}^\top v_i + \epsilon_{ij}$ where $u_{ij}^\top v_i$ are the random effects. Other useful extensions to the proposed paradigm would be on the lines of generalized linear models and Cox regression models.

Author Contributions: Conceptualization, P.X.S. and J.N.; Formal analysis, J.N.; Methodology, J.N. and P.X.S.; Supervision, P.X.S.; Writing—original draft, J.N.; Writing—review & editing, P.X.S. All authors have read and agreed to the published version of the manuscript.

Funding: This work was supported in part by NSF DMS#2113564.

Institutional Review Board Statement: Not applicable.

Informed Consent Statement: Not applicable.

Data Availability Statement: The used data of physical activity counts, BMI and demographic variables (sex and age) are available upon request through a formal data request procedure outlined by the ELEMENT Cohort Study. Contact the corresponding author of this paper for the detail.

Conflicts of Interest: The authors declare no conflict of interest.

Appendix A. Technical Assumptions and Proofs

Appendix A.1. Proof of Lemma 1

It suffices to show that for any $J_1(h, \beta, \gamma)$ in (5) we can always find $\alpha \in \mathcal{R}^n$ such that $J_1(\tilde{h} = \sum_{i=1}^n \alpha_i \mathcal{K}(\cdot, \gamma \circ \vec{z}_i), \gamma, \beta) \leq J_1(h, \beta, \gamma)$ where $\tilde{h}$ is the projection of h onto the linearly spanned space given by $span\{\mathcal{K}(\cdot, \gamma \circ \vec{z}_i), \cdots, \mathcal{K}(\cdot, \gamma \circ \vec{z}_n)\}$. For any h we can write $h = h^\perp + \tilde{h}$ where $h^\perp \in span\{\mathcal{K}(\cdot, \gamma \circ \vec{z}_i), \cdots, \mathcal{K}(\cdot, \gamma \circ \vec{z}_n)\}^\perp$. Since $\mathcal{H}_k$ is a reproducing kernel Hilbert space we can rewrite (5) as follows:

$$J_1(h, \gamma, \beta) = \frac{1}{2n} \sum_{i=1}^n \{y_i - \mathbf{x}_i^\top \beta - <h, \mathcal{K}(\cdot, \gamma \circ \vec{z}_i)>\}^2 + \frac{1}{2}\lambda_1 \|h\|_{\mathcal{H}_k}^2 + \lambda_2 \rho(\gamma; \delta).$$

Since $<h^\perp, \mathcal{K}(\cdot, \gamma \circ \vec{z}_i)>= 0$ for every i, we obtain

$$J_1(h, \gamma, \beta) = \frac{1}{2n} \sum_{i=1}^n \left\{ y_i - \mathbf{x}_i^\top \beta - \sum_{k=1}^n \alpha_k \mathcal{K}(\gamma \circ \vec{z}_i, \gamma \circ \vec{z}_k)) \right\}^2 + \frac{1}{2}\lambda_1 \left\| h^\perp + \tilde{h} \right\|_{\mathcal{H}_k}^2 + \lambda_2 \rho(\gamma; \delta)$$

$$\geq \frac{1}{2n} \sum_{i=1}^n \left\{ y_i - \mathbf{x}_i^\top \beta - \sum_{k=1}^n \alpha_k \mathcal{K}(\gamma \circ \vec{z}_i, \gamma \circ \vec{z}_k)) \right\}^2 + \frac{1}{2}\lambda_1 \|\tilde{h}\|_{\mathcal{H}_k}^2 + \lambda_2 \rho(\gamma; \delta)$$

$$= J_1(\tilde{h}, \gamma, \beta).$$

Appendix A.2. Proof of Lemma 2

The equivalence of forms become clear once we rewrite (6) in the matrix notation. Equation (6) can be written as follows:

$$\min_{\alpha, \beta, \gamma} J_2(\alpha, \beta, \gamma) = \min_{\alpha, \beta, \gamma} \frac{1}{2n} \|\mathbf{Y} - \mathbf{X}\beta - \mathbf{K}(\gamma; \mathbf{Z})\alpha\|_2^2 + \frac{1}{2}\lambda_1 \alpha^\top \mathbf{K}(\gamma; \mathbf{Z})\alpha + \lambda_2 \rho(\gamma; \delta). \quad \text{(A1)}$$

For fixed α, β and λ_1, minimizing the function in (A1) with respect to γ is equivalent to

$$\min_\gamma \left\{ \frac{1}{2n} \left\| \left(\mathbf{Y} - \mathbf{X}\beta - \frac{n}{2}\lambda_1 \alpha\right) - \mathbf{K}(\gamma; \mathbf{Z})\alpha \right\|_2^2 + \lambda_2 \rho(\gamma; \delta) \right\}. \quad \text{(A2)}$$

Appendix A.3. Proof of Theorem 1

With loss of the generality we use the penalty function for sparse group lasso but this proof can easily be modified for other penalty functions. Also, we fix $\lambda_1 = \lambda_2 = \delta = 1$, and consider $\beta \in \mathcal{R}$ as well as set the design matrix $\mathbf{X}$ (or vector in this case) scaled to have norm 1. The case of $\beta \in \mathcal{R}^q$ will follow along similar lines of arguments. Let $\gamma \in D_3$ with $D_3 = \{\gamma : \|\gamma\|_1 \leq \frac{1}{2n}\|\mathbf{Y}\|_2^2\}$. Define $f(\gamma) = \|\mathbf{K}(\gamma; \mathbf{Z})\| = \eta_{max}(\mathbf{K}(\gamma; \mathbf{Z})) \geq 0$, where $\eta_{max}(\mathbf{K}(\gamma; \mathbf{Z}))$ denotes the largest eigenvalue of $\mathbf{K}(\gamma; \mathbf{Z})$ with the operator norm (the norm of $\mathbf{K}(\gamma; \mathbf{Z})$) defined in its usual way $\|\mathbf{K}(\gamma; \mathbf{Z})\| = sup\{\|\mathbf{K}(\gamma; \mathbf{Z})\mathbf{x}\|_2^2 : \|\mathbf{x}\|_2^2 = 1\}$. Since D_3

is compact and $\mathbf{K}(\gamma; Z)$ is continuous with respect to γ it achieves its maximum over D_3. Thus, we define $\eta^\star = sup_{\gamma \in D_3} f(\gamma) \geq 0$. Define $D_2 = \{\beta : | \beta | \leq (1 + \eta^\star) \|\mathbf{Y}\|_2\}$, where the upper bound is denoted by $b^\star = (1 + \eta^\star) \|\mathbf{Y}\|_2 \geq 0$. Moreover, define $D_1 = \{\alpha : \|\alpha\|_2 \leq \sqrt{n}(\|\mathbf{Y}\|_2 + b^\star)\}$.

Since D_1, D_2 and D_3 are compact there exists a $(\alpha^\star, \beta^\star, \gamma^\star)$ such that $J_2(\alpha^\star, \beta^\star, \gamma^\star) \leq J_2(\alpha, \beta, \gamma)$ for all $(\alpha, \beta, \gamma) \in D_1 \times D_2 \times D_3$. Note that $J_2(0,0,0) = \frac{1}{2n}\|\mathbf{Y}\|_2^2$ and $(0,0,0) \in D_1 \times D_2 \times D_3$. We claim that $(\alpha^\star, \beta^\star, \gamma^\star)$ is a global minimizer, which is proved below by contradiction.

Suppose that there exists $(\tilde{\alpha}, \tilde{\beta}, \tilde{\gamma}) \notin D_1 \times D_2 \times D_3$ where $J_2(\tilde{\alpha}, \tilde{\beta}, \tilde{\gamma}) < J_2(\alpha^\star, \beta^\star, \gamma^\star)$. We must have that $\tilde{\gamma} \in D_3$; if not, we have $J_2(\tilde{\alpha}, \tilde{\beta}, \tilde{\gamma}) \geq \|\tilde{\gamma}\|_1 \geq J_2(0,0,0) \geq J_2(\alpha^\star, \beta^\star, \gamma^\star)$. Let $q_1, \cdots, q_n$ be the orthonormal vectors of $\mathbf{K}(\tilde{\gamma}; Z)$ with its associated eigenvalues $\eta_1 \geq \cdots \geq \eta_n \geq 0$. We can write out $\tilde{\alpha}, \mathbf{X}, \mathbf{Y}$ in terms of these basis functions where $\tilde{\alpha} = \sum_{i=1}^{n} < \tilde{\alpha}, q_i > q_i$, $\mathbf{Y} = \sum_{i=1}^{n} < \mathbf{Y}, q_i > q_i$ and $\mathbf{X} = \sum_{i=1}^{n} < \mathbf{X}, q_i > q_i$. Let $C_i^{\tilde{\alpha}} = < \tilde{\alpha}, q_i >$, $C_i^{\mathbf{Y}} = < \mathbf{Y}, q_i >$ and $C_i^{\mathbf{X}} = < \mathbf{X}, q_i >$. It follows that

$$J_2(\tilde{\alpha}, \tilde{\beta}, \tilde{\gamma}) \geq \frac{1}{2n} \left\| \sum_{i=1}^{n} C_i^{\mathbf{Y}} q_i - \sum_{i=1}^{n} C_i^{\mathbf{X}} \tilde{\beta} q_i - \sum_{i=1}^{n} C_i^{\tilde{\alpha}} \eta_i q_i \right\|_2^2 + \frac{1}{2} \sum_{i=1}^{n} (C_i^{\tilde{\alpha}})^2 \eta_i,$$

which is equal to $\frac{1}{2n} \sum_{i=1}^{n} (C_i^{\mathbf{Y}} - C_i^{\mathbf{X}} \tilde{\beta} - C_i^{\tilde{\alpha}} \eta_i)^2 + \frac{1}{2} \sum_{i=1}^{n} (C_i^{\tilde{\alpha}})^2 \eta_i$. We can minimize the above objective function with respect to $C_i^{\tilde{\alpha}}$ and $\tilde{\beta}$. First, note that for any $\eta_i = 0$ we can let $C_i^{\tilde{\alpha}} = 0$ as it will not affect the expression above. It is sufficient to consider $\eta_i > 0$. Taking the first derivative and setting it equal to zero, we obtain the score equations the minimizer must satisfy, for our minimum $\tilde{\beta}$ and $C_i^{\tilde{\alpha}}$

$$\beta = \sum_{i=1}^{n} C_i^{\mathbf{X}} (C_i^{\mathbf{Y}} - C_i^{\tilde{\alpha}} \eta_i) \tag{A3}$$

$$C_i^{\tilde{\alpha}} = \frac{1}{n + \eta_i} (C_i^{\mathbf{Y}} - C_i^{\mathbf{X}} \tilde{\beta}). \tag{A4}$$

In the above derivation we used the fact that $1 = \|\mathbf{X}\|_2^2 = \sum_{i=1}^{n} (C_i^{\mathbf{X}})^2$. Plugging (A4) into (A3), we obtain

$$\beta = \frac{\sum_{i=1}^{n} C_i^{\mathbf{X}} C_i^{\mathbf{Y}} (1 - \frac{\eta_i}{n + \eta_i})}{1 - \sum_{i=1}^{n} (C_i^{\mathbf{X}})^2 \frac{\eta_i}{n + \eta_i}}. \tag{A5}$$

It follows that

$$\beta \leq \frac{\sum_{i=1}^{n} | C_i^{\mathbf{X}} C_i^{\mathbf{Y}} |}{1 - \sum_{i=1}^{n} (C_i^{\mathbf{X}})^2 \frac{\eta^\star}{n + \eta^\star}} \leq \frac{\|\mathbf{X}\|_2 \|\mathbf{Y}\|_2}{\|\mathbf{X}\|_2^2 (1 - \frac{\eta^\star}{n + \eta^\star})} \leq \frac{\|\mathbf{Y}\|_2}{(1 - \frac{\eta^\star}{1 + \eta^\star})} = b^\star.$$

Thus, the β that minimizes J_2 for a given $\gamma \in D_3$ is in D_2. Also, (A4) implies that $| C_i^{\tilde{\alpha}} | \leq (\|\mathbf{Y}\|_2 + \|\mathbf{X}\|_2 \|\beta\|_2)$; consequently, the optimal α for the given $\tilde{\gamma} \in D_3$ and $\beta \in D_2$ that minimizes J_2 satisfies $\|\alpha\|_2 \leq \sqrt{n}(\|\mathbf{Y}\|_2 + b^\star)$. As a result, $\alpha \in D_2$. This suggests that for any $(\tilde{\alpha}, \tilde{\beta}, \tilde{\gamma}) \notin D_1 \times D_2 \times D_3$ we can find an $(\alpha, \beta, \gamma) \in D_1 \times D_2 \times D_3$ such that $J_2(\tilde{\alpha}, \tilde{\beta}, \tilde{\gamma}) \geq J_2(\alpha, \beta, \gamma)$.

Appendix A.4. Proof of Theorem 2

By Lemma 8.4 on page 129 in [32], Assumptions 1, 2, and 3 imply:

$$P\left(\sup_{b \subset \mathcal{B}} \frac{\frac{1}{\sqrt{n}} |\sum_{i=1}^{n} \epsilon_i b(\mathbf{z}_i)|}{\|b\|_{P_n}^{1-\psi}} \geq T \right) \leq c \exp\left(-\frac{T^2}{c^2} \right), \quad T \geq c \tag{A6}$$

where the constant c is dependent on C_1, C_2, C_3, C_4, and ψ. It follows that

$$\sup_{b \in \mathcal{B}} \frac{\frac{1}{\sqrt{n}} |\sum_{i=1}^{n} \epsilon_i b(\mathbf{z}_i)|}{\|b\|_{P_n}^{1-\psi}} = O_p(1). \tag{A7}$$

Therefore, for any $h \in \mathcal{H}_K$ and a scaling map function $\Gamma \in \mathcal{A}$, we obtain

$$\frac{\sqrt{n}(\epsilon, h \circ \Gamma - h_0 \circ \Gamma_0)_n \left(\|h\|_{\mathcal{H}_K}^2 + \|h_0\|_{\mathcal{H}_K}^2 + \|\Gamma\|_{SGL}^2 + \|\Gamma_0\|_{SGL}^2 \right)^{-\psi}}{\|h \circ \Gamma - h_0 \circ \Gamma_0\|_{P_n}^{1-\psi}} = O_p(1). \tag{A8}$$

For our estimators, $\hat{h}$ and $\hat{\Gamma}$, it is easy to see that

$$(\epsilon, \hat{h} \circ \hat{\Gamma} - h_0 \circ \Gamma_0)_n =$$
$$O_p(n^{-\frac{1}{2}}) \left\| \hat{h} \circ \hat{\Gamma} - h_0 \circ \Gamma_0 \right\|_n^{1-\psi} \left(\left\| \hat{h} \right\|_{\mathcal{H}_K}^2 + \|h_0\|_{\mathcal{H}_K}^2 + \|\hat{\Gamma}\|_{SGL}^2 + \|\Gamma_0\|_{SGL}^2 \right)^{\psi}. \tag{A9}$$

From (A9), we obtain the following inequality:

$$\left\| \hat{h} \circ \hat{\Gamma} - h_0 \circ \Gamma_0 \right\|_n^2 + \lambda_1 \left\| \hat{h} \right\|_{\mathcal{H}_K}^2 + \lambda_2 \|\hat{\Gamma}\|_{SGL}^2 \leq$$
$$O_p(n^{-\frac{1}{2}}) \left\| \hat{h} \circ \hat{\Gamma} - h_0 \circ \Gamma_0 \right\|_n^{1-\psi} \left(\left\| \hat{h} \right\|_{\mathcal{H}_K}^2 + \|h_0\|_{\mathcal{H}_K}^2 + \|\hat{\Gamma}\|_{SGL}^2 + \|\Gamma_0\|_{SGL}^2 \right)^{\psi} \tag{A10}$$
$$+ \lambda_1 \|h_0\|_{\mathcal{H}_K}^2 + \lambda_2 \|\Gamma_0\|_{SGL}^2.$$

We require $\lambda_1 = O_p(1)\lambda_2$, namely λ_2 and λ_1 go to zero at the same rate. We will show at the end of the proof what happens if they are not of the same order. Therefore, without loss of generality, we set $\lambda_1 = \lambda_2$, denoted by λ. In what follows, we divide (A10) into two cases.

Case 1: Suppose that

$$O_p(n^{-\frac{1}{2}}) \left\| \hat{h} \circ \hat{\Gamma} - h_0 \circ \Gamma_0 \right\|_n^{1-\psi} \left(\left\| \hat{h} \right\|_{\mathcal{H}_K}^2 + \|h_0\|_{\mathcal{H}_K}^2 + \|\hat{\Gamma}\|_{SGL}^2 + \|\Gamma_0\|_{SGL}^2 \right)^{\psi}$$
$$\geq \lambda \left(\|h_0\|_{\mathcal{H}_K}^2 + \|\Gamma_0\|_{SGL}^2 \right).$$

In this case, we have

$$\left\| \hat{h} \circ \hat{\Gamma} - h_0 \circ \Gamma_0 \right\|_n^2 + \lambda \left(\left\| \hat{h} \right\|_{\mathcal{H}_K}^2 + \|\hat{\Gamma}\|_{SGL}^2 \right) \leq$$
$$O_p(n^{-\frac{1}{2}}) \left\| \hat{h} \circ \hat{\Gamma} - h_0 \circ \Gamma_0 \right\|_n^{1-\psi} \left(\left\| \hat{h} \right\|_{\mathcal{H}_K}^2 + \|h_0\|_{\mathcal{H}_K}^2 + \|\hat{\Gamma}\|_{SGL}^2 + \|\Gamma_0\|_{SGL}^2 \right)^{\psi}. \tag{A11}$$

Above (A11) is further discussed separately in two sub-cases.

Case 1a: If $\|h_0\|_{\mathcal{H}_K}^2 + \|\Gamma_0\|_{SGL}^2 \leq \left\| \hat{h} \right\|_{\mathcal{H}_K}^2 + \|\hat{\Gamma}\|_{SGL}^2$, then we have

$$\left\| \hat{h} \circ \hat{\Gamma} - h_0 \circ \Gamma_0 \right\|_n^2 + \lambda \left(\left\| \hat{h} \right\|_{\mathcal{H}_K}^2 + \|\hat{\Gamma}\|_{SGL}^2 \right) \leq$$
$$O_p(n^{-\frac{1}{2}}) \left\| \hat{h} \circ \hat{\Gamma} - h_0 \circ \Gamma_0 \right\|_n^{1-\psi} \left(\left\| \hat{h} \right\|_{\mathcal{H}_K}^2 + \|\hat{\Gamma}\|_{SGL}^2 \right)^{\psi}. \tag{A12}$$

Therefore,

$$\left(\left\| \hat{h} \right\|_{\mathcal{H}_K}^2 + \|\hat{\Gamma}\|_{SGL}^2 \right)^{\psi} \leq O_p(n^{-\frac{\psi}{2(1-\psi)}}) \left\| \hat{h} \circ \hat{\Gamma} - h_0 \circ \Gamma_0 \right\|_n^{\psi} \lambda^{-\frac{\psi}{1-\psi}}. \tag{A13}$$

It follows that

$$\left\|\hat{h}\circ\hat{\Gamma}-h_0\circ\Gamma_0\right\|_n = O_p(n^{-\frac{1}{2(1-\psi)}})O_p(\lambda^{-\frac{\psi}{1-\psi}}),$$

$$\left\|\hat{h}\right\|_{H_K}^2 + \|\hat{\Gamma}\|_{SGL}^2 = O_p(n^{-\frac{1}{1-\psi}})O_p(\lambda^{-\frac{1+\psi}{1-\psi}}). \tag{A14}$$

Case 1b: If $\|h_0\|_{\mathcal{H}_K}^2 + \|\Gamma_0\|_{SGL}^2 \geq \left\|\hat{h}\right\|_{H_K}^2 + \|\hat{\Gamma}\|_{SGL}^2$, then:

$$\left\|\hat{h}\right\|_{\mathcal{H}_K}^2 + \|\hat{\Gamma}\|_{SGL}^2 = O_p(\|h_0\|_{\mathcal{H}_K}^2 + \|\Gamma_0\|_{SGL}^2)O_p(1).$$

Therefore,

$$\left\|\hat{h}\circ\hat{\Gamma}-h_0\circ\Gamma_0\right\|_n = O_p(n^{-\frac{1}{2(1+\psi)}})\left(\|h_0\|_{\mathcal{H}_K}^2 + \|\Gamma\|_{SGL}^2\right)^{\frac{\psi}{1+\psi}}.$$

Consequently, we obtain

$$\left\|\hat{h}\circ\hat{\Gamma}-h_0\circ\Gamma_0\right\|_n = O_p(n^{-\frac{1}{2(1-\psi)}})O_p(\lambda^{-\frac{\psi}{1-\psi}}),$$

$$\left\|\hat{h}\right\|_{\mathcal{H}_K}^2 + \|\hat{\Gamma}\|_{SGL}^2 = O_p(n^{-\frac{1}{1-\psi}})O_p(\lambda^{-\frac{1+\psi}{1-\psi}}). \tag{A15}$$

Both terms in (A15) are the same rates as those in (A14).

Case 2: Suppose that

$$O_p(n^{-\frac{1}{2}})\left\|\hat{h}\circ\hat{\Gamma}-h_0\circ\Gamma_0\right\|_n^{1-\psi}\left(\left\|\hat{h}\right\|_{\mathcal{H}_K}^2 + \|h_0\|_{\mathcal{H}_K}^2 + \|\hat{\Gamma}\|_{SGL}^2 + \|\Gamma_0\|_{SGL}^2\right)^{\psi}$$

$$\leq \lambda(\|h_0\|_{\mathcal{H}_K}^2 + \|\Gamma_0\|_{SGL}^2).$$

Then, we have

$$\left\|\hat{h}\circ\hat{\Gamma}-h_0\circ\Gamma_0\right\|_n^2 + \lambda\left(\left\|\hat{h}\right\|_{\mathcal{H}_K}^2 + \|\hat{\Gamma}\|_{SGL}^2\right) \leq 2\lambda\left(\|h_0\|_{\mathcal{H}_K}^2 + \|\Gamma_0\|_{SGL}^2\right).$$

This implies that

$$\left\|\hat{h}\circ\hat{\Gamma}-h_0\circ\Gamma_0\right\|_n = O_p(\lambda^{\frac{1}{2}})\left(\|h_0\|_{\mathcal{H}_K}^2 + \|\Gamma_0\|_{SGL}^2\right)^{\frac{1}{2}},$$

$$\left\|\hat{h}\right\|_{\mathcal{H}_K}^2 + \|\hat{\Gamma}\|_{SGL}^2 = O_p(1)\left(\|h_0\|_{\mathcal{H}_K}^2 + \|\Gamma_0\|_{SGL}^2\right). \tag{A16}$$

In order to make (A14) and (A16) have the same rates we first equate the two term $O_p(\lambda^{\frac{1}{2}})\left(\|h\|_{\mathcal{H}_K}^2 + \|\Gamma\|_{SGL}^2\right)^{\frac{1}{2}}$ and $O_p(n^{-\frac{1}{2(1-\psi)}})O_p(\lambda^{-\frac{\psi}{1-\psi}})$, and then solve for a common λ. The solution is given as follows:

$$\lambda^{-1} = n^{\frac{1}{1+\psi}}\left(\|h\|_{\mathcal{H}_K}^2 + \|\Gamma\|_{SGL}^2\right)^{\frac{1-\psi}{1+\psi}}.$$

Under this λ value we obtain that (A14)–(A16) as of the form:

$$\left\|\hat{h}\circ\hat{\Gamma}-h_0\circ\Gamma_0\right\|_n = O_p(n^{-\frac{1}{2(1+\psi)}})\left(\|h_0\|_{\mathcal{H}_K}^2 + \|\Gamma_0\|_{SGL}^2\right)^{\frac{\psi}{1+\psi}}, \tag{A17}$$

$$\left\|\hat{h}\right\|_{\mathcal{H}_K}^2 + \|\hat{\Gamma}\|_{SGL}^2 = O_p(1)\left(\|h_0\|_{\mathcal{H}_K}^2 + \|\Gamma_0\|_{SGL}^2\right). \tag{A18}$$

This completes the proof of Theorem 2.

Now we discuss the situation where the tuning parameters λ_1 and λ_2 are not of the same order. As seen blow, the selection consistency may not be guaranteed. Take Case 2 as an example. Suppose that

$$O_p(n^{-\frac{1}{2}})\left\|\hat{h}\circ\hat{\Gamma}-h_0\circ\Gamma_0\right\|_n^{1-\psi}\left(\left\|\hat{h}\right\|_{\mathcal{H}_\mathcal{K}}^2+\|h_0\|_{\mathcal{H}_\mathcal{K}}^2+\|\hat{\Gamma}\|_{SGL}^2+\|\Gamma_0\|_{SGL}^2\right)^\psi$$
$$\leq\lambda_1\|h_0\|_{\mathcal{H}_\mathcal{K}}^2+\lambda_2\|\Gamma_0\|_{SGL}^2.$$

Let us consider two cases.
Case 2a: If $\lambda_1\|h_0\|_{\mathcal{H}_\mathcal{K}}^2\leq\lambda_2\|\Gamma_0\|_{SGL}^2$, following the same arguments above, we have

$$\left\|\hat{h}\circ\hat{\Gamma}-h_0\circ\Gamma_0\right\|_n=O_p(\lambda_2^{\frac{1}{2}})\|\Gamma_0\|_{SGL},$$
$$\left\|\hat{h}\right\|_{\mathcal{H}_\mathcal{K}}^2=O_p(\frac{\lambda_2}{\lambda_1})\|\Gamma_0\|_{SGL}^2, \tag{A19}$$
$$\|\hat{\Gamma}\|_{SGL}^2=O_p(1)\|\Gamma_0\|_{SGL}^2.$$

Case 2b: If $\lambda_1\|h_0\|_{\mathcal{H}_\mathcal{K}}^2\geq\lambda_2\|\Gamma_0\|_{SGL}^2$, then following the same logic as before:

$$\left\|\hat{h}\circ\hat{\Gamma}-h_0\circ\Gamma_0\right\|_n=O_p(\lambda_1^{\frac{1}{2}})\|h_0\|_{\mathcal{H}_\mathcal{K}},$$
$$\|\hat{\Gamma}\|_{SGL}^2=O_p(\frac{\lambda_1}{\lambda_2})\|h_0\|_{\mathcal{H}_\mathcal{K}}^2, \tag{A20}$$
$$\left\|\hat{h}\right\|_{\mathcal{H}_\mathcal{K}}^2=O_p(1)\|h_0\|_{\mathcal{H}_\mathcal{K}}^2.$$

Both terms involve $O_p(\frac{\lambda_1}{\lambda_2})$ and $O_p(\frac{\lambda_2}{\lambda_1})$, indicating that these two tuning parameters λ_1 and λ_2 should go to zero at the same rates. Moreover, we can think of our estimator $\hat{h}\circ\hat{\Gamma}$ as one operational object. See Appendix B for more details on this, which can further explain the need of one rate for the two penalties.

Appendix A.5. Proof of Corollary 1

For convenience, we present the following lemma proved by [32] (on page 20).

Lemma A1. *(Geer's Lemma) A d dimensional ball of radius R, $B_d(R)$, in $\mathcal{R}^d$ with Euclidean metric can be covered by $(\frac{4R+\delta}{\delta})^d$ balls of radius δ.*

We have shown in the proof of Theorem 1 that the optimal γ vector is restricted to be within a ball of a radius that depends on the norm of $\mathbf{Y}$. For the sake of simplicity let us confine our γ to be within a norm ball of radius 1, $\gamma\in\mathcal{G}=\{\gamma:\|\gamma\|_2^2\leq1\}$. We then confine our set which we called $\mathcal{A}$ to be restricted to those γ, that is $\mathcal{A}=\{\Gamma:\Gamma(\mathbf{z})=\gamma\circ\mathbf{z},\gamma\in\mathcal{G}\}$. Since our $\gamma\in R^s$, we can use above Lemma A1 and cover our set $\mathcal{A}$ with $N_1=\left(\frac{4+\delta}{\delta}\right)^s$ number of functions in the following sense. The ball of radius 1 in R^s can be covered (using the Euclidean metric) by $\{\gamma_1,\cdots\gamma_{N_1}\}$. Since there is a one to one relationship between the functions Γ and γ, take the set $\{\Gamma_1,\ldots,\Gamma_{N_1}\}$ and define the metric between some Γ_j and Γ_k in the set $\mathcal{A}$ as $d(\Gamma_j,\Gamma_k)=\|\gamma_j-\gamma_k\|_2$. Then, the set of functions $\{\Gamma_1,\ldots,\Gamma_{N_1}\}$ is a δ-covering for $\mathcal{A}$ under this metric with entropy $s\,log(\frac{4+\delta}{\delta})$. For each Γ_j we have an induced RKHS, $\mathcal{H}_{\mathcal{K}\circ\Gamma_j}=\{h\circ\Gamma_j:h\in\mathcal{H}_\mathcal{K}\}$ with entropy no larger than that of $\mathcal{H}_\mathcal{K}$, which according to the assumption, has entropy $\leq A\delta^{-2\psi}$ for some $\psi\in(0,1)$ and $A\in\mathcal{R}$. Therefore, the covering number $N_2=N(\delta,\mathcal{H}_{\mathcal{K}\circ\Gamma_j},P_n)\leq\exp\{A\delta^{-2\psi}\}$. This implies that for every Γ_j there exists a set $\{h_{j_1}\circ\Gamma_j,\cdots,h_{j_{N_2}}\circ\Gamma_j\}$ such that for every $h\circ\Gamma_j\in\mathcal{H}_{\mathcal{K}\circ\Gamma_j}$ there exists an integer $i\in\{1,\ldots,N_2\}$ we have $\|h\circ\Gamma_j-h_{j_i}\circ\Gamma_j\|_{P_n}\leq\delta$. Set $\mathcal{B}$ is essentially the union of the different Hilbert spaces of the form $\mathcal{H}_{\mathcal{K} \circ \Gamma}$. Under the setup, a natural estimate of the *delta*-covering number of this set would be approximately of size $N_1 \times N_2$ where functions take the form of $\{h_{1_1} \circ \Gamma_1, \cdots, h_{1_{N_2}} \circ \Gamma_1, \cdots, h_{N_{11}} \circ \Gamma_{N_1}, \cdots, h_{N_{1 N_2}} \circ \Gamma_{N_1}\}$. In addition, we add N_2 functions from the set $\{h_1 \circ \Gamma_0, \cdots, h_{N_2} \circ \Gamma_0\}$ where Γ_0 is the true Γ_0 (or one of the true Γ_0). Since $\mathcal{H}_{\mathcal{K} \circ \Gamma_j}$ is a Hilbert space for every j, if $h \circ \Gamma_j \in \mathcal{H}_{\mathcal{K} \circ \Gamma_j}$ so is $\dfrac{h \circ \Gamma_j}{\|h\|^2_{\mathcal{H}_\mathcal{K}} + \|h_0\|^2_{\mathcal{H}_\mathcal{K}} + \|\Gamma_j\|^2_{SGL} + \|\Gamma_0\|^2_{SGL}}$.

We can simply ignore the denominator and substitute $\dfrac{h \circ \Gamma_j}{\|h\|^2_{\mathcal{H}_\mathcal{K}} + \|h_0\|^2_{\mathcal{H}_\mathcal{K}} + \|\Gamma_j\|^2_{SGL} + \|\Gamma_0\|^2_{SGL}}$ with $\tilde{h} \circ \Gamma_j \in H_{\mathcal{K} \circ \Gamma_j}$ where $\tilde{h} = \dfrac{h}{\|h\|^2_{\mathcal{H}_\mathcal{K}} + \|h_0\|^2_{\mathcal{H}_\mathcal{K}} + \|\Gamma_j\|^2_{SGL} + \|\Gamma_0\|^2_{SGL}}$.

We now prove Corollary 1.

Proof. Set $M = \sup_h < \nabla h(\mathbf{z}), \nabla h(\mathbf{z}) >$ where the inner product is the standard Euclidean inner product. This is for a fixed $\mathbf{z}$, or under the assumption that the gradient is uniformly bounded, we can take the $\sup_{h \in \mathcal{H}_\mathcal{K}, \mathbf{z} \in R^s} < \nabla h(\mathbf{z}), \nabla h(\mathbf{z}) >$. Let $N_1 = \dfrac{4 + \left(\frac{\delta}{3M^{\frac{1}{2}}} \right)^s}{\left(\frac{\delta}{3M^{\frac{1}{2}}} \right)}$ which is the number of balls needed to provide a $\left(\frac{\delta}{3M^{\frac{1}{2}}} \right)$ covering for a norm 1 ball in R^s. Let $N_2 = \exp \left\{ \left(A(\frac{\delta}{3})^{-2\psi} \right) \right\}$ which is the covering number needed to provide a $\frac{\delta}{3}$ cover of our space $\mathcal{H}_\mathcal{K}$. Let:

$$\tilde{h} \circ \hat{\Gamma} - \tilde{h}_0 \circ \Gamma_0 =$$

$$\frac{\hat{h} \circ \hat{\Gamma}}{\left\| \hat{h} \right\|^2_{\mathcal{H}_\mathcal{K}} + \|h_0\|^2_{\mathcal{H}_\mathcal{K}} + \|\hat{\Gamma}\|^2_{SGL} + \|\Gamma_0\|^2_{SGL}} - \frac{h_0 \circ \Gamma_0}{\left\| \hat{h} \right\|^2_{\mathcal{H}_\mathcal{K}} + \|h_0\|^2_{\mathcal{H}_\mathcal{K}} + \|\hat{\Gamma}\|^2_{SGL} + \|\Gamma_0\|^2_{SGL}}$$

be an arbitrary function in the set $\mathcal{B}$. There exists a Γ_j where $j \in \{1, \ldots, N_1\}$ such that $d(\Gamma_j, \hat{\Gamma}) \leq \dfrac{\delta}{3 \max\limits_{i=1,\cdots,n} \|\mathbf{z}_i\|_2 \sqrt{M}}$, and there exists an i where $i \in \{1, \ldots, N_2\}$ such that $\left\| \tilde{h} \circ \Gamma_j - h_{j_i} \circ \Gamma_j \right\|_{P_n} \leq \frac{\delta}{3}$.

Similarly, there exists a $t \in \{1, \ldots, N_2\}$ such that $\left\| \tilde{h}_0 \circ \Gamma_0 - h_t \circ \Gamma_0 \right\|_{P_n} \leq \frac{\delta}{3}$. We construct our approximating function of $\tilde{h} \circ \hat{\Gamma} - \tilde{h}_0 \circ \Gamma_0$ as $h_{j_i} \circ \Gamma_j - h_t \circ \Gamma_0$. We now show that this function is within δ of our arbitrary function $\tilde{h} \circ \hat{\Gamma} - \tilde{h}_0 \circ \Gamma_0$. Applying the mean value theorem for multivariate functions, $\tilde{h} \circ \hat{\Gamma}(\mathbf{z}) = \tilde{h} \circ \Gamma_j(\mathbf{z}) + \nabla \tilde{h}(C(\mathbf{z}))(\hat{\Gamma}(\mathbf{z}) - \Gamma_j(\mathbf{z}))$, we have:

$$\left\| (\tilde{h} \circ \hat{\Gamma} - \tilde{h}_0 \circ \Gamma_0) - (h_{j_i} \circ \Gamma_j - h_t \circ \Gamma_0) \right\|_{P_n}$$

$$\leq \left\| \tilde{h} \circ \hat{\Gamma} - h_{j_i} \circ \Gamma_j \right\|_{P_n} + \left\| \tilde{h}_0 \circ \Gamma_0 - h_t \circ \Gamma_0 \right\|_{P_n}$$

$$\leq \left\| \tilde{h} \circ \hat{\Gamma} - h_{j_i} \circ \Gamma_j \right\|_{P_n} + \frac{\delta}{3}$$

$$= \left\| \tilde{h} \circ \Gamma_j - h_{j_i} \circ \Gamma_j + \nabla \tilde{h}(C(\cdot))(\hat{\Gamma} - \Gamma_j) \right\|_{P_n} + \frac{\delta}{3}$$

where vector $\mathbf{z} \in \mathcal{R}^s$ lies in the segment from $\gamma_j \circ \mathbf{z}$ and $\hat{\gamma} \circ \mathbf{z}$, and $C(\cdot)$ is an unknown function that maps from $\mathcal{R}^s$ into $\mathcal{R}^s$ that allows for the formula to hold. Continuing our chain of inequalities, we obtain:

$$\left\| \tilde{h} \circ \Gamma_j - h_{j_i} \circ \Gamma_j + \nabla \tilde{h}(C(\cdot))(\hat{\Gamma} - \Gamma_j) \right\|_{P_n} + \frac{\delta}{3} \leq$$

$$\left\| \nabla \tilde{h}(C(\cdot))(\hat{\Gamma} - \Gamma_j) \right\|_{P_n} + \frac{\delta}{3} + \frac{\delta}{3} =$$

$$\sqrt{\frac{1}{n} \sum_{i=1}^{n} \left(\nabla \tilde{h}(C(\mathbf{z}_i))(\hat{\Gamma}(\mathbf{z}_i) - \Gamma_j(\mathbf{z}_i)) \right)^2} + \frac{\delta}{3} + \frac{\delta}{3} \leq$$

$$\sqrt{\frac{1}{n} \sum_{i=1}^{n} M \| \hat{\gamma} \circ \mathbf{z_i} - \gamma_j \circ \mathbf{z_i} \|_2^2} + \frac{\delta}{3} + \frac{\delta}{3} \leq$$

$$\sqrt{M \left(\frac{\delta}{3 \max\limits_{i=1,\cdots,n} \| \mathbf{z}_i \|_2 \sqrt{M}} \right)^2 \max\limits_{i=1,\cdots,n} \| \mathbf{z}_i \|_2^2} + \frac{\delta}{3} + \frac{\delta}{3} =$$

$$\frac{\delta}{3} + \frac{\delta}{3} + \frac{\delta}{3} = \delta.$$

Therefore, to provide a δ cover we need $N_1 \times N_2 + N_2$ number of functions or:

$$\exp\left\{ \left(A(\frac{\delta}{3})^{-2\psi} \right) \right\} \left(\frac{4 + \left(\frac{\delta}{3M^{\frac{1}{2}}} \right)}{\left(\frac{\delta}{3M^{\frac{1}{2}}} \right)} \right)^s + \exp\left\{ \left(A\left(\frac{\delta}{3}\right)^{-2\psi} \right) \right\} =$$

$$\exp\{ \tilde{A}\delta^{-2\psi} \} \left(\frac{C + \delta}{\delta} \right)^s + \exp\{ \tilde{A}\delta^{-2\psi} \},$$

where $\tilde{A} = \frac{A}{3^{-2\psi}}$ and $C = 12M^{\frac{1}{2}}$. Taking the log we see the entropy is $\leq \tilde{A}\delta^{-2\psi} + \log\left((\frac{C+\delta}{\delta})^s + 1 \right)$ which is of the same order as $\leq \tilde{A}\delta^{-2\psi}$ (the *log* term is dominated by the first term). Therefore a sufficient (but not necessary) condition for our set $\mathcal{B}$ to have the same entropy as that of the original RKHS $\mathcal{H}_\mathcal{K}$ is for the $\sup_h < \nabla h(\mathbf{z}), \nabla h(\mathbf{z}) >$ to be bounded. Having bounded derivatives is reasonable for any RKHS since every RKHS satisfies the Lipschitz condition of the form:

$$|h(X) - h(Y)| = |< h, \mathcal{K}_X > - < h, \mathcal{K}_Y >| \leq \|h\|_{\mathcal{H}_\mathcal{K}} < \mathcal{K}_X, \mathcal{K}_Y >^{\frac{1}{2}} = \|h\|_{\mathcal{H}_\mathcal{K}} d(X, Y),$$

where the distance metric in $\mathcal{R}^s$ is defined as $d(X, Y)^2 = \mathcal{K}(X, X) - 2\mathcal{K}(X, Y) + \mathcal{K}(Y, Y)$. If we restrict our functions in the RKHS of norm $\leq C$ for some constant C then we have a universal Lipschitz constant C to ensure bounded derivatives. $\square$

Appendix B. Discussion about the FKMR Estimator

We introduce γ as a way of performing variable selection on our vector of FPC features. We want to illustrate this technical trick with some concrete examples and discuss identifiability issues with the resulting estimator. There are two ways of looking at the estimation of the unknown functions h_0 and Γ_0. The first way is to view our feature vector, $\mathbf{z}$, as being related to the dependent variable y through the composite function $h \circ \Gamma$, as explained in Section 4. The second and equivalent way is to view our features as unknown. The true features take the form of $\gamma \circ \mathbf{z}$, where in this case the $\circ$ denotes the Hadamard product. We are given $\mathbf{z}$ and need to estimate the "true" features $\gamma \circ \mathbf{z}$. In addition, we need to estimate the relationship between $\gamma \circ \mathbf{z}$ and y, which is done through the function $h \in \mathcal{H}_\mathcal{K}$.

The first way is to estimate the function $h_0 \circ \Gamma_0$. The function belongs to the RKHS $\mathcal{H}_{\mathcal{K} \circ \Gamma}$. We essentially consider many different function spaces to construct our estimator. The intersection between the function spaces is not necessarily empty, implying that our estimator may not be unique. We proceed this discussion more formally. Let $\mathcal{K} : \mathcal{R}^s \times \mathcal{R}^s \mapsto \mathcal{R}$ be a positive definite function. Let $\Gamma : \mathcal{R}^s \mapsto \mathcal{R}^s$. We define $\mathcal{K} \circ \Gamma : \mathcal{R}^s \times \mathcal{R}^s \mapsto \mathcal{R}$ as the function given by $\mathcal{K} \circ \Gamma(\mathbf{s}, \mathbf{t}) = \mathcal{K}(\Gamma(\mathbf{s}), \Gamma(\mathbf{t}))$. This new function, $\mathcal{K} \circ \Gamma$ is positive definite. There is a relationship between the original RKHS, $\mathcal{H}_{\mathcal{K}}$ and the new RKHS, $\mathcal{H}_{\mathcal{K} \circ \Gamma}$. This results in $\mathcal{H}_{\mathcal{K} \circ \Gamma} = \{h \circ \Gamma : h \in \mathcal{H}_{\mathcal{K}}\}$. For any vector $u \in H_{\mathcal{K} \circ \Gamma}$, we have that $\|u\|_{\mathcal{H}_{\mathcal{K} \circ \Gamma}} = inf\{\|h\|_{\mathcal{H}_{\mathcal{K}}} : u = h \circ \Gamma\}$. In general, $\mathcal{H}_{\mathcal{K} \circ \Gamma} \not\subset \mathcal{H}_{\mathcal{K}}$. In (5), we take the norm with respect to the original space $\mathcal{H}_{\mathcal{K}}$. Our iterative procedure essentially presents the second way in which the true features are unknown, whereas our theoretical arguments are justified through the first way. Given the knowledge of the features (which translates to fixing a γ), we are confined to just one RKHS, $\mathcal{H}_{\mathcal{K}}$. Take the linear kernel, $\mathcal{K}(\mathbf{x}_1, \mathbf{x}_2) = \mathbf{x}_1^{\top} \mathbf{x}_2$ as an example. Suppose the truth is that y is related to a one-dimensional feature $\mathbf{z}_0$ through the following formulation: $y = h_0(z_0) + \varepsilon$ where $h_0 \in \mathcal{H}_{\mathcal{K}_1}$, where $\mathcal{K}_1$ is the kernel that maps from $\mathcal{R} \times \mathcal{R} \mapsto \mathcal{R}$. Therefore, if we knew the feature z_1, we would proceed to optimize (6) using the standard LSKM. However, when each y is associated with a two-dimensional vector $\mathbf{z} = (z_1, z_2)$, where z_2 is a "noisy" feature and unrelated to y. Suppose that *a priori* we do not know this information. Typically we use a model $y = h(z_1, z_2) + \varepsilon$ where $h \in \mathcal{H}_{\mathcal{K}}$, where $\mathcal{K}$ is the kernel that maps from $\mathcal{R}^2 \times \mathcal{R}^2 \mapsto \mathcal{R}$. In this case, we introduce our γ vector (γ_1, γ_2) and formulate $y = h(\gamma_1 z_1, \gamma_2 z_2) + \epsilon$. All functions, h in the space $\mathcal{H}_{\mathcal{K}}$, are of the form $h(\mathbf{z}) = \mathbf{x}^{\top} \mathbf{z}$ for some two-dimensional vector $\mathbf{x} = (x_1, x_2)$. There is a one-to-one relationship between h and $\mathbf{x}$. The true function, h_0, has an associated real number c where $h_1(z_1) = c z_1$. We can recover $h_1 \in \mathcal{H}_{\mathcal{K}_1}$ from our estimation of h and γ if we set $\gamma = (1, 0)$ and $\mathbf{x} = (c, \star)$, where "$\star$" is any real number. Equivalently, we can recover h_1 under $\gamma = (1, 1)$ where $\mathbf{x} = (c, 0)$. There are many functions that may recover the original function in the RKHS corresponding to the linear space kernel. Formulating our problem in the first way, through function composition, we can estimate Γ_0 with the γ being $(1, 0)$ or $(1, 1)$.

We can now see that in the intersection between $\mathcal{H}_{\mathcal{K} \circ \Gamma_1}$ and $\mathcal{H}_{\mathcal{K} \circ \Gamma_2}$, where Γ_1 has associated $\gamma_1 = (1, 0)$ and Γ_2 has associated $\gamma_2 = (1, 1)$, lies our estimate of h_1. In truth, for the linear space RKHS, there is no need to apply our method since $h_0 \in \mathcal{H}_{\mathcal{K}_1}$ can be estimated directly from the larger space $\mathcal{H}_{\mathcal{K}}$ where we set $h(\mathbf{z}) = \mathbf{x}^{\top} \mathbf{z}$ where $\mathbf{x} = (c, 0)$. We can never hope to have variable selection consistency nor can we hope to have identifiability of our estimator for these types of spaces. However, from a goodness-of-fit standpoint, we are able to do just as good a job with many types of function compositions. Our hope is that we can glean some variable selection by penalizing the γ vector with the $\rho(\gamma; \delta)$ term which, going back to the above scenario, should give preference to $\gamma = (1, 0)$ over $\gamma = (1, 1)$. For the RKHS associated with the Gaussian Kernel, the "larger dimensional space", a Gaussian Kernel mapping from higher dimensions, does not necessarily contain the functions from a "lower dimensional space", a Gaussian Kernel mapping from lower dimensions. However through the introduction of the γ transformation of the features, we can recover the equivalent functions of the "lower dimensional space".

References

1. Chandler, J.L.; Brazendale, K.; Beets, M.W.; Mealing, B.A. Classification of Physical Activity Intensities Using a Wrist-worn Accelerometer in 8–12-Year-old Children. *Pediatric Obes.* **2016**, *11*, 120–127. [CrossRef] [PubMed]
2. Chen, K.Y.; Bassett, D.R. The Technology of Accelerometry-based Activity Monitors: Current and Future. *Med. Sci. Sport. Exerc.* **2005**, *37*, S490–S500. [CrossRef] [PubMed]
3. Bai, J.; Di, C.; Xiao, L.; Evenson, K.R.; LaCroix, A.Z.; Crainiceanu, C.M.; Buchner, D.M. An Activity Index for Raw Accelerometry Data and Its Comparison with Other Activity Metrics. *PLoS ONE* **2016**, *11*, e0160644. [CrossRef] [PubMed]
4. John, D.; Freedson, P. ActiGraph and Actical Physical Activity Monitors: A Peek under the Hood. *Med. Sci. Sport. Exerc.* **2012**, *44*, S86–S89. [CrossRef]
5. Kim, Y.; Lee, J.M.; Peters, B.P.; Gaesser, G.A.; Welk, G.J. Examination of Different Accelerometer Cut-points for Assessing Sedentary Behaviors in Children. *PLoS ONE* **2014**, *9*, e90630. [CrossRef]

6. Bai, J.; Sun, Y.; Schrack, J.A.; Crainiceanu, C.M.; Wang, M.C. A Two-stage Model for Wearable Device Data. *Biometrics* **2018**, *74*, 744–752. [CrossRef]
7. Sasaki, J.E.; Hickey, A.M.; Staudenmayer, J.W.; John, D.; Kent, J.A.; Freedson, P.S. Performance of Activity Classification Algorithms in Free-Living Older Adults. *Med. Sci. Sport. Exerc.* **2016**, *48*, 941–950. [CrossRef]
8. Di, C.Z.; Crainiceanu, C.M.; Caffo, B.S.; Punjabi, N.M. Multilevel Functional Principal Component Analysis. *Ann. Appl. Stat.* **2009**, *3*, 458–488. [CrossRef]
9. Goldsmith, J.; Liu, X.; Rundle, A.; Jacobson, J. New Insights into Activity Patterns in Children, Found Using Functional Data Analyses. *Med. Sci. Sport. Exerc.* **2016**, *48*, 1723–1729. [CrossRef]
10. Li, H.; Keadle, S.K.; Staudenmayer, J.; Assaad, H.; Huang, J.Z.; Carroll, R.J. Methods to Assess An Exercise Intervention Trial Based on 3-Level Functional Data. *Biostatistics* **2015**, *16*, 754–771. [CrossRef]
11. Zhang, Y.; Li, H.; Keadle, S.K.; Matthews, C.E.; Carroll, R.J. A Review of Statistical Analyses on Physical Activity Data Collected from Accelerometers. *Stat. Biosci.* **2019**, *11*, 465–476. [CrossRef] [PubMed]
12. Ramsay, J.O.; Silverman, B.W. *Functional Data Analysis*; Springer Series in Statistics; Springer: Berlin/Heidelberg, Germany, 2005.
13. Cardot, H.; Ferraty, F.; Sarda, P. Spline Estimators for the Functional Linear model. *Stat. Sin.* **2003**, *13*, 571–591.
14. Cardot, H.; Ferraty, F.; Sarda, P. Functional Linear Model. *Stat. Probab. Lett.* **1999**, *45*, 11–22. [CrossRef]
15. Zhu, H.; Yao, F.; Zhang, H.H. Structured Functional Additive Regression in Reproducing Kernel Hilbert Spaces. *J. R. Stat. Soc. Ser. (Stat. Methodol.)* **2014**, *76*, 581–603. [CrossRef]
16. Ferraty, F.; Mas, A.; Vieu, P. Nonparametric Regression on Functional Data: Inference and Practical Aspects. *Aust. N. Z. J. Stat.* **2007**, *49*, 267–286. [CrossRef]
17. McLean, M.W.; Hooker, G.; Staicu, A.M.; Scheipl, F.; Ruppert, D. Functional Generalized Additive Models. *J. Comput. Graph. Stat.* **2014**, *23*, 249–269. [CrossRef]
18. Bosq, D. *Linear Processes in Function Spaces*; Lecture Notes in Statistics; Springer: New York, NY, USA, 2000; Volume 149.
19. Hall, P.; Müller, H.G.; Wang, J.L. Properties of Principal Component Methods for Functional and Longitudinal Data Analysis. *Ann. Stat.* **2006**, *34*, 1493–1517. [CrossRef]
20. Hall, P.; Hosseini-Nasab, M. On Properties of Functional Principal Components Analysis. *J. R. Stat. Soc. Ser. (Stat. Methodol.)* **2006**, *68*, 109–126. [CrossRef]
21. Müller, H.G.; Yao, F. Functional Additive Models. *J. Am. Stat. Assoc.* **2008**, *103*, 1534–1544. [CrossRef]
22. Lin, Y.; Zhang, H.H. Component Selection and Smoothing in Multivariate Nonparametric Regression. *Ann. Stat.* **2006**, *34*, 2272–2297. [CrossRef]
23. Liu, D.; Lin, X.; Ghosh, D. Semiparametric Regression of Multidimensional Genetic Pathway Data: Least-Squares Kernel Machines and Linear Mixed Models. *Biometrics* **2007**, *63*, 1079–1088. [CrossRef] [PubMed]
24. Wood, S.N. *Generalized Additive Models: An Introduction with R*; Chapman and Hall: London, UK, 2006.
25. Lin, X.; Zhang, D. Inference in Generalized Additive Mixed Models by Using Smoothing Splines. *J. R. Stat. Soc. Ser. (Stat. Methodol.)* **1999**, *61*, 381–400. [CrossRef]
26. Yuan, M.; Lin, Y. Model Selection and Estimation in Regression with Grouped Variables. *J. R. Stat. Soc. Ser. (Stat. Methodol.)* **2006**, *68*, 49–67. [CrossRef]
27. Simon, N.; Friedman, J.; Hastie, T.; Tibshirani, R. A Sparse-Group Lasso. *J. Comput. Graph. Stat.* **2013**, *22*, 231–245. [CrossRef]
28. Breiman, L. Better Subset Regression Using the Nonnegative Garrote. *Technometrics* **1995**, *37*, 373–384. [CrossRef]
29. Salzo, S.; Villa, S. Convergence Analysis of a Proximal Gauss–Newton Method. *Comput. Optim. Appl.* **2012**, *53*, 557–589. [CrossRef]
30. Naiman, J. Multivariate Functional Kernel Machine Regression and Feature Selection with Applications to Accelerometer Mobile Health Devices. Ph.D. Dissertation, University of Michigan, Ann Arbor, MI, USA, 2020.
31. Peng, H.; Huang, T. Penalized Least Squares for Single Index Models. *J. Stat. Plan. Inference* **2011**, *141*, 1362–1379. [CrossRef]
32. Geer, S.A. *Empirical Processes in M-Estimation*; Cambridge Series in Statistical and Probabilistic Mathematics; Cambridge University Press: Cambridge, UK, 2000.
33. Hainmueller, J.; Hazlett, C. Kernel Regularized Least Squares: Reducing Misspecification Bias with a Flexible and Interpretable Machine Learning Approach. *Political Anal.* **2014**, *22*, 143–168. [CrossRef]
34. Yao, F.; Müller, H.G.; Wang, J.L. Functional Data Analysis for Sparse Longitudinal Data. *J. Am. Stat. Assoc.* **2005**, *100*, 577–590. [CrossRef]
35. Lewis, R.C.; Meeker, J.D.; Peterson, K.E.; Lee, J.M.; Pace, G.G.; Cantoral, A.; Téllez-Rojo, M.M. Predictors of Urinary Bisphenol A and Phthalate Metabolite Concentrations in Mexican Children. *Chemosphere* **2013**, *93*, 2390–2398. [CrossRef]
36. Schrack, J.A.; Zipunnikov, V.; Goldsmith, J.; Bai, J.; Simonsick, E.M.; Crainiceanu, C.; Ferrucci, L. Assessing the Physical Cliff: Detailed Quantification of Age-related Differences in Daily Patterns of Physical Activity. *J. Gerontol. Ser. Biol. Sci. Med. Sci.* **2014**, *69*, 973–979. [CrossRef] [PubMed]
37. Jansen, E.C.; Dunietz, G.L.; Chervin, R.D.; Baylin, A.; Baek, J.; Banker, M.; Song, P.X.K.; Cantoral, A.; Tellez Rojo, M.M.; Peterson, K.E. Adiposity in Adolescents: The Interplay of Sleep Duration and Sleep Variability. *J. Pediatr.* **2018**, *203*, 309–316. [CrossRef] [PubMed]

Article

Comparative Analysis of Social Support in Online Health Communities Using a Word Co-Occurrence Network Analysis Approach

Mengque Liu [1], Xia Zou [1], Jiyin Chen [1] and Shuangge Ma [2,*]

[1] School of Journalism and New Media, Xi'an Jiaotong University, No.28 Xianning West Road, Xi'an 710049, China; mengqueliu@xjtu.edu.cn (M.L.); zx124557896@xjtu.edu.cn (X.Z.); cjyxajd@mail.xjtu.edu.cn (J.C.)

[2] Department of Biostatistics, Yale University, 60 College Street, New Haven, CT 06520, USA

* Correspondence: shuangge.ma@yale.edu

Abstract: Online health communities (OHCs) have become a major source of social support for people with health problems. Members of OHCs interact online with others facing similar health problems and receive multiple types of social support, including but not limited to informational support, emotional support, and companionship. The aim of this study is to examine the differences in social support communication among people with different types of cancers. A novel approach is developed to better understand the types of social support embedded in OHC posts. Our approach, based on the word co-occurrence network analysis, preserves the semantic structures of the texts. Information extraction from the semantic structures is supported by the interplay of quantitative and qualitative analyses of the network structures. Our analysis shows that significant differences in social support exist across cancer types, and evidence for the differences across diseases in terms of communication preferences and language use is also identified. Overall, this study can establish a new venue for extracting and analyzing information, so as to inform social support for clinical care.

Keywords: online health community; social support; network analysis; cancer

Citation: Liu, M.; Zou, X.; Chen, J.; Ma, S. Comparative Analysis of Social Support in Online Health Communities Using a Word Co-Occurrence Network Analysis Approach. *Entropy* **2022**, 24, 174. https://doi.org/10.3390/e24020174

Academic Editors: S. Ejaz Ahmed and Farouk Nathoo

Received: 21 December 2021
Accepted: 22 January 2022
Published: 25 January 2022

Publisher's Note: MDPI stays neutral with regard to jurisdictional claims in published maps and institutional affiliations.

1. Introduction

A cancer diagnosis and treatment can cause significant changes to a person's path in life and affect his/her daily activities, work, relationships, and family roles. Cancer patients (and their surrounding members) often suffer from a high level of psychological stress, which can lead to anxiety and depression. They strongly demand social support, which is broadly defined as resources or aids that are exchanged by members within a specific community. Extensive research [1–3] has reported social support as a complex construction with direct and buffering effects on a person's well-being and psychological adjustment to cancer. For example, studies have suggested the association between social support and cancer progression [4]. In addition, insufficient social support can lead to poor health behaviors, which may result in an increased vulnerability toward cancer and its associated mortality [5]. It has also been identified as a consistent indicator for survival.

According to the Health Information National Trends Survey, the proportion of cancer survivors reporting internet use has increased over time, from 49.5% in 2003 to 76.9% in 2017 [6]. Consistent with that, social support is also increasingly exchanged via computer-mediated communication, which has been referred to as computer-mediated social support. It can be developed among strangers whose only connection is their common affliction or concern about a source of personal discomfort. The anonymous nature of online communities also allows patients to exchange personal concerns and advice without the fear of being judged or recognized [7]. We refer to published studies for more discussions on the advantages of computer-mediated social support [8–10]. Online health communities

(OHCs) are online social networks with a focus on health. OHCs can be categorized as either general-purpose communities or those dedicated to a specific health issue. Many OHCs have their own websites, while others are built on existing social networking services, such as Facebook. Compared to traditional health-related websites that only allow users to retrieve information, OHCs can increase members' ability to interact with peers facing similar health problems and, as a result, better meet their immediate needs for social support. People show emotional support for others in OHCs by offering encouragement, reassurance, compassion, etc. OHCs are helpful in empowering patients through personal participation and providing access to information as well as emotional support.

Understanding how members of these online groups interact with each other and make use of online support resources is of critical interest. A handful of content analyses have been conducted, examining the nature of support messages communicated in OHCs [11]. In several studies that analyzed a variety of cancer support groups, information support was found to be the predominant type of support exchanged [12,13]. Some other studies reported that emotional support was the most frequent type of support message [14,15]. Questions, though, about when and why social support messages in computer-mediated contexts vary systematically remain largely unanswered [16]. Blank et al. [17] and Seale et al. [18] revealed significant gender differences. There is also evidence that the support needs of those who were diagnosed, and their families, vary by disease [12,19,20]. It is noted that these studies are mostly limited to breast cancer and prostate cancer, which are mostly gender-specific. Our literature review suggests that, in general, differences across diseases have not been sufficiently examined—something that is critical for understanding patients' needs related to information, emotional support, and relationship-building in OHCs. Only by understanding patients' more specific perceptions and needs can we further optimize the designs and services of OHCs, especially for cancer survivors, who have complex support needs and require different levels of care [21].

Our objective is to provide a detailed and inductively generated account of cancer-type differences in a large number of postings in online cancer support forums. To this end, a novel approach is applied to better understand the types of social support embedded in OHC posts. Different from some previous studies that relied on a commensurate coding scheme with all posts coded [22], which is not feasible with a large amount of data, our approach, based on a word co-occurrence network analysis technique, can provide a macroscopic field-wide view to extract information from big data, making it possible to process a massive amount of online community data. Some other studies adopted quantitative analysis approaches. For example, Seale et al. [18] conducted a comparative keyword analysis to facilitate an interpretive and qualitative examination focused on the meanings of word clusters associated with keywords. There are limitations, however, such as a lack of relevance of word clusters and an inaccurate expression of text themes. Wang et al. [23] used machine learning techniques to reveal the types of social support embedded in each post of an OHC. Wu et al. [24] proposed a social support classification method, using an LDA (linear discriminant analysis) to extract topic features from data. A significant limitation of this analysis is that a certain amount of human annotation is needed, which can be time-consuming and subjective. In addition, an unbalanced data distribution can affect the accuracy of prediction and performance. In this study, the adopted analysis approach can advance from the aforementioned and other studies and directly overcome their limitations. Text data are organized and analyzed with a network perspective, which is system-oriented. Our analysis can identify patterns and relationships among all the words in a system. It can capture properties of individual words and provide insight on how individual words are tied to a larger web (collection of interconnections).

Overall, this study fits well in the scope of information theory-based research. Specifically, it extracts information by conducting complex text mining, and generates knowledge on a complex system by conducting an advanced network analysis, which can more effectively describe variables by taking a system perspective and modeling interconnections. Although the analytic methods adopted in this article have roots in the existing literature, their "combination" and application to a new domain and new biomedical questions are novel. The most essential merit of this study may come from its data analysis findings, which can reveal the social support needed for multiple deadly cancers and the significant differences across cancer types: this has been suggested in the literature but not well quantified to date. The findings can be valuable for stakeholders at multiple levels including healthcare providers, patients, family members, and others. This study can also serve as a prototype for future social support analyses using state-of-the-art network and information analysis techniques, and noting that the existing social support analysis has mostly been based on less advanced methods.

2. Materials and Methods

2.1. Data Source

Patientslikeme.com (PLM) is the world's largest personalized health network, with a growing community of more than 830,000 users. It was designed to facilitate information-sharing between users within disease-specific communities, with the goal of improving the well-being of all users through knowledge derived from shared, real-world experiences and outcomes. In addition to general social networking service (SNS) tools such as user profiles, comments, and private messages, each community has disease-specific tools that allow patients to track and share relevant information such as symptoms, treatments, and medical data. These features have enabled PLM to play a leading role in empowering patients and facilitating social support exchanges and communication online. We note that PLM is not specific to cancer. However, it may still be one of the best resources for studying cancer social support. Beyond the aforementioned advantages, it also has a close working relationship with various healthcare providers. For example, two-thirds of its users felt that their healthcare providers approved/supported using PLM, and about one-third had printed out their patient profiles for use during healthcare visits [25].

PLM has a representative cancer community of more than 50,000 people with over 50 types of cancers, and it is focused on providing customized, disease-specific services that are closely related to our research goal. Extensive research into patient perspectives has been based on this information source. For example, there have been several evaluations of patient perspectives on diseases as well as patient-reported clinical and treatment experience studies of social support groups [26,27]. Other OHCs, such as Breastcancer.org [28], Google Groups [19], and WebMD [29], have also been utilized as data resources in related research.

A web crawler was designed and used to collect data from the PLM online cancer forums, which were launched in 2011. The original dataset consists of all the public posts and user profile information from February 2011 to September 2020. There are 12,150 posts that were contributed by 1358 users who were cancer patients or family members. All posts were in English. The cancer patients were then filtered (according to tags and conditions), leading to 6262 posts. Most of the posts (87.85%) are related to eight cancers. Our exploration shows that the dominating majority of patients had a single type of cancer, which matches clinical practice. Additional details are presented in Figure 1. Our study is centered around these eight specific cancers.

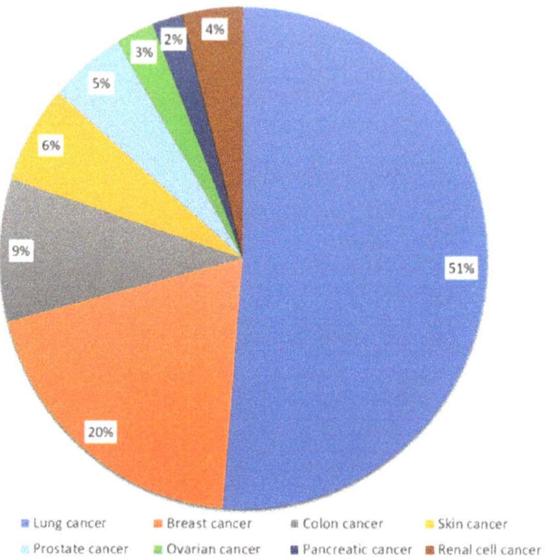

Figure 1. Percentages of posts for the eight types of cancer.

2.2. Method and Procedures

The key steps include the construction of the word co-occurrence network, module detection, social support examination, and interpretation. They are discussed in detail in the following subsections.

Step1: Word Co-Occurrence Network Construction

The posts are split into sentences. For pre-processing, we first conduct tokenization. Stop words that are not informative are removed. Punctuation marks are excluded. Multi-word tokenization is also conducted to expand a raw token into multiple syntactic words. A word co-occurrence network is created with unigram tokens and concatenated multi-word units.

A word co-occurrence network can be expressed as $G = (V, E)$, where V is a set of nodes (where each node represents a word) and E is a set of edges. Edge $e_{ij} \in E$ connects nodes i and j if those two words co-occur within at least one sentence. The number of edges is denoted as $m = |E|$, and $n = |V|$ denotes the number of nodes. The degree of a node i is the number of edges connected to that node, that is, $k_i = | \{ j \in V \,|\{i, j\} \in E\}|$. The weight w_{ij} of edge e_{ij} is defined as the count of joint word occurrence, describing the co-occurrence relationship between the corresponding words in one sentence. The network is undirected by construction. Figure 2 shows a representative word co-occurrence network plotted using the software *Gephi* and containing information on the words and semantic structures. Some important statistical parameters that characterize a network are examined. First, the average shortest-path length ($ASPL$) is the average value of the shortest-path length between any two nodes in the network, which is calculated as:

$$ASPL = \frac{2 \sum_{i>j} d_{ij}}{n(n-1)},$$

where d_{ij} is the shortest-path length between nodes i and j. Second, the clustering coefficient of the network CC is the average of the clustering coefficients of all the nodes in the network defined as:

$$CC = \frac{1}{n} \sum_i \frac{m_i}{k_i(k_i-1)/2},$$

where k_i is the degree of node i, and m_i is the number of edges among the k_i neighbor nodes. For example, for an Erdös–Renyi random network, its average shortest-path length is $ASPL_r \approx \ln(n)/(\ln(2m) - \ln(n))$, and its clustering coefficient is $CC_r \approx 2m/n(n-1)$. A network is said to be a small-world network if $ASPL \approx ASPL_r$ and $CC \approx CC_r$ [30]. Third, degree distribution $p(k)$ is defined as the probability that a randomly chosen node has exactly degree k. For example, if $p(k)$ satisfies the power-law degree distribution, that is, $p(k) \propto k^{-\gamma}$, where γ is a positive constant, then the network is said to be scale-free [31].

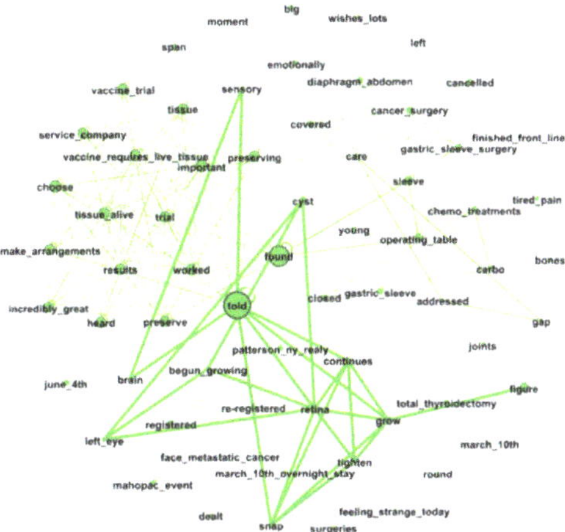

Figure 2. A sample word co-occurrence network for randomly selected posts.

The study of co-occurrence can allow researchers to quantitatively describe the semantic structures of posts. However, significant challenges appear immediately. The word co-occurrence network of posts is usually very hard to visualize, and it is impossible to directly extract meaningful information. As such, there is a strong need to simplify the network, which can reduce complexity, improve visualization, and serve other purposes. One approach is to construct subgraphs, in which most of the useful information contained in the initial graph can be preserved. Here, we achieve this goal via network modules.

Step2: Module Detection

A module is defined as a set of densely connected nodes that are sparsely connected to the other modules in the network. The Louvain algorithm [32], which is based on the optimization of the quality function known as modularity over all possible divisions of a network, is adopted in this analysis and realized using the *Gephi* software. More specifically, this algorithm identifies modules by minimizing:

$$Q(c) = \frac{1}{2M} \sum_i \sum_j \left[w_{ij} - \lambda \frac{\ell_i \ell_j}{2M} \right] \delta_{ij}(c),$$

where c is a partition of nodes, w_{ij} is the edge weight between nodes i and j, λ is a tuning parameter,

$$M = \frac{1}{2} \sum_i \sum_j w_{ij},$$

$$\ell_i = \sum_j w_{ij},$$

and

$$\delta_{ij}(c) = \begin{cases} 1 & if \;\; c(i) = c(j) \\ 0 & otherwise \end{cases}.$$

Here $c(i)$ denotes the module to which node i belongs in the partition c.

The algorithm can unfold a complete hierarchical modular structure for the network, thereby giving access to different resolutions of module detection. In *Gephi*, the resolution parameter, which describes how much between-group edges impact the modularity score, determines the granularity level at which modules are detected [33], with a low-resolution value resulting in more modules. It has been suggested that this algorithm outperforms all other module detection methods in computation time. Moreover, highly satisfied module detection has been observed in practice. For our analysis, module detection of the word co-occurrence network can reduce the size of data, and the analysis of co-occurrences in an individual module can allow researchers to keep track of the semantic structures, which are useful in understanding social support.

Step3: Social Support Quantification and Interpretation

The analysis of word co-occurrences involves clustering words together without breaking their semantic links. In this step, we examine social support by analyzing the semantic structures of the identified modules. As a representative example, Figure 3 presents a module in the word co-occurrence network for ovarian cancer. The words grouped in one module are likely to describe tightly connected topics. For example, most of the words in Figure 3 are related to treatments and medical terminologies. As such, this module can be considered as describing informational support.

- The Taxonomy of Social Support.

Several taxonomies have been developed for the categories of support messages (see for example, [34,35]). Literature on social support suggests that OHCs mainly offer three types of social support: informational support, emotional support, and companionship [11,36]. Informational support is the transmission of facts, suggestions, and/or guidance to community users. Example topics include medication side effects, ways to deal with a symptom, experience with a physician, and medical insurance problems. Emotional support is the expression of understanding, encouragement, empathy, affection, affirmation, caring, and concern. Such support can help reduce stress and anxiety. Companionship consists of chatting, humor, teasing, and discussions of daily life that are not necessarily related to health problems. Examples include diet plans, birthday wishes, holiday plans, and online scrabble games. Companionship helps expand or reinforce a group member's connections.

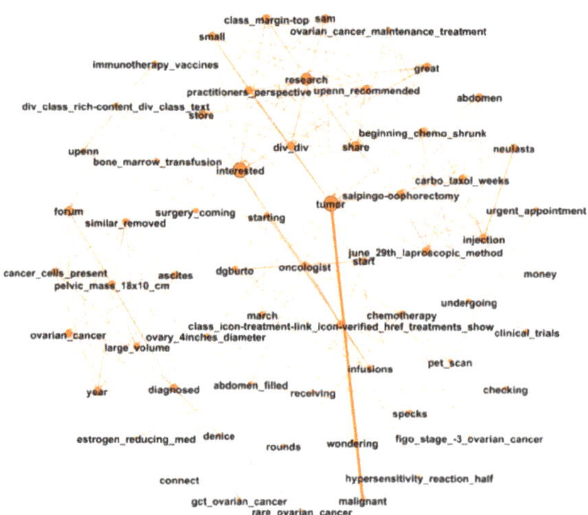

Figure 3. A sample module from the word co-occurrence network for ovarian cancer.

Through the quantitative analysis of semantic structures, the prevalence of specific types of support messages can be revealed. To do this, the first step is to calculate the proportion of edges in each module, which is defined as:

$$P_{C_k} = \frac{\sum_{i \in C_k} \{j \in C_k | \{i,j\} \in E\}}{\sum_{k=1}^{K} \sum_{i \in C_k} \{j \in C_k | \{i,j\} \in E\}}, \quad k = 1, \ldots, K,$$

where K is the number of modules, C_k represents module k, $\sum_{i \in C_k} \{j \in C_k | \{i,j\} \in E\}$ denotes the sum of edges between nodes in C_k. Then, we can compute the proportion of each social support category by summing up the proportions from the individual modules. Exploring communication preferences and language use can also be achieved by taking a closer look at the semantic structures.

3. Results

We apply the analysis approach described above to the data on individual cancers. Pancreatic cancer is highlighted as a representative example.

3.1. Word Co-Occurrence Network

Sentences drawn from the posts were tokened prior to the co-occurrence search, resulting in a list of unique co-occurrence pairs. The word co-occurrence network was then constructed for each cancer. Summary information on the word co-occurrence networks is provided in Table 1. Based on this, an overview of the co-occurrence networks can be provided.

Table 1. Summary of the word co-occurrence networks.

Cancer Type	Sentences	Words	Co-Occurrence Pairs	$ASPL/ASPL_r$	CC/CC_r	γ
Lung cancer	15,690	12,830	196,620	3.167/2.764	0.789/0.001	2.416
Breast cancer	3222	4059	48,559	3.481/2.617	0.821/0.003	2.930
Colon cancer	2746	3524	57,430	3.188/2.344	0.826/0.005	3.017
Basal cell skin cancer	1295	1462	12,124	3.901/2.595	0.831/0.006	3.453
Prostate cancer	751	2005	28,475	3.409/2.272	0.867/0.007	3.056
Ovarian cancer	585	936	10,842	3.592/2.177	0.884/0.012	4.802
Pancreatic cancer	315	729	6749	3.595/2.258	0.861/0.013	3.939
Renal cell cancer	848	1196	9692	4.054/2.544	0.858/0.007	3.414

Compared to a same-scale random network, all the networks have similar average shortest-path lengths and higher clustering coefficients. For example, the average shortest-path length of the pancreatic cancer network is 3.595 (in comparison, an Erdös–Renyi random network has a value of 2.258), and the average clustering coefficient is 0.861 (in comparison, an Erdös–Renyi random network has a value of 0.013). This suggests the presence of the small-world phenomenon in the networks.

In the analysis of degree distribution, it is found that all networks exhibit power-law degree distributions, with the power-law exponent γ ranging between 2.4 and 4.8. Table 1 shows that γ of the ovarian cancer network is the largest, and that of the lung cancer network is the smallest. The scale-free characteristics suggest that the connectivity values of a small number of nodes are quite large (with a large number of connections), rendering them leading roles in the networks. On the other hand, most other nodes have limited connections.

3.2. Module Detection

Take pancreatic cancer as an example. When we visualize its network (Figure 4), words in different modules are represented with different colors. Under the default resolution value of 1.0, there are 72 modules, and the modularity is 0.769. Modules with fewer than five words are removed to improve presentation, leading to 25 modules. Among the remaining modules, the average clustering coefficient is 0.890, suggesting a significant clustering effect. The silhouette for each module is also calculated. The mean silhouette value is 0.649. The silhouette values of the five largest modules are shown in Table 2, which suggest a satisfactory partitioning of the network. The same analysis is also conducted on the other cancers, and the summary of the module detection results is presented in Table 3.

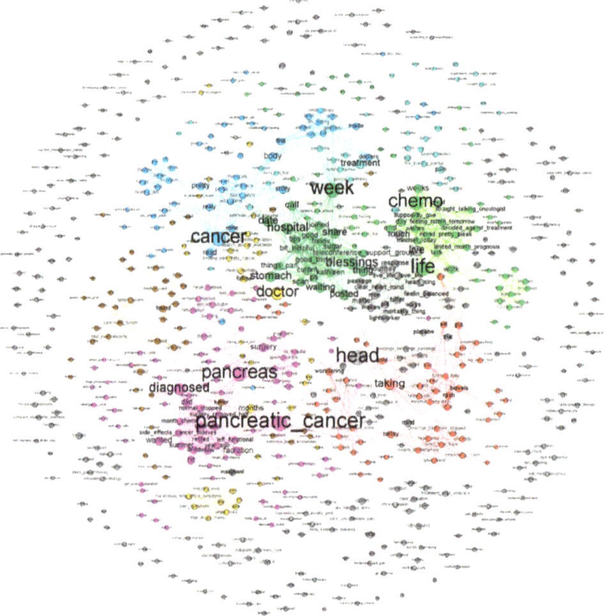

Figure 4. Word co-occurrence network for pancreatic cancer. Different modules are represented using different colors.

Table 2. Information on the five largest modules for pancreatic cancer.

Module ID	% of Edges	Silhouette	Selected Keywords
1	11.35%	0.503	pancreatic cancer; diagnosed; side effects; surgery; left functional
2	7.90%	0.546	chemo; life; oncologist; monitoring; weeks
3	6.32%	0.615	cancer; healthy diet; drinker; fatty tissue; chest cavity
4	6.08%	0.771	doctors; medical cars; scans; sign; pain symptoms
5	4.39%	0.791	treatment; pain; happy; awful; painful

Table 3. Summary of module detection.

Cancer Type	Modularity	CC	Silhouette	Number of Modules
Lung cancer	0.414	0.855	0.426	28
Breast cancer	0.646	0.842	0.433	29
Colon cancer	0.576	0.840	0.506	27
Basal cell skin cancer	0.751	0.857	0.575	29
Prostate cancer	0.685	0.898	0.802	28
Ovarian cancer	0.786	0.903	0.593	27
Pancreatic cancer	0.769	0.890	0.649	25
Renal cell cancer	0.797	0.877	0.503	26

3.3. Social Support Quantification and Interpretation

Summary information for the five largest modules for pancreatic cancer is shown in Table 2. It is observed that the themes of modules 1–4 are mainly concentrated around cancer information, that is, information social support. The keywords of module 5 are mostly associated with the feelings of patients, corresponding to emotional social support. With a similar analysis of the other modules, the proportion of edges in each module is calculated, and the proportions of different social support types after aggregation are obtained. Results are shown in Table 4.

Table 4. Proportions of different social support categories.

Cancer Type	Information Support	Emotional Support	Companionship
Lung cancer	54.94%	13.32%	31.74%
Breast cancer	40.68%	40.45%	18.87%
Colon cancer	58.81%	8.99%	32.20%
Basal cell skin cancer	42.02%	24.19%	33.79%
Prostate cancer	41.15%	36.73%	22.12%
Ovarian cancer	37.22%	36.43%	26.35%
Pancreatic cancer	54.34%	13.13%	32.53%
Renal cell cancer	47.92%	23.61%	28.47%

3.3.1. Differences across Diseases in Types of Social Support

Table 4 shows the proportion of each social support category for each cancer type. Overall, information support (mean 47.14%) and companionship (mean 28.26%) are exchanged most frequently. Sharing is caring, and most posts talk about medical treatments and daily life. The Chi-squared analysis confirms that the overall distribution of social support categories is significantly different across cancer types ($p < 0.001$). Specifically, lung cancer, colon cancer, and pancreatic cancer have the highest percentages (above 50%) of information support. Ovarian and breast cancers have the lowest percentages of information support. Breast cancer has the highest percentage of emotional support (40.45%), followed by prostate cancer (36.73%), ovarian cancer (36.43%), and skin cancer (24.19%). Skin cancer has the highest percentage of companionship (33.79%), while breast cancer (18.87%) and prostate cancer (22.12%) have the lowest.

3.3.2. Differences across Diseases in Communication Preference and Language Use

There is evidence of differences in language use and communication preference across diseases. Four cancers (breast, ovarian, prostate, and skin) have pronounced communication preference and language use patterns. Figure 5 shows the representative network modules, revealing the emotional support of these four cancers. It is observed that breast and ovarian cancer patients mainly talked about their pains and feelings, and their language style was sentimental. In comparison, prostate cancer patients talked more about their thoughts and beliefs, and their language style was calmer and more rational. Figure 6 shows the companionship traits of the four cancers. Skin and breast cancer patients mainly talked about their daily lives, ovarian cancer patients talked more about their family members, and prostate cancer patients talked more broadly. Differences in language use and communication preference mainly exist in the categories of emotional support and companionship. Overall, these findings can reveal several key differences in the use of OHCs across cancer types.

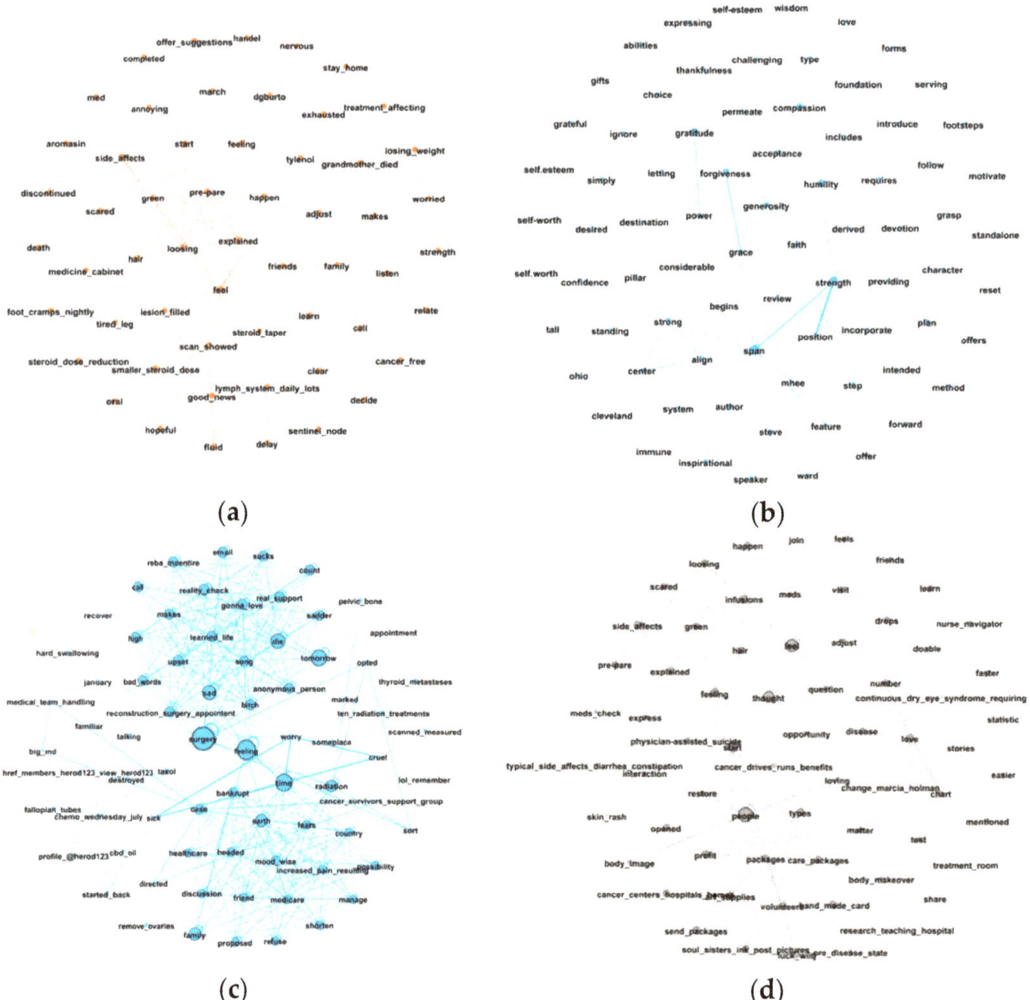

Figure 5. Emotional social support revealed by network modules: (**a**) breast cancer; (**b**) prostate cancer; (**c**) ovarian cancer; (**d**) skin cancer.

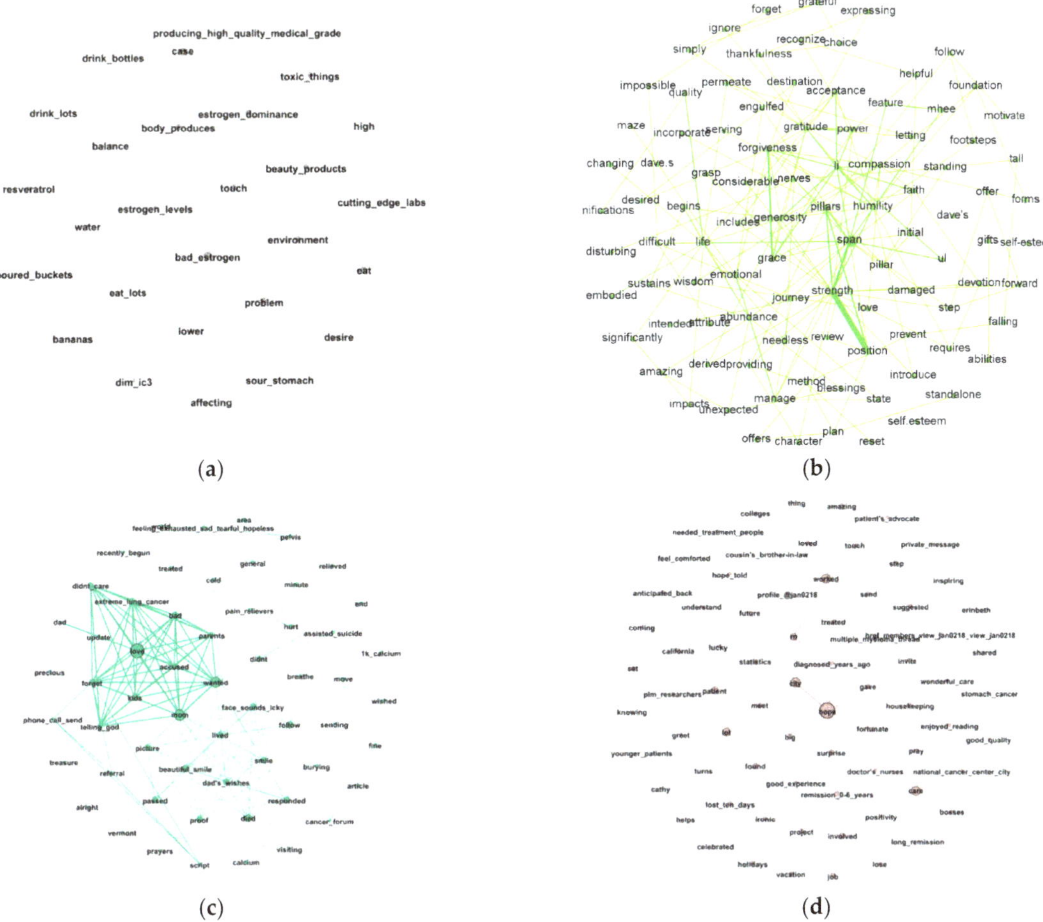

Figure 6. Companionship revealed by network modules: (**a**) breast cancer; (**b**) prostate cancer; (**c**) ovarian cancer; (**d**) skin cancer.

4. Discussion

Our findings are mostly consistent with published research. For example, information support has been identified as the most common type of social support, and published literature has suggested that messages of emotional well-being and medical-related comments are most common on breast cancer sites [17,19,37]. Meanwhile, our research has also added to the existing knowledge of the significant differences between social support categories across cancer types. For example, lung cancer, colon cancer, and pancreatic cancer survivors have been found to mainly utilize OHCs for information-gathering. Notably, prostate cancer survivors also used OHCs as a source of emotional support. Breast, ovarian, prostate, and skin cancer survivors appeared to be in most need of emotional social support. This is likely because people with these cancers had to bear more mental pressure and had a higher risk of also experiencing depression after a new cancer diagnosis [38]. For skin cancer, the high percentage of companionship indicates that the survivors had many daily struggles that led them to seek out support.

Besides adding to existing knowledge by complementing and extending previous research into computer-mediated social support communicated by cancer patients, our analysis has also demonstrated the need for greater recognition of the differences between people with different types of cancer. This knowledge can assist in the design of OHCs. The work can also be a resource for guiding cancer survivors and their families to OHCs that tend to focus more on their specific types of cancer and issues. Similarly, clinicians need to be more aware of the different needs of patients and their families and be able to direct them to online resources that are the most likely to be supportive. In this line, recent studies have shown that the internet has changed the patterns of doctor–patient communication. Social support in OHCs has sometimes played an ambiguous role, making patients behave in a strategic, uncooperative way toward physicians [39,40]. Patient care services have been recommended to enhance the patient–physician relationship. More studies on patients' specific support needs and patient–physician cooperation are needed.

The adopted analysis method can also be used, along with or in replacement of machine learning techniques, in the identification of user roles in OHCs. Further studies on user roles (for example, the differences between lurkers and posters, their specific behaviors, and impact) are also warranted.

Limitations

This study inevitably has limitations. Although PLM is representative and its data has also been examined in other published studies, it is a single OHC and may have a problem of biasedness; although, this has not been observed in existing studies. We have extracted all cancer forum data from PLM. Still, the amount of data for some cancers is limited. This may be true for pancreatic, ovarian, and renal cell cancers. Another data limitation is the possible lack of reliability. Medical information researchers have found that social media sites are identified by limited information [41]. Online users may also be vulnerable to both hidden and overt conflicts of interest, and so they may be incapable of interpreting [42]. In this dataset, there is a lack of information on the duration of diagnosis. As such, we are not able to conduct, for example, a longitudinal analysis to examine temporal trends. Another missed opportunity is that, with a small number of patients with multiple types of cancers, we are not able to provide insights into poly chronic conditions.

There may also be methodological limitations. For example, there is an emphasis on a module-based analysis over individual-message based, which may lead to certain challenges in result interpretation. We have studied the most essential network properties, and it may be of interest to explore more subtle network information.

5. Conclusions

This study has made both domain-specific and methodological contributions to the investigation of OHC use among cancer survivors. There is evidence, some of which confirms and some of which adds to the existing literature, about the significant differences across diseases in terms of social support needs. Specifically, lung cancer, colon cancer, and pancreatic cancer survivors mainly utilized OHCs to meet information support needs. Healthcare providers and physicians are recommended to provide guidance to patients and families on how to gather information and verify its authenticity. Breast, ovarian, prostate, and skin cancer survivors were found to be the most in need of emotional support. For them, targeted patient care can be advice and help to build healthy relationships in a community. Moreover, there is evidence for differences across diseases in language use and communication preference when exchanging social support. For example, skin and breast cancer patients mainly talked about their daily lives, ovarian cancer patients talked more about their family members, and prostate cancer patients talked more about their thoughts and beliefs. Getting familiar with patients' communication preferences can be valuable for establishing the patient–provider bond. With collaboration, liking, and trust, patients are more likely to adhere to treatment especially for long-term medical issues. This work has also introduced a novel method for social support quantification and interpretation, which has multiple advantages over the analyses applied in previous studies.

Author Contributions: Conceptualization, S.M. and M.L.; methodology, S.M. and M.L.; software, M.L.; investigation, X.Z. and J.C.; data curation, X.Z. and J.C.; writing—original draft preparation, M.L.; writing—review and editing, S.M. and M.L.; visualization, M.L.; supervision, S.M.; project administration, S.M.; funding acquisition, M.L. All authors have read and agreed to the published version of the manuscript.

Funding: This research was partly supported by the China Postdoctoral Science Foundation, grant number 2019M663764, the big data visualization technology open sharing platform of Science and Technology Department of Shaanxi Province, grant number 2020PT-029 and National Institutes of Health R03 CA241699.

Institutional Review Board Statement: Not applicable.

Informed Consent Statement: Not applicable.

Data Availability Statement: The analyzed data are in the public domain and accessible to all researchers. However, we do not have the authority to re-distribute data.

Acknowledgments: We thank the editors and reviewers for their kind consideration and careful review.

Conflicts of Interest: The authors declare no conflict of interest.

References

1. Van der Eijk, M.; Faber, M.J.; Aarts, J.W.; Kremer, J.A.; Munneke, M.; Bloem, B.R. Using online health communities to deliver patient-centered care to people with chronic conditions. *J. Med. Internet Res.* **2013**, *15*, e115. [CrossRef] [PubMed]
2. Goh, J.M.; Gao, G.; Agarwal, R. The creation of social value: Can an online health community reduce rural–urban health disparities? *MIS Q.* **2016**, *40*, 247–263. [CrossRef]
3. Merolli, M.; Gray, K.; Martin-Sanchez, F. Health outcomes and related effects of using social media in chronic disease management: A literature review and analysis of affordances. *J. Biomed. Inform.* **2013**, *46*, 957–969. [CrossRef] [PubMed]
4. Nausheen, B.; Gidron, Y.; Peveler, R.; Moss-Morris, R. Social support and cancer progression: A systematic review. *J. Psychosom. Res.* **2009**, *67*, 403–415. [CrossRef] [PubMed]
5. Crispo, A.; Brennan, P.; Jockel, K.H.; Schaffrath-Rosario, A.; Wichmann, H.E.; Nyberg, F.; Simonato, L.; Merletti, F.; Forastiere, F.; Boffetta, P.; et al. The cumulative risk of lung cancer among current, ex-and never-smokers in European men. *Br. J. Cancer* **2004**, *91*, 1280–1286. [CrossRef]
6. Fareed, N.; Swoboda, C.M.; Jonnalagadda, P.; Huerta, T.R. Persistent digital divide in health-related internet use among cancer survivors: Findings from the Health Information National Trends Survey, 2003–2018. *J. Cancer Surviv.* **2021**, *15*, 87–98. [CrossRef] [PubMed]
7. Oh, H.J.; Lee, B. The effect of computer-mediated social support in online communities on patient empowerment and doctor-patient communication. *Health Commun.* **2012**, *27*, 30–41. [CrossRef]
8. Walther, J.B.; Boyd, S. Attraction to computer-mediated social support. *Commun. Technol. Soc. Audience Adopt. Uses* **2002**, *153188*, 2.
9. Wright, K.B.; Miller, C.H. A measure of weak-tie/strong-tie support network preference. *Commun. Monogr.* **2010**, *77*, 500–517. [CrossRef]
10. Wright, K.B.; Johnson, A.; Bernard, D.R.; Averbeck, J. Computer-mediated social support: Promises and pitfalls for individuals coping with health concerns. In *The Routledge Handbook of Health Communication*; Thompson, T.L., Parrott, R., Nussbaum, J.F., Eds.; Routledge: Abingdon, UK, 2011; pp. 349–362.
11. Keating, D.M. Spirituality and support: A descriptive analysis of online social support for depression. *J. Relig. Health* **2013**, *52*, 1014–1028. [CrossRef] [PubMed]
12. Gooden, R.J.; Winefield, H.R. Breast and prostate cancer online discussion boards: A thematic analysis of gender differences and similarities: A thematic analysis of gender differences and similarities. *J. Health Psychol.* **2007**, *12*, 103–114. [CrossRef] [PubMed]
13. Meier, A.; Lyons, E.; Frydman, G.; Forlenza, M.; Rimer, B. How cancer survivors provide support on cancer-related Internet mailing lists. *J. Med. Internet Res.* **2007**, *9*, e12. [CrossRef] [PubMed]
14. Barger, S.D. Perceived emotional support and frequent social contacts are associated with greater knowledge of stroke warning signs: Evidence from two cross-sectional us population surveys. *J. Health Psychol.* **2012**, *17*, 169–178. [CrossRef] [PubMed]
15. Buis, L.R. Emotional and informational support messages in an online hospice support community. *Comput. Inform. Nurs.* **2008**, *26*, 358–367. [CrossRef]
16. Rains, S.A.; Peterson, E.B.; Wright, K.B. Communicating social support in computer-mediated contexts: A meta-analytic review of content analyses examining support messages shared online among individuals coping with illness. *Commun. Monogr.* **2015**, *82*, 403–430. [CrossRef]

17. Blank, T.O.; Adams-Blodnieks, M. The who and the what of usage of two cancer online communities. *Comput. Hum. Behav.* **2007**, *23*, 1249–1257. [CrossRef]
18. Seale, C.; Ziebland, S.; Charteris-Black, J. Gender, cancer experience and internet use: A comparative keyword analysis of interviews and online cancer support groups. *Soc. Sci. Med.* **2006**, *62*, 2577–2590. [CrossRef]
19. Blank, T.O.; Schmidt, S.D.; Vangsness, S.A.; Monteiro, A.K.; Santagata, P.V. Differences among breast and prostate cancer online support groups. *Comput. Hum. Behav.* **2010**, *26*, 1400–1404. [CrossRef]
20. Liu, X.; Sun, M.; Li, J. Research on gender differences in online health communities. *Int. J. Med. Inform.* **2018**, *111*, 172–181. [CrossRef]
21. Brandenbarg, D.; Maass, S.W.; Geerse, O.P.; Stegmann, M.E.; Handberg, C.; Schroevers, M.J.; Duijts, S.F. A systematic review on the prevalence of symptoms of depression, anxiety and distress in long-term cancer survivors: Implications for primary care. *Eur. J. Cancer Care* **2019**, *28*, e13086. [CrossRef]
22. Coursaris, C.K.; Liu, M. An analysis of social support exchanges in online HIV/AIDS self-help groups. *Comput. Hum. Behav.* **2009**, *25*, 911–918. [CrossRef]
23. Wang, X.; Zhao, K.; Street, N. Analyzing and predicting user participations in online health communities: A social support perspective. *J. Med. Internet Res.* **2017**, *19*, e130. [CrossRef]
24. Wu, J.; Hou, S.; Jin, M. Social support and user roles in a Chinese online health community: A LDA based text mining study. In Proceedings of the International Conference on Smart Health, Hong Kong, China, 26–27 June 2017; Springer: Berlin/Heidelberg, Germany, 2017; pp. 169–176.
25. Wicks, P.; Massagli, M.; Frost, J.; Brownstein, C.; Okun, S.; Vaughan, T.; Bradley, R.; Heywood, J. Sharing health data for better outcomes on PatientsLikeMe. *J. Med. Internet Res.* **2010**, *12*, e19. [CrossRef]
26. Eaneff, S.; Wang, V.; Hanger, M.; Levy, M.; Mealy, M.A.; Brandt, A.U.; Eek, D.; Ratchford, J.N.; Nyberg, F.; Goodall, J.; et al. Patient perspectives on neuromyelitis optica spectrum disorders: Data from the PatientsLikeMe online community. *Mult. Scler. Relat. Disord.* **2017**, *17*, 116–122. [CrossRef]
27. Lustria, M.L.A.; Burnett, G.; Cortese, J.; Kazmer, M.; Frost, J.; Kim, J.H.; Ma, J. PatientsLikeMe: ALS patients sharing experiences and personal health information online. *Proc. Am. Soc. Inf. Sci. Technol.* **2009**, *46*, 1–5. [CrossRef]
28. Wang, X.; Zuo, Z.; Zhao, K.; Street, N. Predicting User Participation and Detecting User Role Diffusion in Online Health Communities. In Proceedings of the IEEE 2015 International Conference on Healthcare Informatics, Dallas, TX, USA, 21–23 October 2015; p. 483.
29. Xing, W.; Goggins, S.; Introne, J. Quantifying the effect of informational support on membership retention in online communities through large-scale data analytics. *Comput. Hum. Behav.* **2018**, *86*, 227–234. [CrossRef]
30. Watts, D.J.; Strogatz, S.H. Collective dynamics of 'small-world' networks. *Nature* **1998**, *393*, 440–442. [CrossRef]
31. Barabási, A.L.; Albert, R. Emergence of scaling in random networks. *Science* **1999**, *286*, 509–512. [CrossRef]
32. Blondel, V.D.; Guillaume, J.L.; Lambiotte, R.; Lefebvre, E. Fast unfolding of communities in large networks. *J. Stat. Mech.* **2008**, *10*, P10008. [CrossRef]
33. Lambiotte, R.; Delvenne, J.C.; Barahona, M. Laplacian dynamics and multiscale modular structure in networks. *IEEE Trans. Netw. Sci. Eng.* **2014**, *1*, 76–90. [CrossRef]
34. Schaefer, C.; Coyne, J.C.; Lazarus, R.S. The health-related functions of social support. *J. Behav. Med.* **1981**, *4*, 381–406. [CrossRef] [PubMed]
35. Cutrona, C.E.; Suhr, J.A. Controllability of stressful events and satisfaction with spouse support behaviors. *Commun. Res.* **1992**, *19*, 154–174. [CrossRef]
36. Bambina, A. *Online Social Support: The Interplay of Social Networks and Computer-Mediated Communication*; Cambria Press: Youngstown, NY, USA, 2007; pp. 27–54.
37. Mo, P.K.; Malik, S.H.; Coulson, N.S. Gender differences in computer-mediated communication: A systematic literature review of online health-related support groups. *Patient Educ. Couns.* **2009**, *75*, 16–24. [CrossRef]
38. Alwhaibi, M.; Sambamoorthi, U.; Madhavan, S.; Bias, T.; Kelly, K.; Walkup, J. Cancer type and risk of newly diagnosed depression among elderly medicare beneficiaries with incident breast, colorectal, and prostate cancers. *J. Natl. Compr. Cancer Netw.* **2017**, *15*, 46–55. [CrossRef]
39. Petrič, G.; Atanasova, S.; Kamin, T. Impact of Social Processes in Online Health Communities on Patient Empowerment in Relationship With the Physician: Emergence of Functional and Dysfunctional Empowerment. *J. Med. Internet Res.* **2017**, *19*, e74. [CrossRef]
40. Audrain-Pontevia, A.F.; Menvielle, L. Do online health communities enhance patient-physician relationship? An assessment of the impact of social support and patient empowerment. *Health Serv. Manag. Res.* **2018**, *31*, 154–162. [CrossRef] [PubMed]
41. Moorhead, S.A.; Hazlett, D.E.; Harrison, L.; Carroll, J.K.; Irwin, A.; Hoving, C. A new dimension of health care: Systematic review of the uses, benefits, and limitations of social media for health communication. *J. Med. Internet Res.* **2013**, *15*, e85. [CrossRef]
42. Pirraglia, P.A.; Kravitz, R.L. Social media: New opportunities, new ethical concerns. *J. Gen. Intern. Med.* **2013**, *28*, 165–166. [CrossRef]

Article

Improved Dividend Estimation from Intraday Quotes

Pontus Söderbäck [1], Jörgen Blomvall [1] and Martin Singull [2,*

[1] Department of Management and Engineering, Production Economics, Linköping University, 581 83 Linköping, Sweden; pontus.soderback@liu.se (P.S.); jorgen.blomvall@liu.se (J.B.)
[2] Department of Mathematics, Linköping University, 581 83 Linköping, Sweden
* Correspondence: martin.singull@liu.se

Abstract: Liquid financial markets, such as the options market of the S&P 500 index, create vast amounts of data every day, i.e., so-called intraday data. However, this highly granular data is often reduced to single-time when used to estimate financial quantities. This under-utilization of the data may reduce the quality of the estimates. In this paper, we study the impacts on estimation quality when using intraday data to estimate dividends. The methodology is based on earlier linear regression (ordinary least squares) estimates, which have been adapted to intraday data. Further, the method is also generalized in two aspects. First, the dividends are expressed as present values of future dividends rather than dividend yields. Second, to account for heteroscedasticity, the estimation methodology was formulated as a weighted least squares, where the weights are determined from the market data. This method is compared with a traditional method on out-of-sample S&P 500 European options market data. The results show that estimations based on intraday data have, with statistical significance, a higher quality than the corresponding single-times estimates. Additionally, the two generalizations of the methodology are shown to improve the estimation quality further.

Keywords: big data adaptation; dividend estimation; options markets; weighted least squares

Citation: Söderbäck, P.; Blomvall, J.; Singull, M. Improved Dividend Estimation from Intraday Quotes. *Entropy* **2022**, *24*, 95. https://doi.org/10.3390/e24010095

Academic Editors: S. Ejaz Ahmed and Farouk Nathoo

Received: 6 December 2021
Accepted: 29 December 2021
Published: 7 January 2022

Publisher's Note: MDPI stays neutral with regard to jurisdictional claims in published maps and institutional affiliations.

1. Introduction

This paper presents a method for extracting dividend information from the equity derivatives market using exchange-traded European-typed call and put options. The central methodology in this paper is an extension of the work of Desmettre et al. [1], that is, to formulate a linear regression with a well-known put–call parity. Moreover, we present a novel option position (the sloped asset position), from which it is possible to compute a dividend estimate without specifying an interest rate. Furthermore, throughout the paper, the primary application in mind for the estimates is derivative pricing. This application framing may prima facie seem like an unnecessary limitation, but we argue that the estimates have often-overlooked inherent assumptions that should be aligned with the application. The derivative pricing application follows naturally, whereas other applications require non-trivial adjustments.

One research question is the connection between an asset and its dividends. One of the earliest examples is the asset valuing method: discounted cash flow. The principle idea of that method is that there is a relationship between the price of an asset and its future dividend payments. A related research question is to understand the effect on asset prices of dividend payments. The price of a dividend-paying asset in a frictionless market, ceteris paribus, would drop when a dividend is paid, and the size of the drop would be the size of the dividend payment, see, e.g., Campbell and Beranek [2] and Miller and Modigliani [3]. However, this theory is not supported in empirical studies, and the price generally drops less than the size of the dividend. Campbell and Beranek [2] attribute the differences to tax effects. This idea was elaborated into a formula by Elton and Gruber [4], where the differences between dividend and capital gain taxes were key. Other explanations have been presented, such as transaction costs and behavioral effects. The former was studied by, e.g., Kalay [5] and Boyd and Jagannathan [6], and the latter by Hartzmark and Solomon [7]. Practical imperfections such as a time difference between the ex-dividend date and the payment date can also explain this idea, as was claimed by Wilmott [8]. This paper neither elaborates upon the discounted cash flows method nor provides explanations of imperfect drops in asset prices. Still, these effects must be considered when estimating dividends and evaluating these estimates. We present the implications of the imperfections for estimate interpretation and how to evaluate estimates accordingly.

Dividends are also central in the derivative pricing literature. Dividends have recently started to be seen as an independent asset class according to Filipović and Willems [9], who also provide an overview of this market. This asset class has some interesting properties, but it is not used in this paper. We elaborate on this decision in Section 2.1. Instead, we follow the traditional focus, which has been on modeling the effect on asset price of dividend payments. One of the first to incorporate dividends in derivative pricing was Merton [10], who modeled dividends as continuous adjustments. Another approach is to have discrete adjustments. Discrete adjustments can be applied either as an adjustment of the spot price or an adjustment of the price as time evolves, where the former is sometimes known as an escrowed model (for an overview of this model, see Haug et al. [11], Frishling [12], and Vellekoop and Nieuwenhuis [13]). The use of discrete adjustments is limited since these models have drawbacks. The former contains the possibility of arbitrage opportunities and logical flaws, while the main problem with the latter is its complexity, which often leads to costly methods, see a more elaborated discussion in Haug et al. [11] and Vellekoop and Nieuwenhuis [13]. These problems can be avoided by following Merton [10] and modeling the dividend as a constant continuous yield for each period of maturity, even though that is a poor representation of reality. For example, that method has been applied to models based on stochastic differential equations, such as Carr and Madan [14], Duffie et al. [15], and Carr et al. [16]; implied volatility models such as Gatheral and Jacquier [17]; and local volatility models such as Derman and Kani [18], Derman and Kani [19], and Geng et al. [20].

Regardless of the method, a critical concept is making pricing consistent, which is a critique against the escrowed approach. Even so, estimating dividends, either as a yield or as a present value of future dividends, from market data is not well-studied in the literature. The estimation method that we propose does naturally handle consistent pricing. Furthermore, we study the difference between estimating yield and present value and find that the latter is preferable, regardless of the choice of pricing model. We explain this performance difference in the inherent connection between the dividend yield and the price of the underlying asset.

Although there has been little effort to estimate the dividends for the derivative pricing perspective, it has received more attention in other fields. For example, dividends have long been of interest in studies, such as Fama and French [21], on how dividend yields predict stock returns. Fama and French [21] were not the first to take an intererest in this topic; for an overview of preceding work see their paper and for succeeding papers see Golez [22]. The aims of these papers are of limited relevance in this current study, and relevance is how dividends are estimated. Earlier papers used historical (realized) dividends, but Golez [22] claims that using those could decrease predictability and argued further that inferring dividend yields from the derivatives market is beneficial. Bilson et al. [23] complemented the work of Golez [22] by introducing a novel approach to dividend growth rates implied by market data. Important to note is that Fama and French [21], Golez [22], and Bilson et al. [23] had other aims than to develop a dividend estimation methodology.

Our focus, i.e., on estimation methodology, is not common, but another similar exception is the linear regression (ordinary least squares) methodology presented by Desmettre et al. [1]. In this paper, we generalize the work of Desmettre et al. [1]. The work of Desmettre et al. [1] and our paper can be seen as a parallel to recent work in interest rate estimation by Azzone and Baviera [24] and Blomvall et al. [25]. The estimation methodologies are similar for interest rates and dividends, but the latter contains additional nuances that must be considered. Papers that have estimated dividend quantities from data, such as Golez [22], Bilson et al. [23], and Desmettre et al. [1], have all based their estimates on data from a single time. We expand the data —in the same way as Blomvall et al. [25] does —to use intraday data and find that it provides more stable estimates, i.e., less sensitive to market noise. Moreover, intraday data introduces a coupling to market dynamics that must be considered via a slight reformulation of the regression developed by Desmettre et al. [1]. Additionally, we also present a generalization in the form of weighted least squares formulations.

The estimation methodology is one part of this paper. Furthermore, Desmettre et al. [1] argue that their method and results are limited to markets that meet specific conditions, e.g., the French and German equity markets. This paper presents another interpretation of the quantities, enabling us to evaluate the dividend estimates for more markets, e.g., the US S&P 500 equity market. However, our method is not applicable when used along with equity shares since our methodology relies on relationships between European-typed options. One key of the result evaluations is that the sloped asset position is introduced, acting as an independent method. This position is analogous to the box position used in interest-rate estimation.

The remaining section of this paper is arranged as follows. First, we start with the modeling of dividends, where different estimation methods are also presented. We continue by discussing our data set: the raw data used in the studies and the processing that we performed on the data. In the subsequent section, we present our evaluation methodology, numerical results, and related discussions. Finally, the paper ends with a conclusion and summary of the results found in the paper.

2. Dividend Modeling and Estimation

A dividend payment is a way to distribute value from companies to their shareholders. The basic dynamic that we utilize in our methodology is that the asset price drops when the asset pays a dividend. To obtain a forward-looking estimate, we use the derivatives market. To schematically exemplify this, assume that we have a—highly theoretical—situation with two European-styled call options with identical contract specifications, i.e., identical time to maturity, strike price, and underlying asset, but one option has an underlying asset that pays a dividend while the underlying asset of the other option does not. The option with the dividend-paying asset has a lower price than the other since its payoff at expiry is smaller. In this highly theoretical—but unrealistic—setting, we could infer the dividend effect from the price difference between the options. It is possible to achieve a similar inference in a realistic setting by utilizing the derivatives market.

The idea is simple, but the interpretation of the estimated quantity—even in the idealistic setting of the above example—is rather complex. First, in the example above, the difference between the two option prices is not the dividend, since the option owner is not entitled to the dividend in either case. The difference is instead dependent on how the asset price reacts to dividend payments. This insight is—according to us—not sufficiently pronounced in the dividend estimation literature. Nevertheless, it is of significant importance when interpreting estimates.

The difference between drops and dividends has been empirically studied for shares, and different explanations have been proposed. This research question has not been closed, but when we look into options with an index as the underlying, we argue that other additional effects may also be present. The most apparent difference between a single share and an index is that the latter is neither a traded asset nor pays dividends. The value of an index, i.e., the quoted index value, is not a traded price but a computation from the index constituents' prices according to an index methodology. From this computation, it follows that the index quote should, ideally, experience a drop that reflects the constituent's equity price drop and weight. There might also be details in the index methodology that further complicate the situation. For example, the S&P 500 index quote is not adjusted for standard cash dividends but extra cash dividends. Hence, in theory, the type of dividend payment is reflected differently in the index quotes. These complications make the estimate interpretation more complex than for a single share.

To summarize, the crucial insight is that the effect seen in the market is not only due to the dividend, but is a result of a mixture of the dividend and its imperfections. Despite the importance of this insight, it has received little attention. Desmettre et al. [1] present a related argumentation, but they limit the imperfections to the tax situation. We instead argue that the estimates should not perfectly reflect the realized dividends but rather a latent quantity. This claim is similar to the claim of Desmettre et al. [1], but the difference is that we do not see tax as the sole imperfection. Nevertheless, throughout this paper, we do not explicitly clarify this point repeatedly and refer to the quantity simply as a dividend to increase readability.

The initial theoretical situation—with two different behaviors, i.e., prices, of the same underlying asset—is impossible to replicate in reality. The key to making the idea usable in reality is creating derivative positions related to the underlying asset. In the following two sections, we first discuss data and different relationships that can be used, and then we discuss how to infer estimates from the relationships.

2.1. Market Dividend Relations

The aim of this paper is to create derivative relationships, or positions, of—exchange-traded contracts—from which it is possible to infer the dividend. The relationships that we use should fulfill two properties. First, the data quality of the position should be high, and, second, the position should not require complex modeling of, e.g., the underlying asset, but rather suffice with few assumptions. The former is not a strict definition, but we regard quality as a synonym to liquidity in this paper, i.e., high liquidity is high quality. The reason that we want high-quality data is to ensure that the data quality does not limit the estimates. The limitation of the second property comes from an interpretation of the estimates. The drawback of complex modeling is that the dividend is strongly coupled to the specific model. Estimating the dividend from such a model requires a calibration of the other model parameters. In essence, this coupling makes the dividend an additional model parameter in the calibration process, and, hence, the dividend is affected by the other parameters. This may be a valid method for the calibration of the model, but the dividend is not transferable to other models or applications. Thus, the market contracts we considered in this study were limited by the two properties: high-quality data and non-complex modeling.

The traditional market, when inferring dividend,s has been the equity derivatives market, such as equity futures or equity options. An alternative that may seem attractive is the dividend derivatives market, because of its close connection to dividends, and also because it was used to infer dividend information by van Binsbergen et al. [26]. The market is interesting, but we see three drawbacks of using this market for dividend estimation. First and most important, the underlying of these derivatives is a dividend point index, which is computed from realized dividends. Therefore, the inherent information in the dividend derivatives is linked to realized dividends rather than the effect dividends have on the equity index. This discrepancy makes the dividend derivatives market ill-suited for our estimation since we want to estimate the effect of the asset rather than the dividend. Second, van Binsbergen et al. [26] introduced a model that makes the corresponding estimates less tractable and violates our second property. Additionally, the method used has been questioned by Tunaru [27], who argues that van Binsbergen et al. [26] fail to recognize that dividend derivatives are part of an incomplete market and, thus, that results pbtained using them are invalid. Third, the asset class is not well developed in most markets, and its liquidity is low.

To avoid both illogical approaches and poor liquidity, we use the equity market. In theory, a wide range of derivatives could be used, from plain vanilla to exotic contracts. However, we exclude contracts in the latter category since they require pricing models or have low illiquidity. To conclude, in this study, we considered futures contracts and plain vanilla call and put options to infer dividend information without introducing models and using liquid market data.

The literature for estimating dividends from market data has had two prevailing contract types: futures contracts and plain vanilla European options. The relationships that typically relate to these contracts are the future-basis and the put–call parity. Variations of these positions have been presented, but the common denominator is that they can be constructed almost exclusively and uniquely with exchange-traded contracts. The only component of the relationships that is not directly market observable is the spot interest rates, which match the periods of maturity of the contracts. These unobservable interest rates must be computed from market data. This computation and the corresponding contracts are undesirable because of their increased complexity and reduced tractability.

2.1.1. General Notation

This paper works with two dividend formulations: a yield formulation and a present value formulation. We let $\delta(t;T)$ denote the dividend yield estimated at time t for the period $[t, T]$ and $D(t; T_1, T_2)$ denote the estimated present value at time t of dividends paid in the period $[T_1, T_2]$. To simplify the notation—when the start of the period coincides with the time of estimation—we also introduce $D(t; T) \equiv D(t; t, T)$. Another key component in the dividend estimation is the continuously compounded interest rate. We use the formulation used by Blomvall et al. [25] and decompose the interest rate into two terms, one risk-less interest and an additional spread. We denote the risk-less interest rate and the spread at time t for the period $[t, T]$ as $r_0(t; T)$ and $s(t; T)$, respectively.

This paper considers options with S&P 500 as their underlying, where the standard S&P 500-option contract is of the European-type. The choice of European-typed options was not made inadvertently. Our method does not hold, in general, for American-typed options. European call and put options are used, but they are always considered in a pair as synthetic forward positions. A synthetic forward position is created from a call–put option-pair, i.e., two options with the same underlying, the same time to maturity, and the same strike price. A long (short) synthetic forward position is equivalent to a long (short) call option position and a short (long) put option position. The name synthetic forward position stems from the payoff, which is similar to a standard forward contract, i.e., linear in the price of the underlying. The payoffs are similar, but there are differences between a standard forward contract and the synthetic forward position. The former is unique for each time of maturity, and, upon entering, the two parties agree on a forward price that marks the contract to the market, i.e., no money is transferred upon entering. The synthetic forward position, on the contrary, is not unique for each time of maturity, and it is possible to specify the strike prices. Thus a money transfer can be necessary to mark the contract to the market.

The quote of the S&P 500 index is computed and presented as a unique value, but the market prices for tradable financial assets are only precise down to a bid–ask spread. Despite this market feature, we formulated all the relationships with a unique price in the remainder of this section. Details are discussed in Section 3, but the unique prices used were mid-prices, i.e., the arithmetic means of the bid and ask prices.

2.1.2. Future-Basis

The future-basis is the relationship between a future and spot price for a futures contract. This position is, in essence, used by Andersen and Brotherton-Ratcliffe [28], and it is also described in various textbooks and practitioner-geared literature, such as Wilmott [8] (p. 1040). Let $S(t)$ denote the spot price at time t, and $F(t; T)$ the future price at time t with a time of maturity T, then the future-basis can be written as

$$F(t; T) = S(t)e^{(r_0(t;T) + s(t;T) - \delta(t;T))(T-t)} \tag{1}$$

and

$$F(t;T) = S(t)e^{(r_0(t;T)+s(t;T))(T-t)} - D(t;T), \tag{2}$$

where the former holds for the dividend yield and the latter for a present value of dividends. These relationships can be rewritten as dividend estimates:

$$\hat{\delta}(t;T) = \frac{1}{T-t}\ln\left[\frac{F(t;T)}{S(t)}\right] - (r_0(t;T)+s(t;T)) \tag{3}$$

and

$$\hat{D}(t;T) = S(t)e^{(r(t;T)+s(t;T))(T-t)} - F(t;T). \tag{4}$$

One clear advantage of basing dividend estimates on the future-basis is that the estimates are uniquely specified. On the other hand, we see three drawbacks to basing the estimation of the future market. First, the interest rate must be determined, and potential misspecification affects the dividend estimate. Second, the liquidity of the futures contract is only high for short times to maturity, and, hence, estimations corresponding to longer times to maturity are challenging. Third, the uniqueness of the estimate comes with a drawback. To rely on a single contract for an estimate makes it fragile to noise in the futures price. The second and third drawbacks can be resolved using the options market, e.g., via the put–call parity. Moreover, all three drawbacks can be removed entirely with the sloped asset position, but at the cost of the non-unique estimates. It is also possible to mitigate the first and third problem using a suitable estimation method, which is discussed in Section 2.2.1.

2.1.3. Put–Call Parity

The put–call parity is a relationship that relates the price of a European call option, the price of a European put option, and the price of their underlying asset. The put–call parity does not hold for American call and put options since American-typed options can be exercised early, i.e., prior to maturity. This optionality provides the American-typed options a premium that violates the parity. However, Kragt [29] presents a methodology to estimate these premiums simultaneously with the dividend component, which is outside the scope of this paper. To base dividend estimates on the put–call parity is not a novelty. Additional examples are van Binsbergen et al. [30], Hull [31], and Desmettre et al. [1], where the second formulates the parity with a dividend yield and the other two with the present value of the dividends. We let $c(t;K,T)$ and $p(t;K,T)$ denote the European call and put option prices, respectively, at time t of options, with strike price K, and time of maturity T. The put–call parity formulated with a yield and a present value can be written as:

$$c(t;K,T) - p(t;K,T) = S(t)e^{-\hat{\delta}(t;T)(T-t)} - Ke^{-(r_0(t;T)+s(t;T))(T-t)}, \tag{5}$$

and

$$c(t;K,T) - p(t;K,T) = S(t) - D(t;T) - Ke^{-(r_0(t;T)+s(t;T))(T-t)}, \tag{6}$$

respectively. The left-hand sides of the two relationships can be identified as synthetic forward positions, which we denote as $f(t;K,T) \equiv c(t;K,T) - p(t;K,T)$. From the parities and fixed t, T, and K, it is possible to find direct formulas of the dividend estimates:

$$\hat{\delta}(t;T) = -\frac{1}{T-t}\ln\left[\frac{c(t;K,T) - p(t;K,T) + Ke^{-(r_0(t;T)+s(t;T))(T-t)}}{S(t)}\right] \tag{7}$$

$$= -\frac{1}{T-t}\ln\left[\frac{f(t;K,T) + Ke^{-(r_0(t;T)+s(t;T))(T-t)}}{S(t)}\right], \tag{8}$$

and

$$\hat{D}(t;T) = S(t) - Ke^{-(r_0(t;T)+s(t;T))(T-t)} - c(t;K,T) + p(t;K,T) \tag{9}$$

$$= S(t) - Ke^{-(r_0(t;T)+s(t;T))(T-t)} - f(t;K,T), \tag{10}$$

for the yield and present value, respectively. All estimates for a given time of maturity should, in theory, be the same irrespective of the strike prices. This unity is not true in practice, and the estimates differ for different strike prices. These multiple estimates mitigate the fragility of a single contract but at the cost of non-uniqueness. If a single-valued estimate is necessary, we require an aggregation method. Additionally, the options market is more liquid than the futures market for most times of maturity. The exception is short times to maturity, where the futures market is more liquid than the options market. The third drawback of the future-basis (the need for an interest rate) is also present for the put–call parity. One option to remove the need is to utilize a new option position—the sloped asset position.

2.1.4. Sloped Asset Position

Ronn and Ronn [32] presented an option position, the box-position, from which a market-implied interest rate could be estimated without specifying a dividend. The position has been used in the literature, e.g., van Binsbergen et al. [33] and Blomvall et al. [25]. The box-position is constructed by combining two put–call parities or the equivalent of two synthetic forward positions. We build upon the same logic but choose the number of synthetic forward contracts differently. Let $K_1 \in \mathbb{R}^+$ and $K_2 \in \mathbb{R}^+$, and let the new position consists of one long position in a synthetic forward with the strike price K_1 and K_1/K_2 short synthetic forward positions with the strike price K_2. We refer to this position as the sloped asset position, where the name stems from the payoff of the position. From (5) and (6), we can write two relationships (see Appendix A for details):

$$\frac{f(t;K_1,T)K_2 - f(t;K_2,T)K_1}{K_2 - K_1} = S_t e^{-\delta(T-t)} \tag{11}$$

and

$$\frac{f(t;K_1,T)K_2 - f(t;K_2,T)K_1}{K_2 - K_1} = S_t - D(t;T), \tag{12}$$

respectively. We note that the left-hand sides are the same but that the right-hand sides differ, and we introduce the concept of an adjusted spot price, to simplify the notation, thus:

$$S^*(t;K_1,K_2,T) := \frac{f(t;K_1,T)K_2 - f(t;K_2,T)K_1}{K_2 - K_1}. \tag{13}$$

It is possible to reformulate (11) and (12) with the adjusted spot price into:

$$\hat{\delta}(t;T) = -\frac{1}{T-t} \ln\left[\frac{S^*(t;K_1,K_2,T)}{S(t)}\right], \tag{14}$$

and

$$\hat{D}(t;T) = S(t) - S^*(t;K_1,K_2,T), \tag{15}$$

respectively. The advantage of the position is twofold. First, it is less exposed against noise since it—similar to the put–call parity—is not based on a single data point. Second, contrary to the put–call parity and the future-basis, it does not need an interest rate specification. The reduced noise exposure comes with two drawbacks since it is possible to construct many positions. First, similar to the put–call parity, the estimates must be aggregated if a single-value is wanted. Second, the method is unfeasible for some data sets that the other relationships could manage. For example, with a data set consisting of n option pairs for a given time of maturity (i.e., n synthetic forward contracts), it is possible to construct $n(n-1)/2$ different sloped asset positions and thus equally as many estimates. This quadratic relationship makes the position computationally unfeasible for data sizes that are feasible for the future-basis and the put–call parity. A solution to this infeasibility problem is to limit the data set, but we have chosen not to limit it, because it is difficult to make such a limitation generally and systematically.

2.2. Estimation Methods

The three relationships: the future-basis, the put–call parity, and the sloped asset position could all be used to estimate a dividend quantity, either a yield or a present value, for specific times, t, and times of maturity, T. The estimation method aims to produce a single estimate for each date, but we have multiple times for every date. Furthermore, the future-basis implies a unique estimate for each time of maturity, while multiple estimates can be inferred from the other two relations. For practical applications, multi-valued estimates do not suffice, and a necessary element in the estimation method is aggregation.

A straightforward approach to produce a single estimate is to limit the data. In doing this, the aim of the method is met, but the drawback is that the technique probably introduces additional noise in the estimates, which comes from the fact that the chosen data points can imply biased estimates. An alternative could be to select data points such that the noise is reduced. The disadvantage of such an alternative is twofold. First, it is challenging to design a method that makes this selection possible. Second, it is a strong assumption that a few data points are representative of the whole market. Therefore, we can adjust the estimates instead of adjusting the (input) data. A technique that would consider all available data points to aggregate the estimates, could, e.g., be a mean or a median computation. The drawbacks of this approach are that it requires that the interest rate is specified exogenously, and that the weights given to specific estimates are arbitrary. For example, in the case of the median, all of the weight is put on a single estimate. To mitigate these drawbacks, we followed the method used by Desmettre et al. [1] and formulated the put–call parity as a linear regression model. Similar formulations have also been used by van Binsbergen et al. [33], Azzone and Baviera [24], and Blomvall et al. [25] for interest rate estimation methodologies.

The regression used by Desmettre et al. [1] is the foundation of our work, but we present three expansions. First, instead of limiting the data used to data from a single time (single-time data), we use data from a whole day (intraday data). Second, we formulate two regressions with different modeling of the dividend: one where the dividend is formulated as a yield and one where it is formulated as a present value. Third, we generalize the regression from an ordinary to a weighted least squares model. It would also be possible to formulate a regression from the future-basis, since we use intraday data. We elaborate slightly in the next section, but we do not see it as an appropriate approach, primarily because of the drawbacks presented in Section 2.1.2.

2.2.1. Linear Regression

We formulated one linear regression model for each time of maturity and each put–call parity formulation, (5) and (6). The first regression model was formulated with a dividend yield, and the second used a present-value formulation. In contrast to Desmettre et al. [1], we used intraday data rather than data from a single time. Further, Desmettre et al. [1] correctly point out that by estimating the dividend with regression, the interest rate is estimated simultaneously, eliminating the need for a separate interest rate estimate. Therefore, it may seem strange to reintroduce the estimation need by formulating the interest rate as a sum of an interest rate and an interest spread, but the reintroduction is necessary due to the fact that we use intraday data. The interest rate for a single time is constant, but it is not, in general, constant across a whole day. Consequently, to formulate the regression with a fixed interest will inevitably involve an approximation. To make a more realistic and suitable formulation, we model the spread as a constant and keep intraday dynamics for the total interest rate. The rationale is the same as that used by Blomvall et al. [25], i.e., that the spread is more stable intraday than the risk-less component.

In the formulation, we let N^d denote the number of days we estimated the dividends and let d denote the day $d \in \{1, \ldots, N^d\}$. Moreover, we let κ denote a pair of one (intraday) time, t, and one strike price, K, $\kappa = (t, K)$. For a day d and a time of maturity T, we collected pairs in a set $\mathscr{H}^{d,T}$ and enumerated the pairs as $1, \ldots, N^{d,T}$, where $N^{d,T} = \left| \mathscr{H}^{d,T} \right|$. (The operator, $|\cdot|$, denotes the cardinality of the set.) (The order is unimportant, and the pair with index i is thus $\kappa_i = (t_i, K_i)$). We also introduced $\tau^{d,T}$, which is the time to maturity computed at the beginning of the day. To compute the time from the beginning of the day, we followed the interest rate market convention. The put–call parity, Equation (5), can then be written as

$$f_i^T = S(t_i)e^{-\delta^{d,T}\tau^{d,T}} - K_i e^{-r_0^T(t_i)\tau^{d,T}} e^{-s^{d,T}\tau^{d,T}}, \quad \forall i = 1, \ldots, N^{d,T}, \tag{16}$$

where $f_i^T \equiv f(t_i; K_i, T)$. We introduced $X_{1,i}^{d,T} := -K_i e^{-r_0^T(t_i)\tau^{d,T}}$ and $X_{2,i} := S(t_i)$ to simplify the notation. (Note that $X_{2,i}$ neither depends on the day nor the time of maturity.) We wanted to estimate $e^{-\delta^{d,T}\tau^{d,T}}$ and $e^{-s^{d,T}\tau^{d,T}}$, and denoted the corresponding regression coefficients as $\gamma_1^{d,T}$ and $\gamma_2^{d,T}$. Thus, it is possible to write the linear regression as

$$f_i^T = X_{1,i}^{d,T}\gamma_1 + X_{2,i}\gamma_2^{d,T}, \quad \forall i = 1, \ldots, N^{d,T}. \tag{17}$$

It is possible to write a similar regression, based on (6) by introducing $h_i^T := f_i^T - S(t_i)$ and a regression constant, γ_0,

$$h_i^T = \gamma_0 + X_{i,1}^{d,T}\gamma_1, \quad \forall i = 1, \ldots, N^{d,T}. \tag{18}$$

To summarize, we can write the financial quantity estimates from the regression estimates as

$$\hat{D}^{d,T} = -\gamma_0, \tag{19}$$

$$\hat{\delta}^{d,T} = -\frac{1}{\tau^{d,T}} \ln\left[\hat{\gamma}_2^{d,T}\right], \tag{20}$$

$$\hat{s}^{d,T} = -\frac{1}{\tau^{d,T}} \ln\left[\hat{\gamma}_1^{d,T}\right], \tag{21}$$

where $\hat{D}^{d,T}$ is the estimate of the present value for the time of maturity T. The interest rate spread estimate, $\hat{s}^{d,T}$, can be estimated from the regression models (17) and (18), but the estimates are not generally equal, with the exception of the single-time data.

It is possible to see the differences between using intraday data and single-time data in the regressions. The interest rate, $r_0^T(t_i)$, is fixed when single-time data is used. This fixed interest rate makes the decomposed interest rate form redundant since the sum of $r_0(t_i)^T + s$ is a constant. Further, the spot price, $S(t_i)$, is also fixed, making it possible to convert the dividend yield estimate to a present value dividend estimate, and vice versa, without loss or distortion of the estimates. This perfect conversion makes the two different dividend formulations redundant. These redundancies are not present when intraday data is used, since neither $r_0(t_i)$ nor $S(t_i)$ is constant, making the decomposed interest rate necessary and the dual regression formulations interesting.

Finally, from (17) and (18), it is easy to see that the future-basis regressions, based on (1) and (2), would follow. The formulation is made possible by the utilization of intraday data rather than a single-time data. Despite the analogue to the put–call parity, the regression has one shortcoming compared to its put–call parity counterpart. The regression coefficient for the dividend yield-formulated regression is the sum of the dividend yield and spread, $e^{(s(t;T)-\delta(t;T))(T-t)}$. Hence, the future-basis could only be used for present value estimates.

This shortcoming and the previously mentioned drawbacks are why we do not consider this regression in this paper.

2.2.2. Linear Regression—Weighted Least Squares

In an ordinary least squares formulation all of the data are considered equally important. This implicit assumption is likely to be incorrect since the quality of data points is likely different. The ordinary least squares formulation does not adjust for this difference in data quality and thus has a drawback. One approach to counteract this behavior is to value some data points more and some less. To formulate this mathematically rigorous method, we followed the idea in Blomvall et al. [25] and use weighted least squares. Considering the models (17) and (18), we can formulate the weighted least squares

$$
\min_{\gamma=(\gamma_1,\gamma_2)} \sum_{i=1}^{N^{d,T}} w_i^{d,T} \left(f_i - X_{1,i}^{d,T}\gamma_1 - X_{2,i}^{d,T}\gamma_2 \right)^2,
\tag{22}
$$

and

$$
\min_{\gamma=(\gamma_0,\gamma_1)} \sum_{i=1}^{N^{d,T}} w_i^{d,T} \left(h_i^{d,T} - \gamma_0 - X_{i,1}^{d,T}\gamma_1 \right)^2,
\tag{23}
$$

where $w_1^{d,T}, \ldots, w_{N^{d,T}}^{d,T}$ are non-negative weights. Note that if each weight is chosen as a positive constant, i.e., $0 < w = w_1^{d,T} = \ldots = w_{N^{d,T}}^{d,T}$, we obtain the ordinary least squares estimator, only if $w = 1$, the same sum of square errors, is the same. The crux with these formulations is to determine weights. The key idea of the weights is to choose them such that the resulting estimator has good properties. An essential property for the ordinary least squares estimator is that if the residuals are independent and homoscedastic (same finite variance), the estimator is the BLUE (best linear unbiased estimator). The residuals from the regressions (17) and (18) likely do not fulfill the homoscedasticity, and an ordinary least squares estimator is not the BLUE. One reason is that the liquidity of the data varies between strike prices, where illiquidity typically leads to higher variance.

The heteroscedasticity can be counteracted, and it is possible to achieve the BLUE with a specific weighting scheme. According to Aitken [34] (the result can also be found in textbooks such as Zwanzig and Liero [35]), if the weights are chosen to be inversely proportional to the variances, the estimator receives the BLUE property. In addition to the statistical properties, Blomvall et al. [25] pointed out that weights chosen inversely to the residuals also have an economic rationale. The residuals can be interpreted as a measure of the repricing capabilities of the linear models, where smaller residuals indicate accurate repricing. Nevertheless, the appealing theoretical property has a practical drawback since the variances are unknown and need to be estimated. Estimating the variance for residuals is non-trivial, since we only have a single residual if we fix the time, strike price, and time of maturity, i.e., we do not have repeated estimates of a quantity. To mitigate this problem, we make the same assumption as Blomvall et al. [25] that the variance is constant intraday, i.e., for a fixed strike price and time of maturity. Hence, the variances can be estimated from different intraday times. The weights are computed with the same four-step processes used by Blomvall et al. [25].

First, an ordinary least squares estimate is computed, and the (raw) residuals are determined, which we, for each strike price and time of maturity, denote as e_i, $\forall i = 1, \ldots, N^{d,T}$. Second, the residuals, e_i, are grouped into (index) groups according to their strike prices, $\mathcal{R}_K^{d,T} = \{i | i \in \{1, \ldots, N^{d,T}\}$ and $K_i^{d,T} = K\}$. Third, a variance is estimated for each group, where:

$$\mu_K^{d,T} = \frac{1}{\left| \mathcal{R}_K^{d,T} \right|} \sum_{i \in \mathcal{R}_K^{d,T}} e_i, \tag{24}$$

$$v_K^{d,T} = \frac{1}{\left| \mathcal{R}_K^{d,T} \right| - 1} \sum_{i \in \mathcal{R}_K^{d,T}} \left(e_i - \mu_K^{d,T} \right)^2, \tag{25}$$

denote the estimated mean and variance, respectively, for the group associated with the date, d, the time of maturity, T, and the strike price, K. Finally, the weights in (22) and (23) can be determined from the auxiliary weights $\tilde{w}_K^{d,T} = 1/v_K^{d,T}$, as

$$w_i^{d,T} = \tilde{w}_{K_i^{d,T}}^{d,T}, \quad \forall i = 1, \ldots, N^{d,T}. \tag{26}$$

3. Data

The data set used in this paper is the same data set used by Blomvall et al. [25]. All the data have been collected from the data provider Thomson Reuters Refinitiv Eikon, and the data set consists of three types of intraday data. First, quotes of the S&P 500-index. Second, bid and ask quotes of European call and put options with the S&P 500-index as their underlying. Third, payer and receiver quotes of fix rates of USD denoted by overnight index swaps contracts with the federal funds rate as the reference rate.

The tick data is collected for all dates in the period from 1 March 2020 to 31 January 2021 between 9 a.m.–4 p.m. The European options are all the available monthly options for the given dates, i.e., all options expiring on the third Friday of each month. The USD overnight index swaps fix rates have a maturity between 1 and 10 years. The data set consists of 6 million S&P 500 index quotes, 110 million bid and ask quotes of the USD overnight index swaps, and 54 billion option prices.

Although granular, the data set must be processed to be useful in the paper. The collected data has four inherent problems. First, we collected tick data, but it is difficult to use because of its irregularities. The data is transformed to a more usable form where the level of granularity is preserved. Second, the quotes of the fix rates of overnight index swaps are not directly usable since (17) and (18) require continuous spot rates, and thus a transformation is needed. Third, in Section 2, all regressions were formulated with a unique price, but in the data set, the prices are only precise down to a bid–ask spread. Earlier, we mentioned that the price used is the mid-price, and below, we discuss this issue. Fourth, we discuss how to identify and remove unrealistic data points.

3.1. Synthetic Forward and Sloped Asset Positions

In Section 2, we used synthetic forward and sloped asset positions with unique prices. Neither of these positions is traded in the market, rather only the options are. Hence, neither synthetic forward nor sloped asset positions have quoted bid or ask prices. To circumnavigate the missing prices of these positions, we first computed their bid and ask prices. The mid-prices were then computed from these bid and ask prices. The bid and ask prices were created by artificially replicating the market prices of entering such positions.

Let c, p, and f, respectively, denote the price of a call option, put option, and synthetic forward; and let a and b denote the ask and bid price, respectively. The payments of the bid and ask positions can be summarized as $f_a = c_a - p_b$ and $f_b = c_b - p_a$. We compute the mid-price of the synthetic forward as

$$f_m = \frac{f_a + f_b}{2} = \frac{(c_a - p_b) + (c_b - p_a)}{2} = c_m - p_m. \tag{27}$$

It is thus necessary to have four prices—bid and ask prices for both the call and put options—to compute the mid-price of the synthetic forward position. Hence, if one or more quotes are missing, the mid-quote is not computable and thus not used in the regression.

We likewise compute mid-prices of the sloped asset position by first computing the bid and ask prices; the argument is analogous to the synthetic forward position. We compute the price of entering the position in two directions, and the mid-price is the average. Let $\phi_{i,j}$ denote the sloped asset position, which is going long in a synthetic forward with strike price K_i, and short in K_i/K_j synthetic forwards with strike price K_j. The cost of entering such a position is the ask price, f_i^a reduced by the bid price, f_j^b for each of the K_i/K_j contracts. The ask price of this contract can thus be written as $\phi_{i,j}^a = f_i^a - K_i/K_j f_j^b$. Alternatively, we receive f_i^b and must pay f_j^a for each of the K_i/K_j contracts, and the bid price of the slope position is thus calculated $\phi_{i,j}^b = f_i^b - K_i/K_j f_j^a$. The mid-price of the slope position is

$$\phi_{i,j}^m = \frac{\phi_{i,j}^a + \phi_{i,j}^b}{2} = \frac{\left(f_i^a - K_i/K_j f_j^b\right) + \left(f_i^b - K_i/K_j f_j^a\right)}{2} \tag{28}$$

$$= \frac{\left(f_i^a + f_i^b\right) - K_i/K_j \left(f_j^b + f_j^a\right)}{2} = f_i^m - K_i/K_j f_j^m. \tag{29}$$

We see that the data of synthetic forward contracts is sufficient to express both the put–call parity and the sloped asset position.

3.2. Transformation and Cleaning of Option Tick Data

Our data management has two aims. First, to make the data appropriate for the (estimation) method, and second, to clean the data from the artifacts. While the former is a necessity, the second can lead to the validity of the method being questioned; hence, the data cleaning is moderate. In this paper, we address two features of this data set: a sudden and temporary downward spike in bid quotes and a lack of bid quotes for out-of-the-money options. We classify the former as a data artifact that normal market dynamics cannot explain, while the second has a natural explanation.

The first problem (downward spikes) is illustrated in Figure 1. We deem the drops of approximately $1200 to be non-realistic and deem further that those spikes have been created in the data collection. We cannot explain why only the bid quotes are affected by this effect. The problem of the downward spikes is easily solved by removing them from the data set. The crux is to determine which of the bid quotes are artifacts and which that are not. In Figure 1, the artifact is evident, but there could be other cases where the spikes are not as obvious. A rough description is that the drops are more pronounced for (deep) in-the-money options than out-of-the-money options, since the former options naturally have higher prices. However, the silver lining is that the effect of the data is less severe, since out-of-the-money options have a lower price; thus, the drops cannot be as big. Therefore, we limited the data cleaning to in-the-money options, since they are more affected and easier to find than out-of-the-money options. The data cleaning procedure that we used was to discard all call (put) options with strike prices greater (less) than the spot price as well as all quotes smaller than $1.

The second problem (missing bid quotes) is not as trivial or obvious as the spikes. Many (deep) out-of-the-money options in the data set lack bid quotes but have corresponding ask quotes. We attribute this data property to the tick size of the market, i.e., the minimum amount that a quote can be changed. This amount may be greater than the fair price of some options, and any (positive) bid quote would thus be overpriced. If the only possible price is an overprice, the only sensible action is not to quote. The ask prices do not suffer from the same dynamics, since it is natural to ask for a higher price than a fair price. The drawback of the estimation method with missing bid quotes is that it decreases the data set significantly. As noted above, a single synthetic forward price mid-price requires both a bid and ask quote for one call and one put option. To reduce the data waste, we recreated the bid quotes.

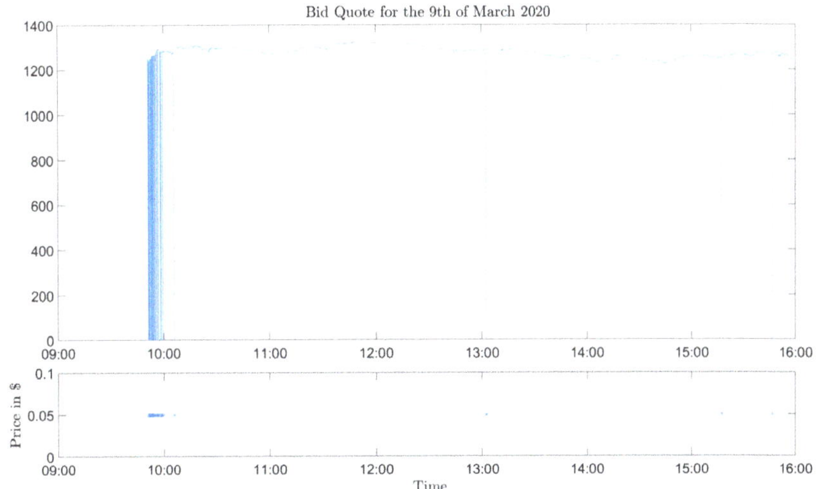

Figure 1. The two panels illustrate the bid quotes of a call option on the 9 March 2020. The strike price of the option is $1500 (in-the-money), and its expiration is the 20 March 2020. The two panels illustrate the same data, but the lower panel focuses on smaller values and thus has a smaller y-axis than the upper panel.

The strict natural lower limit of plain vanilla option quotes is zero. A price of zero means that someone, essentially, gives away a contract for free with only positive (including zero) payoffs, which is an arbitrage opportunity and, thus, a non-realistic scenario. However, the bid and ask prices are not used individually, but rather only in pairs, to compute mid-prices. Therefore, we argue that when the fair option price is within one tick from zero, it is a valid approximation to set the bid price to zero. A zero bid price is too low, the subsequent mid-price is too low, and a bias is introduced in the mid-prices. In order not to introduce biases, we only replaced missing bid quotes for some call and put options, which are essentially options with small prices. Let S denote the intraday median spot price, K the strike price, and let $a^{d,T} := a\sigma\sqrt{\tau^{d,T}}$ where $a > 0$ and $\sigma > 0$. We replaced missing bid quotes for call and put options if $S(1 + a^{d,T}) < K$ and $S(1 - a^{d,T}) > K$, respectively. (The economic interpretation is that bid quotes are only replaced for options with a strike price at least a standard deviations, σ, from the current spot price of the underlying asset, i.e., deep out-of-money options.)

Data cleaning is the first step in the data transformation process, where the second is to process the data into a better-suited format. The collected option data is tick data of values and timestamps, which has a precision of one second. One alternative would be to transform the tick data set into a set where the data points are spaced with a fixed time unit, e.g., a second. In essence, the idea is to use the most recent tick quoted in the market for every time unit, using the most recent tick in the case of tick data to second data. The assumption is that, as long as new ticks have not reached the market, i.e., no new information has reached the market, the old ticks are still valid. There are two practical benefits of such an approach. First, it is easy to work with such data. Second, the data utilization is high. Furthermore, if no new information has reached the market, that implies that the market's dividend and interest-rate beliefs are unchanged. (The converse is not true, changed prices are not synonyms for changes in the markets in terms of the dividend or interest rates beliefs, but it can signify a myriad of factors).

The drawback of such an approach is the risk of amplifying noise. All market information carries some noise, and repeating individual data points would assign higher confidence or weight to arbitrary points and consequently amplify noise in these points. In order not to indirectly assign higher weights to certain points and instead to keep the data utilization high, we are only interested in times where at least one quote of at least one option has changed from the previous time. In this method, the prices of the options should correspond to the index quote for the same times. The transformation that we propose is a two-step process. First, the tick data set is transformed into a set with a specific frequency, i.e., the time between data points, e.g., 1 second, which is the frequency used in this paper. Second, this data set is transformed into the final data set, where only the data points that have changed are kept. Small schematic examples of mock tick data, fixed time unit data, and the final data are presented in Tables 1–3, respectively. Note, before the first tick of the day, the quote is written as not available (N/A). The value from a tick prevails until a new tick comes or the day ends (4 pm). The transformation is performed for all options and all fixed rates. The second transformation is from the one-second data to a data set that only contains seconds that coincide with ticks. Table 3 presents a continuation of the example in Table 2. Note that this transformation is not a reversal of the first transformation. The first transformation was made for individual options' bid and ask quotes, and the second considers all the options' bid and ask quotes (for a given day and time of maturity) simultaneously.

Table 1. The two panels schematically exemplify mock tick market data of two assets. The marker N.U. indicates Not Updated.

Time	Bid Quote (1)	Ask Quote (1)	Time	Bid Quote (2)	Ask Quote (2)
09:01:02	100	102	09:02:00	200	N.U.
09:02:30	N.U.	103	09:02:59	N.U.	213
09:03:15	99	N.U.	09:04:10	199	N.U.
⋮	⋮	⋮	⋮	⋮	⋮

Table 2. The two panels schematically exemplify one-second data that have been derived from Table 1. The left and right panels are derived from the left and right panels in Table 1, respectively. The bold quotes indicate that those quotes were ticks and not repeats of an earlier tick. Bold times indicate that, at that time, at least one of the quotes (bid and ask) was a tick.

Time	Bid Quote (1)	Ask Quote (1)	Time	Bid Quote (2)	Ask Quote (2)
09:01:02	100	102	09:02:00	200	N/A
⋮	⋮	⋮	⋮	⋮	⋮
09:02:29	100	102	09:02:58	200	N/A
09:02:30	100	103	09:02:59	200	213
⋮	⋮	⋮	⋮	⋮	⋮
09:03:14	100	103	09:04:09	200	213
09:03:15	99	103	09:04:10	199	213
⋮	⋮	⋮	⋮	⋮	⋮

Table 3. This table schematically exemplifies one-second data, which combines the two panels in Table 2. The bold numbers indicate that those numbers were ticks in the tick data. (Note that every row has at least one bolded number) The difference between the panels in Table 2 and this table is that times that lack a tick have been removed from this table.

Time	Bid Quote (1)	Ask Quote (1)	Bid Quote (2)	Ask Quote (2)
09:01:02	100	102	N/A	N/A
09:02:00	100	102	200	N/A
09:02:30	100	103	200	N/A
09:02:59	100	103	200	213
09:03:15	99	103	200	213
09:04:10	99	103	199	213
⋮	⋮	⋮	⋮	⋮

3.3. Overnight Index Swap Implied Spot-Rates

The regression formulations (17) and (18) require continuous spot interest rates. In this paper, we follow the arguments in Blomvall et al. [25] and base these rates on OIS contracts. A specific interest rate is not critical since we estimate a spread over this rate, and most rates are stable intraday. From that point of view, we could have used interest rates from a data provider, such as Thomson Reuters Eikon Refinitiv.

However, the interest rate data must match the frequency of the option data, and thus we must compute them. We use the technique proposed by Blomvall [36], which produces a complete forward-term structure of daily forward interest rates. In this paper, only the spot rates that correspond to the options times of maturity are of interest, and these rates are computed from the forward rates.

4. Results and Discussion

This results and discussion section consists of four parts. First, we present the characteristics of the estimates in plots, which are the foundation of the next part in the section. In the second part, in-sample results are presented, that is, results where the data set has not been divided into training and test sets. The in-sample results answer some questions, but the validity of the results can be partly questioned since the results could be the effect of over-fitting. The third part presents the methodology for performing out-of-sample testing, i.e., the data partitioning and evaluation methods. The results include both some basic statistics and a statistical Diebold–Mariano test. Throughout the section, we discuss and highlight results when presented, but one question spans multiple parts—the difference in estimating yield and present value, and hence, it is discussed in the fourth, and final, part of this section.

In addition to the question of the difference between yield and present value, two additional questions are discussed in this section. First, various regressions for dividend estimation have been presented, which can be grouped according to two properties: the weighting scheme and the type of data. The regressions have been formulated generally to handle intraday data, which is similar to the approach used by Blomvall et al. [25]. Contrariwise, in Desmettre et al. [1] and other methodologically similar approaches for interest rate estimation, single-time data is used, see Blomvall et al. [25] for an overview of the latter. From these regressions we make two comparisons. First, we compare the single-time and the intraday dataset. Second, we also study the differences between the weighted least squares and the ordinary least squares models.

4.1. Characteristics of Estimates

The characterization of the estimates is divided into two parts. First, we have three illustrations of the estimates, both for intraday data and single-time data. The data set is not partitioned in this section, but rather all the data for each time has been used. Second, we start by presenting some surf plots in Figure 2. The surf plots provide an overview, but it is difficult to see any small differences. The line plots in Figures 3 and 4 complement the surf plots.

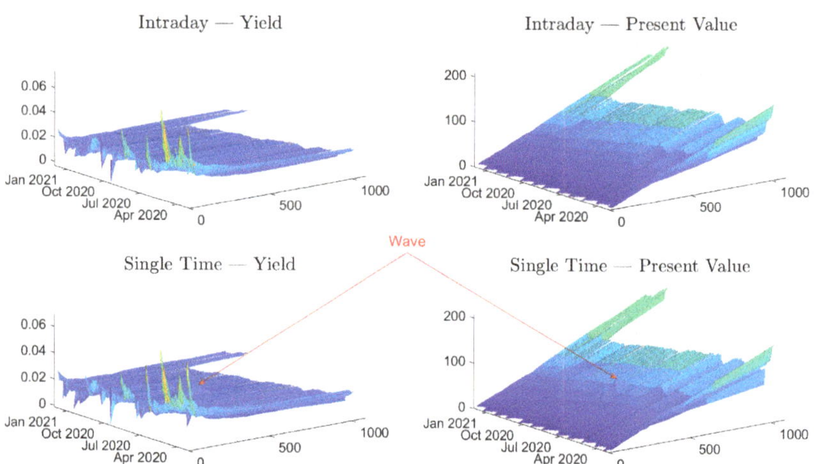

Figure 2. The figure shows the surface plots of the ordinary least squares dividend estimates, which can be grouped by two properties. First, the two upper and two lower panels are computed with intraday and single-time data, respectively. Second, the left and right panels are computed as dividend yields and the present value of dividends, respectively. The z-axes of all four panels indicate the estimated values. The x- and y-axes corresponds to the date and time to maturity (measured in days), respectively, to which the estimates correspond.

The overall illustration in the surf plots and the line plots is that the present value estimates have a downward sloping trend as time progresses. These trends, for the longer maturities, are supported by the mean values in Table 4, which indicates that the mean daily changes are negative. These slopes are expected since the present value—for a specific time of maturity—naturally decreases when ex-dividend dates are passed. On the other hand, the yield estimates do not form a slope. Instead, the main effect is that they converge for longer times of maturity. This convergence can be interpreted as expected dividends, in dollars, being stable over the years.

Table 4. This table shows statistics for the daily differences in market-matched present value dividend estimates for four series of estimates with a constant time of maturity. The numbers (382, 473, 655, 1019) in the first column—TTM—are the times to maturity on the 2 March 2020 (i.e., the first date in the data set.) for the series.

	Single-Time			Intraday		
TTM	**Mean**	**Std**	**Autocorr**	**Mean**	**Std**	**Autocorr**
382	-0.24581	1.7569	-0.1491	-0.23592	1.1137	0.33601
473	-0.25063	1.9994	-0.053606	-0.23916	1.387	0.31847
655	-0.26461	3.3015	-0.17999	-0.25003	2.1288	0.28339
1019	-0.24773	5.7669	-0.32063	-0.23198	3.5832	0.12227

We can observe that both the yield and present value approaches are stable as time evolves, but the estimates vary substantially for different maturities. We can also see an additional effect: estimates drift off shortly before the expiration date. The problem effect is easily observable for the yields in Figures 2 and 4. The effect is also observable for the present value data, but the scale of the plot masks the effect. In most cases, the drift is positive, but we can observe some negative estimates. A negative yield or a negative present value can be interpreted as a cash flow that lifts (negative drop) the price. It is an improbable market dynamic, and, since we experience these negative estimates adjacent to other spurious estimates, we argue that these are not to be taken at face value. Instead, the estimates in these regions should rather be seen as indications of artifacts of the estimation method. A similar effect was reported by Blomvall et al. [25] for interest rate spread estimates, and we follow their argument and explain this effect with low option data quality. Finally, we can see that these spurious values are more pronounced for the single-time data than for the intraday data. We only consider option pairs with a time to maturity exceeding five days to reduce the impact on the results of these spurious values.

Figure 3. The figure consists of two panels, where the upper and lower panels illustrate the present value dividend estimates that are market-matched and stripped, respectively. The plots illustrate a series of dividend estimates with fixed times of maturity, where the x-axis is the date. The series in each panel consists of either estimates determined by intraday data or by single-time data recorded at 3 p.m. The legend of the upper panels indicates the type of data, and the number is the number of days to maturity at the first date. The legend of the lower panel contain a period, which is the number of days to maturity for the two contracts that have created the stripped dividend.

The above observation illustrates potential data problems, and provides some insights into how the market reacted during the period. The core idea of the paper is not to understand and study the market dynamics. Nevertheless, the illustrations indicate shifts in the market, which are too significant to leave without comment. The comments are not detailed but instead focus on the holistic picture. In Figure 4, we can see that around April 2020, the estimates behave differently than for the other period, which is a period when the global pandemic started to affect the markets. It is possible to see that the estimates of present value squeezed together, i.e., the difference in estimates of longer and shorter times of maturity reduced. The yield estimates were also affected, but rather with in the opposite direction. The difference between the long and short times of maturity increased. We can also observe that the S&P 500 index quote also experienced a downturn. The effects on the estimates can prima facie seem contrary, but both behaviors have the same underlying reason. During this period, many companies cut, either partly or entirely, their future dividends but kept dividends that were closer in time (e.g., announced dividends), and the market anticipated further cuts for future dividends. For the present value dividend estimates, the effect was direct. Present value dividend estimates corresponding to longer maturities were reduced more than the corresponding short times to maturity. This phenomenon is natural, since both the realized and anticipated dividend cuts were more pronounced for longer times to maturity. A similar effect would have been seen in the yields if the S&P 500 quote had been constant, but the downturn of the S&P 500 offset the effect of lower yields, especially for short times of maturity, and resulted in higher yield estimates for shorter times to maturity. We can see that the estimates have captured these market dynamics and could potentially be a good measure of how the market predicted large dividend cuts.

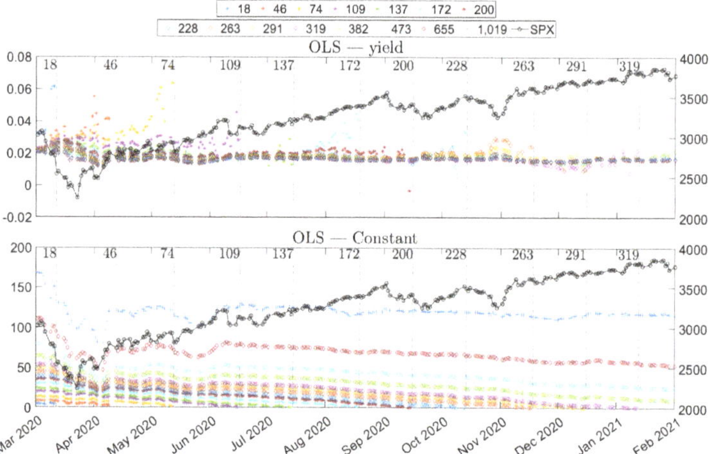

Figure 4. This figure consists of two panels which share the same labels. The upper panel illustrates dividend yield estimates, and the lower panel illustrates present value dividend estimates. The data illustrated in these two panels are part of the data in Figure 2. However, in these two panels, we plot dividend estimates that have a constant time of maturity (one of the dimensions of the surf plots has been removed). The times of maturity that are illustrated are those that were present in the market as of 2 March 2020 (i.e., the first date in our data set). The legends show the time to maturity (measured in days), corresponding to the times of maturity at the first date.

The interconnection between the yield and the quote of the underlying asset is interesting in two regards. First, it gives rise to counter-intuitive behavior. Second, and more important, from the view of assumptions, if a yield is constant intraday, this would imply highly fluctuating present value dividend estimates during the day. Such behavior seems unlikely from market participants. This view is complemented by Vellekoop and Nieuwenhuis [13], who claim that market makers prefer to specify fixed cash amounts rather than yields. We take this as an indication that the constant dividend yield formulation has inherent problems. In the upcoming sections, we present results that support the fact that yield estimates perform worse than their present value counterparts, and that these differences in performance can be related to the variability of the underlying asset.

4.2. In-Sample

We have two types of in-sample results. First, we elaborate on whether the yield or the present value should be used. (This question is also discussed in the next section, where an out-of-sample analysis is performed). Second, we examine the difference between the intraday and single-time data.

An interpretation of the linear models is that it is a pricing method for synthetic forward positions, and an obvious performance measure between the models is to compare the residuals. The residuals are informative but difficult to compare. Therefore, we compare the mean squared errors rather than the residuals themselves in the analyses.

4.2.1. Yield and Present Value Comparison

We consider two different regression models, (17) and (18), where the former is formulated with a dividend yield and the latter with the present values of dividends. The results of the regressions are shown in Table 5, and we can observe that the present value dividend formulation (18) outperforms the yield formulation (17), i.e., the former has a lower mean squared error than the latter.

Table 5. This table consists of mean squared errors (MSE) for in-sample dividend ordinary least squares (OLS) estimates based on intraday and single-time (recorded at 3 p.m.) data. The table shows the mean squared errors for the yield and present value dividend formulation. The MSE for the single-time data is—by construction—equal for both dividend formulations; hence, these are written on the same row.

Dividend Type	Data	MSE
Yield	Intraday	3.2104
Present value	Intraday	3.1806
Yield and Present value	Single-time	3.2114

In Table 5, only the ordinary least squares results are presented, not the corresponding weighted least squares results. The reason is that the ordinary least squares estimator by construction produces a lower mean squared error than the weighted least squares estimator, cf. (22) and (23). Therefore, to have a meaningful comparison, we will compare the ordinary and weighted least squares estimators out-of-sample in the next section.

It is possible to see a difference in predictability between the yield and present value formulation, but it is not easy to relate the two quantities and determine the magnitude of the difference. The yield is transformed into a present value to enable a comparison between the estimates. We represent the yield implied present value with D_y. The key in the transformation is that both estimates can be interpreted as a spot price adjustments. The yield and present value adjusted spot prices can be written as $S(t)e^{-\delta(t;T)(T-t)}$ and $S(t) - D(t;T)$, respectively. By equating these two adjusted spot prices, we write the yield-implied present value as

$$D_y(t;T) = S(t)\left(1 - e^{-\delta(t;T)(T-t)}\right), \tag{30}$$

which can be rewritten into conversions between yield and present value. We can see the results of this conversion in Table 6. The differences between the yield implied and the estimated present value are not big, but there are differences. The statistics in the table do not present any clear differences between the two estimates. The summarized picture shows that the differences are close and symmetric around zero, since the mean values are close to zero with a low standard deviation, while the skewness and kurtosis indicate that there are extreme points. The skewness shows that the implied present value dividend is higher than the present value dividend in eight of eleven ranges and in total. Furthermore, the high kurtosis shows differences notably more extreme than a couple of standard deviations from the mean. The only visible trends in the data are that the standard deviations and the absolute differences seem to increase with longer times to maturity. These results are thus inconclusive as to whether there is a difference or if the estimations only are noisier for longer times to maturity. We continue this discussion around the out-of-sample tests.

Table 6. This table presents the statistics of the differences between the present value estimates and the implied present value quantity, D_y. The second column—Abs. Mean—shows the values of the mean of the absolute value of the differences. The last row—All TTM—shows the statistics for all differences. The other rows show groups of differences corresponding to the times to maturity (days) in the range.

TTM Range [Days]	Abs. Mean	Mean	Std. Dev.	Skewness	Kurtosis
$[0, 100)$	0.0314	−0.00567	0.0587	−1.6364	27.3525
$[100, 200)$	0.0746	−0.00391	0.1244	−0.6311	16.1054
$[200, 300)$	0.1385	−0.00397	0.2241	−0.4728	11.2769
$[300, 400)$	0.1636	0.00575	0.2648	−0.3781	12.5705
$[400, 500)$	0.2032	0.01219	0.3175	−0.2575	11.8386
$[500, 600)$	0.1918	0.03577	0.2692	0.1181	6.7503
$[600, 700)$	0.3390	0.02911	0.5131	−0.3406	9.1445
$[700, 800)$	0.2405	0.05065	0.3331	0.2678	6.0934
$[800, 900)$	0.2991	0.06862	0.4279	0.0755	6.7246
$[900, 1000)$	0.4294	0.05606	0.5963	−0.0068	5.9678
$[1000, 1100]$	0.6999	0.02285	1.0509	−0.5342	7.0079
All TTM	0.1363	0.00642	0.2727	-0.4462	37.0126

4.2.2. Intraday and Single-Time Data

Blomvall et al. [25] concluded that intraday data produce more stable and higher quality estimates than data recorded from a single time. Therefore, we undertook a similar analysis and performed linear regressions where only data points recorded at 3 p.m. were used. First, Table 5 shows mean squared errors for both the intraday and the single-time data set, but the mean squared errors are not directly comparable since the errors are computed from different data sets. Consequently, we do not make any such comparisons in-sample but postpone them to the out-of-sample analysis.

It is possible to consider the surf plots in Figure 2 again. The differences between the intraday and single-time data seem small, but it is possible to observe a wave for both estimate types, which indicates that some estimates differ from adjacent estimates. These estimate differences are more visible in the upper panel of Figure 3 than in Figure 2. We can see that around the period of March–May of 2020, the single-time estimates seems to be more volatile than the intraday estimates. Further, later in the studied period, there are occasional single estimates that are considerably different from their adjacent estimates.

We can study the estimates that correspond to the times of maturity of the options market, which we refer to as market-matched dividend estimates. We want two properties when computing the statistics of the estimates: to estimate the same quantity every day and to have large sample sizes, i.e., long times series of estimates. The latter property is achieved by limiting the data set, such that only the times of maturity that are present in the market for the whole period of study are included. The first property is impossible to achieve completely since the market changes with time. In a period, $[t, T]$, the value of the dividends changes because the ex-dividend dates are passed as t evolves. Additionally, the estimated quantity may also change since the beliefs of future dividends change. By studying the present values of the dividends estimates between two maturities that have not passed in the period, the impact of passed ex-dividends dates is removed. We refer to these differences as stripped dividend estimates. We use the notation introduced in Section 2.1.1, where $D(t; \tau, T)$ is the present value of dividends within the ex-dividend date in the period $[\tau, T]$. We can then measure some statistics of these stripped dividends, an analysis that is similar to the analysis conducted by Desmettre et al. [1].

The market-matched and stripped dividend estimates are similar, but they have some differences in their interpretations. Statistics of the market-matched estimates can be seen in Table 4, and statistics for the stripped dividends estimates are presented in Table 7. Further, the market-matched and stripped dividends are presented in the lower panel of Figure 3. The stripped dividends estimates do not have the downward slopes that the market-matched dividend estimates have. The line plots of Figure 3 are flat, and the means of Table 7 are approximately zero. The reason for the slope is that the ex-dividend dates are passed for the market-matched dividend estimates, but since the stripped dividends are further in the future, no ex-dividend dates have passed.

Table 7. This table shows the statistics for the daily differences of the stripped present value dividend estimates for four series of estimates with a constant time of maturity. The intervals (e.g., 473–382) in the first column—Tenors (TTM)—indicate time to maturity intervals on 2 March 2020 (the first date in the data set) that the stripped dividend estimates correspond to.

Tenors (TTM)	Single-Time			Intraday		
	Mean	**Std**	**Autocorr**	**Mean**	**Std**	**Autocorr**
473–382	−0.0048	1.3296	−0.4810	−0.0032	0.6636	−0.2873
655–473	−0.0140	2.8049	−0.4661	−0.0109	1.1899	−0.0508
1019–655	0.0169	4.2557	−0.4981	0.0181	2.0258	−0.1932

The means are similar for both data sets, but the single-time data estimates have higher volatility values than the intraday data estimates. Furthermore, the standard deviation (volatility) values are similar between the market-matched and stripped dividends, which is surprising. The stripped dividends are estimates of fewer dividends than the market-matched dividends, and additionally, those dividends are shared with the market-matched estimates. Therefore, a natural assumption is that the dispersion of the former would be smaller. A possible explanation is that the future dividends are uncertain. Another explanation is that there is noise in the estimates, which may be because options with longer times to maturity are less liquid than options with shorter times to maturity.

Furthermore, the auto-correlation of the daily differences holds interesting information. We can see in Figure 3 that there are some upward spikes for individual days, i.e., it goes up one day and then comes back to a similar level the following day. This pattern is a clear sign of the noise in the estimates. We can contrast the single-time data plots with the plots of the intraday data, which lack clear spikes. We measured the auto-correlation to see how much the estimates were affected and presented the results in Tables 4 and 7. We noted that the market-matched dividend estimates had a lower auto-correlation since these estimates had a downward trend. This downward trend reduced the information in the auto-correlation, and, thus, the auto-correlation values of the stripped dividends are better indicators of the noise for each method. We can see in Table 7 that the auto-correlation is negative for both the intraday and the single-time data, but the auto-correlations are smaller (more negative) for the latter. The negative sign indicates that both types of estimates are affected by noise, and further, the differences between the auto-correlations indicate that intraday estimates contain less noise than the single-time estimates. Further, it is impossible to make statements concerning the noise level in the market match contra the stripped dividend estimates since the market-matched dividends have a natural downward slope, which thus increases the auto-correlation of the daily differences.

4.3. Out-of-Sample

The in-sample results indicate that the present value of the dividends performs better than the yield estimates. However, these results can be questioned, since the performance may be a result of over-fitting. In this section, we perform an out-of-sample analysis. The analysis is a two-step approach. First, we discuss how to partition the data into two sets:

the training and test sets. The former was used for estimating, while the latter was used for evaluating the estimates. Second, we present the evaluation method.

4.3.1. Partitioning the Data Set

In order to make an out-of-sample analysis, the data set needed be divided into two parts. The data consisted of all (business) dates from 1 March 2020 to 1 February 2021. Each date had some times of maturity, and linear regressions were performed for each time of maturity. The partitions into in- and out-of-sample sets were performed on each such unit, since there was neither data sharing between the dates nor the times of maturity. The data set used for estimation consisted of three data types: the spot price of the underlying (i.e., quotes of the S&P 500 index), spot interest rates, and synthetic forward mid-prices. The regressions use both different times and different strike prices. We argued in Section 3 that information reaches the market over time and that the times are important. For each (intraday) time, a single and unique S&P 500 quote and a single unique spot rate exist. This uniqueness creates the need for these points to be used both in- and out-of-sample. On the other hand, the synthetic forward prices can be partitioned into two sets.

The partition is performed with two principles. First, we want a wide range of strike prices in-sample since they are important for making good estimates. Second, we want to have a greater portion in-sample than out-of-sample. The set of synthetic forward positions is divided into an in- and out-of-sample set according to two criteria. The first criterion is that a synthetic forward is included if its strike price is below a lower limit, $\ell \in \mathbb{R}^+$, or above an upper limit, $u \in \mathbb{R}^+$. The second criterion is that of the synthetic forwards not included in-sample by the first criterion, every k:th is placed in the out-of-sample set, while the remaining are placed in-sample, where $k \in \mathbb{N}^+$, i.e., a strictly positive integer. It would be improper to make the first inclusion criterion static, since the index value changes during the studied period, and thus the limits of in- and out-of-money change. Therefore, rather than assigning static values to ℓ and u, we assign values relative to the index value for each day. The index value was not constant intraday, and we computed the index daily reference value as the median of all intraday index quotes and denoted it with $\hat{S}$, and we defined $\ell = \ell'\hat{S}$ and $u = u'\hat{S}$, where $\ell' \in \mathbb{R}^+$ and $u' \in \mathbb{R}^+$.

The in-sample and out-of-sample data are from the same data set, but their roles are not equal. The in-sample should, in essence, be the data used for estimation. The out-of-sample, on the other hand, was used as a reference, and we could have been more selective when forming this set, and, e.g., used additional filters. One rough measure of the quality of prices is the size of the bid–ask spread, where a wide spread indicates a less reliable price and a narrow spread a more reliable price. The idea is to remove options with too wide spreads, an idea which was used by Blomvall et al. [25] and Azzone and Baviera [24]. The crux is to characterize a typical and reasonable spread. One natural dynamic to keep in mind is that options with higher prices have wider spreads than options with lower prices, if the spreads are measured in an absolute dollar amount. This relationship means that in-sample options have wider spreads than out-of-sample options, and options with longer maturity times have larger spreads than options close to expiry. However, this dynamic is not a big problem in practice. The latter is not a problem, since each time of maturity is managed independently. The former is slightly more challenging, but since the out-of-sample is a subset in which deep in- and out-of-sample options have been excluded, the potential impact is limited. Further, the spreads can also vary between days, and thus they are not suitable to use as a fixed cutoff value. Instead, a reference is computed for each date to account for this variability.

The additional filter handles call and put options separately and are applied for each date and time to maturity. Let, $N^{d,T}$ denote the number of option pairs for date d, with time of maturity T, and let $\Delta c_i = c_a^i - c_b^i$ and $\Delta p_i = p_a^i - p_b^i$ denote the spread of the ith call option and put option, respectively. The scaled median of the spreads is computed as $m_c = (1 + b_c) \underset{i \in \{1,\dots,N^{d,T}\}}{\text{median}} \Delta c_i$ and $m_p = (1 + b_p) \underset{i \in \{1,\dots,N^{d,T}\}}{\text{median}} \Delta p_i$, where $b_c \in \mathbb{R}^+$ and $b_p \in \mathbb{R}^+$. We kept an option if its spread was below the scaled median. Note that a complete option pair was required to compute the synthetic forward price, and, hence, if the one option in the pair was removed, the other one became useless. The parameters used to generate all out-of-sample results are presented in Table 8.

Table 8. This table shows the parameters used to partition the data set into in- and out-of-sample data sets.

Parameter	Value
ℓ'	0.75
u'	1.25
k	10
b_c	0.30
b_p	0.30

4.3.2. Evaluation Method

It is critical to choose how to evaluate an estimate. One approach would be to follow the path used by Desmettre et al. [1]. They estimated the dividends for individual shares and compared the results of their estimates with the realized dividends, but we argue that this approach has some intrinsic drawbacks. First, Desmettre et al. [1] discuss a difference in their estimates of the market consensus of the present value dividends and the actual dividends. They used market data for specific markets with a tax setting that they argued was suitable. This favorable tax setting is not present in the US market, and, in Section 2.1, we argue that we do not measure the dividends but rather how the index is affected by them. Second, there is also a practical problem with index data. The index does not pay dividends but rather its constituents, which results in considerably more dividend payments, and the payments must be scaled with the weight of its constituents. All these technical details make the method error-prone and thus not suitable for use. To summarize, even in idealistic conditions, it is not generally valid to compare dividend estimates with their realized counterparts.

Another natural approach would be to use the linear models and the predicted errors, which, in essence, is how well the linear models reprice the out-of-sample options. The advantage of the prediction errors is that they are easily computable and allow an easy model comparison. The primary disadvantage is that the linear regressions of the put–call parity also include estimations of the interest rate spread. Hence, prediction errors are affected by both the quality of the dividends and the interest rate spreads estimates. The results are, thus, in a strict sense, a measure of linear model performance, but not necessarily of the dividend. Consequently, we base our estimate on another approach: utilizing the sloped asset position. The limitation of using this position as an estimator is the vast amount of combinations. A potential solution to this limitation is to limit the data, but the drawback is figuring out how to make such a limitation systematically. However, in the out-of-sample testing, the data set was, by construction, small enough to use sloped asset positions. The sloped asset position makes it possible to test the estimates isolated from the potential effects of the interest rate. Furthermore, we use the adjusted share price formulation, S^*, to compare yield and present value since the two types are not directly comparable. The regressions were run in-sample, and they were then compared with the help of the out-of-sample data.

The mean squared errors of the residuals is one method of measuring and comparing the different methods. It was difficult to argue if the difference between methods was big or small. Therefore, we complemented the mean squared error with a statistical test on the out-of-sample data. We use the version of the Diebold–Mariano test that was used by Blomvall et al. [25]. This test is a version of the original test presented by Diebold and Mariano [37]. The test consists of four steps. First, we partitioned the data into in- and out-of-sample sets. Second, the regression was performed (in-sample). Third, the linear models were evaluated on the out-of-sample data to measure the errors. Fourth, we performed the Diebold–Mariano test from the errors. We denoted the errors for the two regressions, which we compared using $s_{i,1}$ and $s_{i,2}$, respectively, where $i = 1, \ldots, n$. Let $d_i = s_{i,1}i^2 - s_{i,2}^2$, $\forall i = 1, \ldots, n$ denote the loss differentials, and let

$$\bar{d} = \frac{1}{n} \sum_{i=1}^{n} d_i,$$

denote the mean of the loss differentials, and the autocovariance with lag k be

$$\gamma_k = \frac{1}{n} \sum_{i=k+1}^{n} \left(d_i - \bar{d}\right)\left(d_{i-k} - \bar{d}\right). \tag{31}$$

The Diebold–Mariano statistic was formulated as

$$DM = \frac{\bar{d}}{\sqrt{\frac{1}{n}\left(\gamma_0 + 2\sum_{k=1}^{h-1} \gamma_k\right)}}, \tag{32}$$

where $h \in \mathbb{N}^+$, i.e., a strictly positive integer, and we chose $h = n^{1/3} + 1$. The Diebold–Mariano test statistic follows a standard normal, $N(0,1)$, given the null hypothesis H_0: $E[d_i] = \mu = 0$. We computed the errors, $s_{i,1}$ and $s_{i,2}$, in two ways. First, we used the prediction errors of the linear models of the out-of-sample data. Second, we also used the adjusted spot price that was implied by the sloped asset position.

Further, the Diebold–Mariano test only determines if there is a (significant) difference between methods, but the test does not quantify this difference. However, Blomvall et al. [25] present one measure, $\sqrt{2/\pi}\sqrt{\bar{d}}$, that can be interpreted as the average improvement between estimates. The measure has the same unit as the errors, $s_{i,1}$ and $s_{i,2}$.

4.3.3. Results

The results are divided between the mean squared errors presented in Table 9 and the Diebold–Mariano tests presented in Table 10. The Diebold–Mariano tests cover the comparisons between yield and present value, ordinary and weighted least squares formulations, and single-time and intraday data. Table 9 shows that the out-of-sample results are consistent with the in-sample results, since the present value outperformed the yield formulation. Moreover, the weighted least squares formulation performed better than the ordinary least squares formulation. The mean squared errors indicate the performance of the different models, but they do not quantify the significance or even if the difference is significant.

Table 9. This table shows the out-of-sample mean squared errors (MSE) of the ordinary least squares (OLS) the weighted least squares (WLS) formulations and the differences between these two regressions. This table presents the results of two different measures: dividend yield and present value dividend; and two evaluation methods: regression residuals and difference to the sloped asset position.

Dividend Formulation	Data	MSE		
		OLS (MSE)	WLS (MSE)	OLS − WLS
Present Value	Intraday Prediction	0.2826	0.2800	0.0026
Yield	Intraday Prediction	0.3130	0.3105	0.0025
Present Value	Intraday Sloped	32.609	32.597	0.0118
Yield	Intraday Sloped	32.656	32.646	0.0103

Table 10. This table presents the Diebold–Mariano test statistics for the comparison between the different estimation methods. The first three columns show information about the method; the type of dividend formulation: yield or present value (PV); the regression form: ordinary (OLS) or weighted least squares (WLS); the type of data used: either single-time (Single) or intraday (I-day) data; and the errors that can be based on prediction or sloped asset positions. The Diebold–Mariano test compares a pair of methods, and each row in the table is one such comparison, and the compared methods are indicated with "reference method" vs. "alternative method". For example, in the first row, the yield and present value formulation are compared. The fifth and sixth rows contain the mean of the differential and the Diebold–Mariano test statistics, where a positive or negative sign indicates that the alternative method is better or worse, respectively, than the reference method.

Dividend Formulation	Regression Formulation	Data	$\bar{d}$	$\sqrt{2/\pi\bar{d}}$ [USD]	DM
Yield vs. PV	OLS	I-day Predicition	0.0304	0.1391	22.710
Yield vs. PV	WLS	I-day Predicition	0.0305	0.1394	23.220
Yield vs. PV	OLS	I-day Sloped	0.0472	0.1733	37.096
Yield vs. PV	WLS	I-day Sloped	0.0487	0.1761	37.998
PV	OLS vs. WLS	I-day Prediction	0.0025	0.0395	4.2099
PV	OLS vs. WLS	I-day Sloped	0.0590	0.1937	23.408
PV	OLS	Single vs. I-day Sloped	0.4660	0.5447	79.258

To see the statistical significance between the models, we discuss the Diebold–Mariano results in this section. The Diebold–Mariano test results are presented in Table 10. That table presents the test statistic, and all the comparisons show significant differences. Further, the fifth column, $\sqrt{2/\pi}\sqrt{\bar{d}}$, is a measure of the differences between the methods. The statistical test and the values of the measures yield the same results, which can be summarized in three points. First, the present value dividend formulation is significantly better than the yield formulation, and the improvements are between 13.91 to 17.61 cents. Second, the weighted least squares formulation is significantly better than the ordinary least squares formulation, and the improvements are between 3.95 to 19.37 cents. Third, basing dividend estimates on intraday data is significantly better than single time data, and the improvement is 54.57 cents. These quantitative results align with the earlier qualitative results.

4.4. Performance Difference between Yield and Present Value

We have seen that the present value formulation has a superior performance to the yield formulation both in-sample and out-of-sample for intraday data. If single-time data is used, there is no difference between the two formulations. The methods are similar in assumptions but with a crucial difference. The dividend quantity is assumed constant in both regressions, but a constant dividend yield implies different adjustments to the spot price, which is incompatible with the market participants' perception.

We tested if this variability in the adjustment can explain the inferior performance. We performed a regression that related the difference between the methods and the intraday variability of the spot price to each other. It is possible to create many variability measures, but there are two features that we would like the measure to have. First, the absolute quote changes are less interesting than the relative changes, i.e., the changes should be related to spot price. Second, we want the regression to be easily computable and tractable.

It is also possible to create several measures of dividend differences. We chose to measure the intraday variability of the spot price as the intraday range of the spot price divided by the median spot price. First, we introduced times $t'_j, \, j = 1, \ldots, M^d$, which were the times when the spot price of the index was recorded, and M^d was the number of such times for day d. The variability of the spot price for a day d was then written as

$$\Delta S_d = \frac{\max\limits_{j \in \{1,\dots,M^d\}} S(t_j) - \min\limits_{j \in \{1,\dots,M^d\}} S(t_j)}{\underset{j \in \{1,\dots,M^d\}}{\text{median}}\ S(t_j)}, \tag{33}$$

and the difference between the dividends of the two as

$$Y_d(T) = \frac{1}{T} \left| \hat{D}^{d,T} - D_y^{d,T} \right|. \tag{34}$$

where

$$\hat{D}_y^{d,T} = \underset{j \in \{1,\dots,M^d\}}{\text{median}}\ S(t_j)\left(1 - \hat{\gamma}_2^{d,T}\right). \tag{35}$$

The values of $\hat{D}_y^{d,T}$ were aggregated into a single value, Y_d for each date as the mean of $Y_d(T)$. The regression can then be formulated as

$$Y_d = \beta \Delta S_d, \tag{36}$$

and the results are presented in Table 11 and Figure 5. We can see that the t-statistic indicates that the coefficient is significantly different from zero, indicating that the spot price variability partly explains the difference. Furthermore, we can see from Figure 5 that the spot price variability is probably not the sole explanation, but it is possible to conclude that increased variability increases the difference between the two dividend formulations.

Table 11. This plot is the result of regressing and understanding the problem with estimating the dividend yield.

	Estimate	**SE**	**t-Statistic**
β	4.0059	0.1920	20.8637

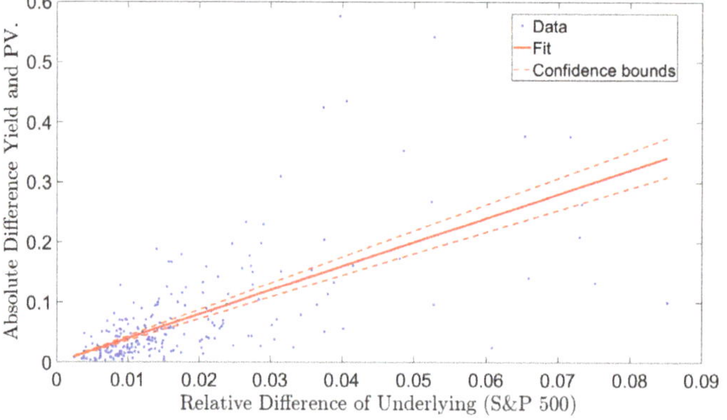

Figure 5. This figure illustrates the regression results between the variability of the underlying and the difference between estimating a dividend yield and a present value of a dividend. The slope coefficient is 4.0059, which means that the greater variability predicts a bigger difference between the yield and the present value estimate.

4.5. Conclusions

This paper has made both practical and theoretical contributions to the literature in this area. The practical contribution is that we have expanded and generalized the regression method presented by Desmettre et al. [1] in two regards. First, we have generalized the regression from an ordinary least squares formulation to a weighted least squares formulation. Second, the regression has been reformulated to utilize intraday data rather than being limited to data recorded at a single time. We have proven that both of these changes improve the quality of the dividend estimates with statistical significance. The latter improved the estimation more than the former. Additionally, one key component of this analysis is the new European option position (the sloped asset position) that we have introduced. This position makes it possible to evaluate dividend estimates independent of interest rate estimates.

The main theoretical contribution is that we have proven that the present value dividend formulation performs significantly better than the yield formulation. We have also proposed an explanation for this phenomenon. We propose that worse performance is caused by the inherent connection between the yield and the spot price. We have also contributed theoretically with the clarification of the interpretation of the dividend. These realizations could affect, e.g., the dividend adjustments in derivative pricing.

Author Contributions: Conceptualization: P.S., J.B. and M.S.; methodology: P.S., J.B. and M.S.; software: P.S.; validation: P.S.; formal analysis: P.S., J.B. and M.S.; investigation: P.S.; resources: P.S.; data curation: P.S. and J.B.; writing–original draft preparation: P.S.; writing–review and editing: P.S., J.B. and M.S.; visualization: P.S.; supervision: J.B. and M.S.; project administration: P.S.; and funding acquisition, N/A. All authors have read and agreed to the published version of the manuscript, please see the following link: CRediT taxonomy (accessed on 5 December 2021) for explanation of terms. Authorship has been limited to those who have contributed substantially to the work reported.

Funding: This research received no external funding.

Data Availability Statement: All the data have been collected from the data provider Thomson Reuters Refinitiv Eikon.

Acknowledgments: The authors would like to thank Jonas Ekblom, Johan Hagenbjörk and the anonymous reviewers for several valuable and helpful suggestions and comments to improve the presentation of the paper.

Conflicts of Interest: The authors declare no conflict of interest.

Appendix A

The derivation of the *sloped asset* position can be made either with a dividend expressed as a yield or as a present value of future dividends. Here, we derive both versions. Let, f_i be a synthetic forward with strike price K_i, i.e., the options in the option pair used to construct the synthetic forward have the strike price K_i. Furthermore, let δ and D denote the dividend yield and present value of the dividends, respectively. Furthermore, let r denote the continuous interest rate. However, the rate is not necessary to prove the relation. Finally, let T denote the time to maturity when entering the contracts, and the time the contract is entered into is t.

The sloped asset position consists of a long position in f_1 and $\frac{K_1}{K_2}$ short positions in f_2. The former has a payoff that can be written as $g_1(s) = s - K_1$, while the short positions provide a payoff of $g_2(s) = -\frac{K_1}{K_2}(s - K_2) = \frac{K_1}{K_2}(K_2 - s)$. The total payoff at expiration for the complete position is:

$$g(s) = g_1(s) + g_2(s) = (s - K_1) + \left(\frac{K_1}{K_2}(K_2 - s)\right) = s\left(1 - \frac{K_1}{K_2}\right).$$

The payoff can be interpreted as a fractional position, either long or short, in the shares. The price of this contract upon entering is the share price adjusted for dividends (scaled with the factor), i.e., the following:

$$f_1 - \frac{K_1}{K_2} f_2 = S^* \left(1 - \frac{K_1}{K_2} \right) = S e^{-\delta(T-t)} \left(1 - \frac{K_1}{K_2} \right) \tag{A1}$$

$$f_1 - \frac{K_1}{K_2} f_2 = S^* \left(1 - \frac{K_1}{K_2} \right) = (S - D) \left(1 - \frac{K_1}{K_2} \right). \tag{A2}$$

It is possible to find the dividend yields and the present values of the dividends directly from the expressions by rearranging them thus:

$$\delta = -\frac{1}{T-t} \ln[S^*/S] = -\frac{1}{T-t} \ln \left[\frac{1}{S} \frac{f_1 K_2 - f_2 K_1}{K_2 - K_1} \right], \tag{A3}$$

$$D = S - S^* = S - \frac{f_1 K_2 - f_2 K_1}{K_2 - K_1}. \tag{A4}$$

References

1. Desmettre, S.; Grün, S.; Seifried, F.T. Estimating discrete dividends by no-arbitrage. *Quant. Financ.* **2017**, *17*, 261–274. [CrossRef]
2. Campbell, J.A.; Beranek, W. Stock Price Behavior on Ex-Dividend Dates. *J. Financ.* **1955**, *10*, 425–429. [CrossRef]
3. Miller, M.H.; Modigliani, F. Dividend Policy, Growth, and the Valuation of Shares. *J. Bus.* **1961**, *34*, 411–433. [CrossRef]
4. Elton, E.J.; Gruber, M.J. Marginal Stockholder Tax Rates and the Clientele Effect. *Rev. Econ. Stat.* **1970**, *52*, 68–74. [CrossRef]
5. Kalay, A. The Ex-Dividend Day Behavior of Stock Prices: A Re-Examination of the Clientele Effect. *J. Financ.* **1982**, *37*, 1059–1070. [CrossRef]
6. Boyd, J.H.; Jagannathan, R. Ex-Dividend Price Behavior of Common Stocks. *Rev. Financ. Stud.* **1994**, *7*, 711–741. [CrossRef]
7. Hartzmark, S.M.; Solomon, D.H. The Dividend Disconnect. *J. Financ.* **2019**, *74*, 2153–2199. [CrossRef]
8. Wilmott, P. *Paul Wilmott on Quantitative Finance 3 Volume Set*, 2nd ed.; Wiley: Hoboken, NJ, USA, 2006.
9. Filipović, D.; Willems, S. A term structure model for dividends and interest rates. *Math. Financ.* **2020**, *30*, 1461–1496. [CrossRef]
10. Merton, R.C. Theory of Rational Option Pricing. *Bell J. Econ. Manag. Sci.* **1973**, *4*, 141–183. [CrossRef]
11. Haug, E.; Haug, J.; Lewis, A.L. Back to Basics: A New Approach to the Discrete Dividend Problem. *Wilmott Mag.* **2003**, *9*, 37–47.
12. Frishling, V. A discrete question. *Risk* **2002**, *15*, 115–116.
13. Vellekoop, M.H.; Nieuwenhuis, J.W. Efficient Pricing of Derivatives on Assets with Discrete Dividends. *Appl. Math. Financ.* **2006**, *13*, 265–284. [CrossRef]
14. Carr, P.; Madan, D.B. Option Valuation Using the Fast Fourier Transform. *J. Comput. Financ.* **1999**, *2*, 61–73. [CrossRef]
15. Duffie, D.; Pan, J.; Singleton, K. Transform analysis and asset pricing for affine jump-diffusions. *Econometrica* **2000**, *68*, 1343–1376.
16. Carr, P.; Geman, H.; Madan, D.B.; Yor, M. Stochastic Volatility for Lévy Processes. *Math. Financ.* **2003**, *13*, 345–382. [CrossRef]
17. Gatheral, J.; Jacquier, A. Arbitrage-free SVI volatility surfaces. *Quant. Financ.* **2014**, *14*, 59–71. [CrossRef]
18. Derman, E.; Kani, I. The Volatility Smile and Its Implied Tree. In *GS Quantitative Strategies Research Notes*; 1994. Available online: http://emanuelderman.com/the-volatility-smile-and-its-implied-tree/ (accessed on 5 December 2021).
19. Derman, E.; Kani, I. Stochastic Implied Trees: Arbitrage Pricing with Stochastic Term and Strike Structure of Volatility. *Int. J. Theor. Appl. Finan.* **1998**, *01*. [CrossRef]
20. Geng, J.; Navon, I.M.; Chen, X. Non-parametric calibration of the local volatility surface for European options using a second-order Tikhonov regularization. *Quant. Financ.* **2014**, *14*, 73–85. [CrossRef]
21. Fama, E.F.; French, K.R. Dividend yields and expected stock returns. *J. Financ. Econ.* **1988**, *22*, 3–25. [CrossRef]
22. Golez, B. Expected Returns and Dividend Growth Rates Implied by Derivative Markets. *Rev. Financ. Stud.* **2014**, *27*, 790–822. [CrossRef]
23. Bilson, J.F.; Kang, S.B.; Luo, H. The term structure of implied dividend yields and expected returns. *Econ. Lett.* **2015**, *128*, 9–13. [CrossRef]
24. Azzone, M.; Baviera, R. Synthetic forwards and cost of funding in the equity derivative market. *Financ. Res. Lett.* **2021**, *41*, 101841. [CrossRef]
25. Blomvall, J.; Söderbäck, P.; Singull, M. Weighted Least Squares Estimation of the Risk-Free Rate from Derivative Prices. *submitted*.
26. van Binsbergen, J.H.; Hueskes, W.; Koijen, R.; Vrugt, E. Equity yields. *J. Financ. Econ.* **2013**, *110*, 503–519. [CrossRef]
27. Tunaru, R.S. Dividend derivatives. *Quant. Financ.* **2018**, *18*, 63–81. [CrossRef]
28. Andersen, L.; Brotherton-Ratcliffe, R. The equity option volatility smile: An implicit finite-difference approach. *J. Comput. Financ.* **1997**, *1*, 5–38. [CrossRef]
29. Kragt, J. Option Implied Dividends. 2017. Available online: https://papers.ssrn.com/sol3/papers.cfm?abstract_id=2980275 (accessed on 5 December 2021).
30. van Binsbergen, J.H.; Brandt, M.; Koijen, R. On the Timing and Pricing of Dividends. *Am. Econ. Rev.* **2012**, *102*, 1596–1618. [CrossRef]
31. Hull, J.C. *Options, Futures and Other Derivatives: Global Edition: QMUL*, 9th ed.; Pearson Education Limited: Edinburgh, UK, 2013.
32. Ronn, A.; Ronn, E. The box spread arbitrage conditions: Theory, tests, and investment strategies. *Rev. Financ. Stud.* **1989**, *2*, 91–108. [CrossRef]

33. van Binsbergen, J.H.; Diamond, W.F.; Grotteria, M. Risk-free interest rates. *J. Financ. Econ.* **2022**, *143*, 1–29. [CrossRef]
34. Aitken, A.C. IV.—On Least Squares and Linear Combination of Observations. *Proc. R. Soc. Edinb.* **1936**, *55*, 42–48. [CrossRef]
35. Zwanzig, S.; Liero, H. *Introduction to the Theory of Statistical Inference*, 1st ed.; Chapman and Hall/CRC: Boca Raton, FL, USA, 2011. [CrossRef]
36. Blomvall, J. Measurement of interest rates using a convex optimization model. *Eur. J. Oper. Res.* **2017**, *256*, 308–316. [CrossRef]
37. Diebold, F.X.; Mariano, R.S. Comparing Predictive Accuracy. *J. Bus. Econ. Stat.* **1995**. *13*, 253–263. [CrossRef]

Article

Sparse Estimation Strategies in Linear Mixed Effect Models for High-Dimensional Data Application

Eugene A. Opoku [1,*], Syed Ejaz Ahmed [2] and Farouk S. Nathoo [1]

[1] Department of Mathematics and Statistics, University of Victoria, Victoria, BC V8P 5C2, Canada; nathoo@uvic.ca
[2] Department of Mathematics and Statistics, Brock University, St. Catharines, ON L2S 3A1, Canada; sahmed5@brocku.ca
* Correspondence: eopoku@uvic.ca

Abstract: In a host of business applications, biomedical and epidemiological studies, the problem of multicollinearity among predictor variables is a frequent issue in longitudinal data analysis for linear mixed models (LMM). We consider an efficient estimation strategy for high-dimensional data application, where the dimensions of the parameters are larger than the number of observations. In this paper, we are interested in estimating the fixed effects parameters of the LMM when it is assumed that some prior information is available in the form of linear restrictions on the parameters. We propose the pretest and shrinkage estimation strategies using the ridge full model as the base estimator. We establish the asymptotic distributional bias and risks of the suggested estimators and investigate their relative performance with respect to the ridge full model estimator. Furthermore, we compare the numerical performance of the LASSO-type estimators with the pretest and shrinkage ridge estimators. The methodology is investigated using simulation studies and then demonstrated on an application exploring how effective brain connectivity in the default mode network (DMN) may be related to genetics within the context of Alzheimer's disease.

Keywords: linear mixed model; ridge estimation; pretest and shrinkage estimation; multicollinearity; asymptotic bias and risk; LASSO estimation; high-dimensional data

Citation: Opoku, E.A.; Ahmed, S.E.; Nathoo, F.S. Sparse Estimation Strategies in Linear Mixed Effect Models for High-Dimensional Data Application. *Entropy* **2021**, *23*, 1348. https://doi.org/10.3390/e23101348

Academic Editor: Matteo Convertino

Received: 9 September 2021
Accepted: 12 October 2021
Published: 15 October 2021

Publisher's Note: MDPI stays neutral with regard to jurisdictional claims in published maps and institutional affiliations.

1. Introduction

In many fields such as bio-informatics, physical biology, and epidemiology, the response of interest is represented by repeated measures of some variables of interest that are collected over a specified time period for different independent subjects or individuals. These types of data are commonly encountered in medical research where the responses are subject to various time-dependent and time-constant effects such as pre- and post-treatment types, gender effect, and baseline measures, among others. A widely-used statistical tool in the analysis and modeling of longitudinal and repeated measures data is the linear mixed effects model (LMM) [1,2]. This model provides an effective and flexible way to describe the means and the covariance structures of a response variable after accounting for within subject correlation.

The rapid growth in the size and scope of longitudinal data has created a need for innovative statistical strategies in longitudinal data analysis. Classical methods are based on the assumption that the number of predictors is less than the number of observations. However, there is an increasing demand for efficient prediction strategies for analysis of high-dimensional data, where the number of observed data elements (sample size) are smaller than the number of predictors in a linear model context. Existing techniques that deal with high-dimensional data mostly rely on various penalized estimators. Due to the trade-off between model complexity and model prediction, the statistical inference of model selection becomes an extremely important and challenging problem in high-dimensional data analysis.

Over the years, many penalized regularization approaches have been developed to do variable selection and estimation simultaneously. Among them, the least absolute shrinkage and selection operator (LASSO) is commonly used [3]. It is a useful estimation technique in part due to its convexity and computational efficiency. The LASSO approach is based on an ℓ_1 penalty for regularization of regression parameters. Ref. [4] provides a comprehensive summary of the consistency properties of the LASSO approach. Related penalized likelihood methods have been extensively studied in the literature, see for example [5–10]. The penalized likelihood methods have a close connection to Bayesian procedures. Thus, the LASSO estimate corresponds to a Bayes method that puts a Laplacian (double-exponential) prior on the regression coefficients [11,12].

In this paper, our interest lies in estimating the fixed effect parameters of the LMM using a ridge estimation technique when it is assumed that some prior information is available in the form of potential linear restrictions on the parameters. One possible source of prior information is using a Bayesian approach. An alternative source of prior information may be obtained from previous studies or expert knowledge that search for or assume sparsity patterns.

We consider the problem of fixed effect parameter estimation for LMMs when there exist many predictors relative to the sample size. These predictors may be classified into two groups: sparse and non-sparse. Thus, there are two choices to be considered: a full model with all predictors, and a sub-model that contains only non-sparse predictors. When the sub-model based on available subspace information is true (i.e., the assumed restriction holds), it then provides more efficient statistical inferences than those based on a full model. In contrast, if the sub-model is not true, the estimates could become biased and inefficient. The consequences of incorporating subspace information therefore depend on the quality or reliability of the information being incorporated into the estimation procedure. One way to deal with uncertain subspace information is to use a pretest estimation strategy. The validity of the information is tested before incorporation into a final estimator. Another approach is shrinkage estimation, which shrinks the full model estimator to the sub-model estimator by utilizing subspace information. Besides these estimation strategies, there is a growing literature on simultaneous model selection and estimation. These approaches are known as penalty strategies. By shrinking some regression coefficients toward zero, the penalty methods simultaneously select a sub-model and estimate its regression parameters. Several authors have investigated the pretest, shrinkage, and penalty estimation strategies in partial linear model, Poisson regression model, and Weibull censored regression model [13–15].

To formulate the problem, we suppose that the vector of the fixed effects parameter β in the LMM can be partitioned into two sub-vectors $\beta = (\beta_1', \beta_2')'$, where β_1 is the coefficient vector of non-sparse predictors and β_2 is the coefficient vector of sparse predictors. Our interest lies in the estimation of β_1 when β_2 is close to zero. To deal with this problem in the context of low dimensional data, ref. [16] propose an improved estimation strategy using sub-model selection and post-estimation for the LMM. Within this framework, linear shrinkage and shrinkage pretest estimation strategies are developed, which combine full model and sub-model estimators in an effective way as a trade-off between bias and variance. Ref. [17] extend this study by using a likelihood ratio test to develop James–Stein shrinkage and pretest estimation methods based on LMM for longitudinal data. In addition, the non-penalty estimators are compared with several penalty estimators (LASSO, adaptive LASSO and Elastic Net) for best performance.

In most real data situations, there is also the problem of multicollinearity among predictor variables for high-dimensional data. Various biased estimation techniques such as shrinkage estimation, partial least squares estimation [18] and Liu estimators [19] have been implemented to deal with this problem, but the widely used technique is ridge estimation [20]. The ridge estimator overcomes the weakness of the least squares estimator with a smaller mean squared error. To overcome and combat multicollinearity, ref. [21] propose pretest and Stein-type ridge regression estimators for linear and partially linear models. Furthermore, ref. [22] also develop shrinkage estimation based on Liu regression to overcome multicollinearity in linear models.

Our primary focus is on the estimation and prediction problem for linear mixed effect models when there are many potential predictors that have a weak or no influence on the response of interest. This method simultaneously controls overfitting using general least square estimation with a roughness penalty. We propose pretest and shrinkage estimation strategies using the ridge estimation technique as a base estimator and numerically compare their performance with the LASSO and adaptive LASSO estimators. Our proposed estimation strategy is applied to both high-dimensional and low-dimensional data.

The rest of this article is organized as follows. In Section 2, we present the linear mixed effect model and the proposed estimation techniques. We introduce the full and sub-model estimators based on ridge estimation. Thereafter, we construct the pretest and shrinkage ridge estimators. Section 3 provides the asymptotic bias and risk of these estimators. A Monte Carlo simulation is used to evaluate the performance of the estimators including a comparison with the lasso-type estimators, and the results are reported in Section 4. Section 5 presents a demonstration of the proposed methodology on a high-dimensional resting-state effective brain connectivity and genetic data. We also illustrate the proposed estimation methods in an application to a low-dimensional Amsterdam growth and health study. Section 6 presents a discussion with recommendations.

2. Model and Estimation Strategies

In this section, we present the linear mixed effect model and the proposed estimation strategies.

2.1. Linear Mixed Model

Suppose that we have a sample of N subjects. For the ith subject, we collect the response variable y_{ij} for the jth time, where $i = 1\ldots, n; j = 1\ldots, n_i$ and $N = \sum_{i=1}^{n} n_i$. Let $Y_i = (y_{i1}, \ldots y_{in_i})'$ denotes the $n_i \times 1$ vector of responses from the ith subject. Let $X_i = (x_{i1}, \ldots, x_{in_i})'$ and $Z_i = (z_{i1}, \ldots, z_{in_i})'$ be $n_i \times$ p and $n_i \times$ q known fixed-effects and random-effect design matrix for the ith subject of full rank p and q, respectively. The linear mixed effect model [1] for a vector of repeated responses Y_i on the ith subject is assumed to have the form

$$Y_i = X_i \beta + Z_i a_i + \epsilon_i, \tag{1}$$

where $\beta = (\beta_1, \ldots, \beta_p)'$ is the p $\times$ 1 vector of unknown fixed-effect parameters or regression coefficients, a_i is the q $\times$ 1 vector of unobservable random effects for the ith subject, assumed to come from a multivariate normal distribution with zero mean and a covariance matrix G, where G is an unknown $q \times q$ covariance matrix and ϵ_i denotes $n_i \times 1$ vector of error terms assumed to be normally distributed with zero mean, covariance matrix $\sigma^2 I_{n_i}$. Further, ϵ_i are assumed to be independent of the random effects a_i.

The marginal distribution for the response y_i is normal with mean $X_i \beta$ and covariance matrix $Cov(Y_i) = Z_i \sigma_i^2 Z_i^T + \sigma^2 I_n$. By stacking the vectors, the mixed model can be can be expressed as $Y = X\beta + Za + \epsilon$. From the Equation (1), the distribution of the model follows $Y \sim \mathcal{N}_n(X\beta, V)$, where $E(Y) = X\beta$ with covariance, $V = \sum_{i=1}^{n} Z_i \sigma_i^2 Z_i^T + \sigma^2 I_n$.

2.2. Ridge Full Model and Sub-Model Estimator

The generalized least square estimator (GLS) is defined as $\hat{\beta}^{GLS} = (X^T V^{-1} X)^{-1} X^T V^{-1} Y$ and the ridge full model estimator can be obtained by introducing a penalized regression so that $\hat{\beta} = \arg\min_{\beta} \left\{ (Y - X\beta)^T V^{-1} (Y - X\beta) + k\beta^T \beta \right\}$ and

$$\hat{\beta}^{Ridge} = (X^T V^{-1} X + kI)^{-1} X^T V^{-1} Y, \text{ where } \hat{\beta}^{Ridge} \text{ is the ridge full model estimator}$$

and $k \in [0, \infty)$ is the tuning parameter. If k = 0, $\hat{\beta}^{Ridge}$ is the GLS estimator and $\hat{\beta}^{Ridge} = 0$ for k is sufficiently large. We select the value of k using cross validation.

We let $\mathbf{X} = (\mathbf{X}_1, \mathbf{X}_2)$, where $\mathbf{X}_1$ is an $n \times p_1$ sub-matrix containing the non-sparse predictors and $\mathbf{X}_2$ is an $n \times p_2$ sub-matrix that contains the sparse predictors. Accordingly, $\beta = (\beta_1, \beta_2)$ where β_1 and β_2 have dimensions p_1 and p_2, respectively, with $p_1 + p_2 = p$, $p_i \geq 0$ for $i = 1, 2$.

A sub-model is defined as $Y = X\beta + Za + \epsilon$ subject to $\beta^T\beta \leq \phi$ and $\beta_2 = 0$ which corresponds to $Y = X_1\beta_1 + Za + \epsilon$ subject to $\beta_1^T\beta_1 \leq \phi$. The sub-model estimator $\hat{\beta}_1^{\text{RSM}}$ of β_1 has the form $\hat{\beta}_1^{\text{RSM}} = (\mathbf{X}_1^T\mathbf{V}^{-1}\mathbf{X}_1 + k\mathbf{I})^{-1}\mathbf{X}_1^T\mathbf{V}^{-1}\mathbf{Y}$. We denote $\hat{\beta}_1^{\text{RFM}}$ as the full model ridge estimator of β_1 and given as

$$\hat{\beta}_1^{\text{RFM}} = (\mathbf{X}_1^T\mathbf{V}^{-1/2}\mathbf{M}_{X_2}\mathbf{V}^{-1/2}\mathbf{X}_1 + k\mathbf{I})^{-1}\mathbf{X}_1^T\mathbf{V}^{-1/2}\mathbf{M}_{X_2}\mathbf{V}^{-1/2}\mathbf{Y}, \text{ where } \mathbf{M}_{X_2} = \mathbf{I} - \mathbf{P} = \mathbf{I} - \mathbf{V}^{-1/2}\mathbf{X}_2(\mathbf{X}_2\mathbf{V}^{-1}\mathbf{X}_2)^{-1}\mathbf{X}_2^T\mathbf{V}^{-1/2}.$$

2.3. Pretest Ridge Estimation Strategy

Generally, the sub-model estimator will be more efficient than the full model estimator if the information embodied in the imposed linear restrictions is valid, thus β_2 is close to zero. However, if the information is not valid the sub-model estimator is likely to be more biased and may have a higher risk than the full model estimator. There is, therefore, some doubt as to whether or not to impose the restrictions on the model's parameter. It is in response to this uncertainty that a statistical test may be used to determine the validity of the proposed restrictions. Accordingly, the procedure to follow in practice is pretest the validity of the restrictions and if the outcome of the pretest suggests that they are correct then the model parameters are estimated incorporating the restrictions. If the pretest rejects the restrictions then the parameters are estimated from the sample information alone. This motivates the consideration of the pretest estimation strategy for the LMM.

The pretest estimator is a combination of the full model estimator $\hat{\beta}_1^{\text{RFM}}$, and sub-model estimator $\hat{\beta}_1^{\text{RSM}}$, through an indicator function $I(L_n \leq d_{n,\alpha})$, where L_n is an appropriate test statistic to test $H_0 : \beta_2 = 0$ versus $H_A : \beta_2 \neq 0$. Moreover, $d_{n,\alpha}$ is an α level critical value based on distribution of L_n under H_0. We define test statistics based on the log-likelihood ratio test as $L_n = 2\{\ell^*(\hat{\beta}^{\text{RFM}} \mid \mathbf{Y}) - \ell^*(\hat{\beta}^{\text{RSM}} \mid \mathbf{Y})\}$.

Under H_0, the test statistic L_n follows asymptotic chi-square distribution with p_2 degrees of freedom. The pretest test ridge estimator $\hat{\beta}_1^{\text{RPT}}$ of β_1 is then defined by

$$\hat{\beta}_1^{\text{RPT}} = \hat{\beta}_1^{\text{RFM}} - (\hat{\beta}_1^{\text{RFM}} - \hat{\beta}_1^{\text{RSM}})I(L_n \leq d_{n,\alpha}), \quad p_2 \geq 1.$$

2.4. Shrinkage Ridge Estimation Strategy

The pre-test estimator is a discontinuous function of the sub-model $\hat{\beta}_1^{\text{RSM}}$ and full model $\hat{\beta}_1^{\text{RFM}}$, which depends on the hard threshold $(d_{n,\alpha} = \chi^2_{p_2,\alpha})$. We address this limitation by defining the shrinkage ridge estimator based on soft thresholding. The shrinkage ridge estimator (RSE) of β_1, denoted as $\hat{\beta}_1^{\text{RSE}}$, is defined as

$$\hat{\beta}_1^{\text{RSE}} = \hat{\beta}_1^{\text{RSM}} + (\hat{\beta}_1^{\text{RFM}} - \hat{\beta}_1^{\text{RSM}})(1 - (p_2 - 2)L_n^{-1}), \quad p_2 \geq 3.$$

Here, $\hat{\beta}_1^{\text{RSE}}$ is the linear combination of the full model $\hat{\beta}_1^{\text{RFM}}$ and sub-model $\hat{\beta}_1^{\text{RSM}}$ estimates. If $L_n \leq (p_2 - 2)$, then a relatively large weight is placed on $\hat{\beta}_1^{\text{RSM}}$ otherwise, more weight is on $\hat{\beta}_1^{\text{RFM}}$. A setback with $\hat{\beta}_1^{\text{RSE}}$ is that it is not a convex combination of $\hat{\beta}_1^{\text{RFM}}$ and $\hat{\beta}_1^{\text{RSM}}$. This can cause over-shrinkage, which gives the estimator opposite sign of $\hat{\beta}_1^{\text{RFM}}$. This could happen if $(p_2 - 2)L_n^{-1}$ is larger than one. To counter this, we use the positive-part shrinkage ridge estimator (RPS) defined as

$$\hat{\beta}_1^{\text{RPS}} = \hat{\beta}_1^{\text{RSM}} + (\hat{\beta}_1^{\text{RFM}} - \hat{\beta}_1^{\text{RSM}})(1 - (p_2 - 2)L_n^{-1})^+, p_2 \geq 3$$

where $(1 - (p_2 - 2)L_n^{-1})^+ = \max(0, 1 - (p_2 - 2)L_n^{-1})$. The RPS estimator will control possible over-shrinking in the RSE estimator.

3. Asymptotic Results

In this section, we derive the asymptotic distributional bias and risk of the estimators considered in Section 2. We examine the properties of the estimators for increasing n and as β_2 approaches the null vector under the sequence of local alternatives defined as

$$K_n : \beta_2 = \beta_{2(n)} = \frac{\kappa}{\sqrt{n}}, \tag{2}$$

where $\kappa = (\kappa_1, \kappa_2 \ldots, \kappa_{p_2})' \in \mathbb{R}^{p_2}$ is a fixed vector. The vector $\frac{\kappa}{\sqrt{n}}$ is a measure of how far local alternatives K_n differ from the subspace information $\beta_2 = 0$. In order to evaluate the performance of the estimators, we define the asymptotic distributional bias of the estimator $\hat{\beta}_1^*$ as

$$\mathrm{ADB}(\hat{\beta}_1^*) = \lim_{n \to \infty} E\{\sqrt{n}(\hat{\beta}_1^* - \beta_1)\},$$

In order to compute the risk functions, we first compute the asymptotic covariance of the estimators. The asymptotic covariance of an estimator $\hat{\beta}_1^*$ is expressed as

$$\mathrm{Cov}(\hat{\beta}_1^*) = \lim_{n \to \infty} E\{n(\hat{\beta}_1^* - \beta_1)(\hat{\beta}_1^* - \beta_1)^{\mathrm{T}}\}.$$

Following the asymptotic covariance matrix, we define the asymptotic risk of an estimator $\hat{\beta}_1^*$ as $\mathrm{R}(\hat{\beta}_1^*) = \mathrm{tr}\left(\mathbf{Q}\mathrm{Cov}(\hat{\beta}_1^*)\right)$. $\mathbf{Q}$ is a positive definite matrix of weights with dimensions of $p \times p$. We set $\mathbf{Q} = \mathbf{I}$ in this study.

Assumption 1. *We make the following two regularity conditions to establish the asymptotic properties of the estimators.*

1. $\frac{1}{n} \max_{1 \leq i \leq n} x_i^T [X^T V^{-1} X]^{-1} x_i \to 0$ *as* $n \to \infty$, *where* x_i^T *is the ith row of* $\mathbf{X}$.

2. $B_n = n^{-1}[X^T V^{-1} X]^{-1} \to B$, *for some finite* $B = \begin{pmatrix} B_{11} & B_{12} \\ B_{21} & B_{22} \end{pmatrix}$.

Theorem 1. *For $k < \infty$, If $k/\sqrt{n} \to \lambda_o$ and B is non-singular, the distribution of the full model ridge estimator, $\hat{\beta}_n^{RFM}$ is*

$$\sqrt{n}(\hat{\beta}_n^{RFM} - \beta) \xrightarrow{D} \mathcal{N}(-\lambda_o B^{-1}\beta, B^{-1}),$$

where $\xrightarrow{D}$ denotes convergence in distribution.

Proof. See Theorem 2 in [23]. $\square$

Proposition 1. *Assuming the above assumption 1 together with Theorem 1 hold, under the local alternatives K_n, we have*

$$\begin{pmatrix} \varphi_1 \\ \varphi_3 \end{pmatrix} \xrightarrow{D} \mathcal{N}\left[\begin{pmatrix} -\mu_{11.2} \\ \delta \end{pmatrix}, \begin{pmatrix} B_{11.2}^{-1} & \Phi \\ \Phi & \Phi \end{pmatrix}\right],$$

$$\begin{pmatrix} \varphi_3 \\ \varphi_2 \end{pmatrix} \xrightarrow{D} \mathcal{N}\left[\begin{pmatrix} \delta \\ -\gamma \end{pmatrix}, \begin{pmatrix} \Phi & 0 \\ 0 & B_{11}^{-1} \end{pmatrix}\right],$$

where $\varphi_1 = \sqrt{n}(\hat{\beta}_1^{RFM} - \beta_1)$, $\varphi_2 = \sqrt{n}(\hat{\beta}_1^{RSM} - \beta_1)$, $\varphi_3 = \sqrt{n}(\hat{\beta}_1^{RFM} - \hat{\beta}_1^{RSM})$, $\gamma = \mu_{11.2} + \delta$, $\delta = B_{11}^{-1}B_{12}\kappa$, $\Phi = B_{11}^{-1}B_{12}B_{22.1}^{-1}B_{21}B_{11}^{-1}$, $B_{22.1} = B_{22} - B_{21}B_{11}^{-1}B_{12}$, $\mu = -\lambda_o B^{-1}\beta = \begin{pmatrix} \mu_1 \\ \mu_2 \end{pmatrix}$ and $\mu_{11.2} = \mu_1 - B_{12}B_{22}^{-1}((\beta_2 - \kappa) - \mu_2)$.

Proof. See Appendix A □

Theorem 2. *Under the condition of Theorem 1 and the local alternatives K_n, the ADBs of the proposed estimators are*

$$ADB(\hat{\beta}_1^{RFM}) = -\mu_{11.2},$$
$$ADB(\hat{\beta}_1^{RSM}) = -\mu_{11.2} - B_{11}^{-1}B_{12}\delta = -\gamma,$$
$$ADB(\hat{\beta}_1^{RPT}) = -\mu_{11.2} - \delta H_{p_2+2}(\chi_{p_2,\alpha}^2; \Delta),$$
$$ADB(\hat{\beta}_1^{RSE}) = -\mu_{11.2} - (p_2 - 2)\delta E(\chi_{p_2+2}^{-2}(\Delta)),$$
$$ADB(\hat{\beta}_1^{RPS}) = -\mu_{11.2} - \delta H_{p_2+2}(\chi_{p_2-2}^2; \Delta)\} - (p_2 - 2)\delta E\{\chi_{p_2+2}^{-2}(\Delta)I(\chi_{p_2+2}^{-2} > p_2 - 2)\},$$

where $\Delta = \kappa^T B_{22.1}^{-1}\kappa$, $B_{22.1} = B_{22} - B_{21}B_{11}^{-1}B_{12}$, and $H_v(x; \Delta)$ is the cumulative distribution function of the non-central chi-squared distribution with non-centrality parameter Δ and v degrees of freedom, and $E(\chi_v^{-2j}(\Delta))$ is the expected value of the inverse of a non-central χ^2 distribution with v degrees of freedom and non-centrality parameter Δ,

$$E(\chi_v^{-2j}(\Delta)) = \int_0^\infty x^{-2j}dH_v(x, \Delta).$$

Proof. See Appendix B.1 □

Since the ADBs of the estimators are in non-scalar form, we define the following asymptotic quadratic bias (AQDB) of $\hat{\beta}_1^*$ by

$$AQDB(\hat{\beta}_1^*) = \left(ADB(\hat{\beta}_1^*)\right)' B_{11.2}\left(ADB(\hat{\beta}_1^*)\right),$$

where $\mathbf{B}_{11.2} = \mathbf{B}_{11} - \mathbf{B}_{12}\mathbf{B}_{22}^{-1}\mathbf{B}_{21}$.

Corollary 1. *Suppose Theorem 2 holds. Then, under $\{K_n\}$, the AQDBs of the estimators are*

$$AQDB(\hat{\beta}_1^{RFM}) = \mu_{11.2}^T B_{11.2}\mu_{11.2},$$
$$AQDB(\hat{\beta}_1^{RSM}) = \gamma^T B_{11.2}\gamma,$$
$$AQDB(\hat{\beta}_1^{RPT}) = \mu_{11.2}^T B_{11.2}\mu_{11.2} + \mu_{11.2}^T B_{11.2}\delta H_{p_2+2}(\chi_{p_2}^2; \Delta)$$
$$+ \delta^T B_{11.2}\mu_{11.2}H_{p_2+2}(\chi_{p_2}^2; \Delta) + \delta^T B_{11.2}\delta H_{p_2+2}^2(\chi_{p_2}^2; \Delta),$$
$$AQDB(\hat{\beta}_1^{RSE}) = \mu_{11.2}^T B_{11.2}\mu_{11.2} + (p_2 - 2)\mu_{11.2}^T B_{11.2}\delta E(\chi_{p_2+2}^{-2}(\Delta))$$
$$+ (p_2 - 2)\delta^T B_{11.2}\mu_{11.2}E(\chi_{p_2+2}^{-2}(\Delta)) + (p_2 - 2)^2\delta^T B_{11.2}\delta\left(E(\chi_{p_2+2}^{-2}(\Delta))\right)^2,$$
$$AQDB(\hat{\beta}_1^{RPS}) = \mu_{11.2}^T B_{11.2}\mu_{11.2} + (\delta^T B_{11.2}\mu_{11.2} + \mu_{11.2}^T B_{11.2}\delta)[H_{p_2+2}(p_2 - 2; \Delta)$$
$$+ (p_2 - 2)E\{\chi_{p_2+2}^{-2}(\Delta)I(\chi_{p_2+2}^{-2}(\Delta) > p_2 - 2)\}] + \delta^T B_{11.2}\delta\Big[H_{p_2+2}(p_2 - 2; \Delta)$$
$$+ (p_2 - 2)E\{\chi_{p_2+2}^{-2}(\Delta)I(\chi_{p_2+2}^{-2}(\Delta) > p_2 - 2)\}\Big]^2.$$

When $\mathbf{B}_{11.2} = \mathbf{0}$, the AQDB of all estimators are equivalent, and the estimators are therefore asymptotically unbiased. If we assume that $\mathbf{B}_{11.2} \neq \mathbf{0}$, the results for the bias of the estimators can be summarized as follows:

1. The AQDB of $\hat{\beta}_1^{RSM}$ is an unbounded function of $\gamma^T \mathbf{B}_{11.2}\gamma$.

2. The AQDB of $\hat{\beta}_1^{RPT}$ starts from $\mu_{11.2}^T \mathbf{B}_{11.2}\mu_{11.2}$ at $\Delta = 0$, and when Δ increases, it increases to the maximum and then decreases to zero.
3. The characteristics of $\hat{\beta}_1^{RSE}$ and $\hat{\beta}_1^{RPS}$ are similar to $\hat{\beta}_1^{RPT}$. The AQDB of $\hat{\beta}_1^{RSE}$ and $\hat{\beta}_1^{RPS}$ similarly start from $\mu_{11.2}^T \mathbf{B}_{11.2}\mu_{11.2}$ at $\Delta = 0$, and increase to a point, and then decrease towards zero, since $E\{\chi_{p_2+2}^{-2}(\Delta)\}$ is a non-increasing on of Δ.

Theorem 3. *Suppose Theorem 1 holds and under the local alternatives K_n, the covariance matrices of the estimators are*

$$Cov(\hat{\beta}_1^{RFM}) = \mathbf{B}_{11.2}^{-1} + \mu_{11.2}\mu_{11.2}^T,$$

$$Cov(\hat{\beta}_1^{RSM}) = \mathbf{B}_{11}^{-1} + \gamma\gamma^T,$$

$$Cov(\hat{\beta}_1^{RPT}) = \mathbf{B}_{11.2}^{-1} + \mu_{11.2}\mu_{11.2}^T + 2\mu_{11.2}^T \delta H_{p_2+2}(\chi_{p_2}^2; \Delta) - \Phi H_{p_2+2}(\chi_{p_2}^2; \Delta)$$
$$+ \delta\delta^T [2H_{p_2+2}(\chi_{p_2}^2; \Delta) - H_{p_2+4}(\chi_{p_2}^2; \Delta)],$$

$$Cov(\hat{\beta}_1^{RSE}) = \mathbf{B}_{11.2}^{-1} + \mu_{11.2}\mu_{11.2}^T + 2(p_2 - 2)\mu_{11.2}^T \delta E\left(\chi_{p_2+2}^{-2}(\Delta)\right)$$
$$- (p_2 - 2)\Phi\left\{2E\left(\chi_{p_2+2}^{-2}(\Delta)\right) - (p_2 - 2)E\left(\chi_{p_2+2}^{-4}(\Delta)\right)\right\}$$
$$+ (p_2 - 2)\delta\delta^T\left\{- 2E\left(\chi_{p_2+4}^{-2}(\Delta)\right) + 2E(\chi_{p_2+2}^{-2}(\Delta)) + (p_2 - 2)E\left(\chi_{p_2+4}^{-4}(\Delta)\right)\right\},$$

$$Cov(\hat{\beta}_1^{RPS}) = Cov(\hat{\beta}_1^{RSE}) + 2\delta\mu_{11.2}^T E\left(\left\{1 - (p_2 - 2)\chi_{p_2+2}^{-2}(\Delta)\right\}I\left(\chi_{p_2+2}^2(\Delta) \leq p_2 - 2\right)\right)$$

$$- 2\Phi E\left(\left\{1 - (p_2 - 2)\chi_{p_2+2}^{-2}(\Delta)\right\}I\left(\chi_{p_2+2}^2(\Delta) \leq p_2 - 2\right)\right)$$

$$- 2\delta\delta^T E(\{1 - (p_2 - 2)\chi_{p_2+4}^{-2}(\Delta)\}I(\chi_{p_2+4}^2(\Delta) \leq p_2 - 2))$$

$$+ 2\delta\delta^T E\left(\left\{1 - (p_2 - 2)\chi_{p_2+2}^{-2}(\Delta)\right\}I\left(\chi_{p_2+2}^2(\Delta) \leq p_2 - 2\right)\right)$$

$$- (p_2 - 2)^2\Phi E\left(\chi_{p_2+2}^{-4}(\Delta)I\left(\chi_{p_2+2,\alpha}^2(\Delta) \leq p_2 - 2\right)\right)$$

$$- (p_2 - 2)^2\delta\delta^T E\left(\chi_{p_2+2,\alpha}^{-4}(\Delta)I\left(\chi_{p_2+2,\alpha}^2(\Delta) \leq p_2 - 2\right)\right)$$

$$+ \Phi H_{p_2+2}(p_2 - 2; \Delta) + \delta\delta^T H_{p_2+4}(p_2 - 2; \Delta).$$

Proof. See Appendix B.2. $\square$

Corollary 2. *Under the local alternatives (K_n) and from Theorem 3, the risk of the estimators are obtained as*

$$R\big[\hat{\beta}_1^{RFM}\big] = tr\Big(QB_{11.2}^{-1}\Big) + \mu_{11.2}^T Q\mu_{11.2},$$

$$R\big[\hat{\beta}_1^{RSM}\big] = tr\Big(QB_{11}^{-1}\Big) + \gamma^T Q\gamma,$$

$$R\big[\hat{\beta}_1^{RPT}\big] = tr\Big(QB_{11.2}^{-1}\Big) + \mu_{11.2}^T Q\mu_{11.2} + 2\mu_{11.2}^T Q\delta H_{p_2+2}\Big(\chi_{p_2}^2;\Delta\Big)$$
$$- tr(Q\Phi) H_{p_2+2}\Big(\chi_{p_2}^2;\Delta\Big) + \delta Q\delta^T\Big[2H_{p_2+2}(\chi_{p_2}^2;\Delta) - H_{p_2+4}(\chi_{p_2}^2;\Delta)\Big],$$

$$R\big[\hat{\beta}_1^{RSE}\big] = tr\Big(QB_{11.2}^{-1}\Big) + \mu_{11.2}^T Q\mu_{11.2} + 2(p_2-2)\mu_{11.2}^T Q\delta E\Big(\chi_{p_2+2}^{-2}(\Delta)\Big)$$
$$- (p_2-2)tr(Q\Phi)\Big[E\Big(\chi_{p_2+2}^{-2}(\Delta)\Big) - (p_2-2)E\Big(\chi_{p_2+2}^{-4}(\Delta)\Big)\Big]$$
$$+ (p_2-2)\delta^T Q\delta\Big[2E\Big(\chi_{p_2+2}^{-2}(\Delta)\Big) - 2E\Big(\chi_{p_2+4}^{-2}(\Delta)\Big) - (p_2-2)E\Big(\chi_{p_2+4}^{-4}(\Delta)\Big)\Big],$$

$$R\big[\hat{\beta}_1^{RPS}\big] = R\big[\hat{\beta}_1^{RSE}\big] + 2\delta Q\mu_{11.2}^T E\bigg(\Big\{1 - (p_2-2)\chi_{p_2+2}^{-2}(\Delta)\Big\}I\Big(\chi_{p_2+2}^2(\Delta) \le p_2-2\Big)\bigg)$$
$$- 2tr(Q\Phi)E\bigg(\Big\{1 - (p_2-2)\chi_{p_2+2}^{-2}(\Delta)\Big\}I\Big(\chi_{p_2+2}^2(\Delta) \le p_2-2\Big)\bigg)$$
$$- 2\delta^T Q\delta E\big(\{1 - (p_2-2)\chi_{p_2+4}^{-2}(\Delta)\}I(\chi_{p_2+4}^2(\Delta) \le p_2-2)\big)$$
$$+ 2\delta^T Q\delta E\bigg(\Big\{1 - (p_2-2)\chi_{p_2+2}^{-2}(\Delta)\Big\}I\Big(\chi_{p_2+2}^2(\Delta) \le p_2-2\Big)\bigg)$$
$$- (p_2-2)^2 tr(Q\Phi)E\bigg(\chi_{p_2+2}^{-4}(\Delta)I\Big(\chi_{p_2+2}^2(\Delta) \le p_2-2\Big)\bigg)$$
$$- (p_2-2)^2\delta^T Q\delta E\bigg(\chi_{p_2+2}^{-4}(\Delta)I\Big(\chi_{p_2+2}^2(\Delta) \le p_2-2\Big)\bigg)$$
$$+ tr(Q\Phi)H_{p_2+2}\big(p_2-2;\Delta\big) + \delta^T Q\delta H_{p_2+4}\big(p_2-2;\Delta\big).$$

From Theorem 2, when $\mathbf{B}_{12} = 0$, the risks of estimators $\hat{\beta}_1^{RSM}, \hat{\beta}_1^{RPT}, \hat{\beta}_1^{RSE}$, and $\hat{\beta}_1^{RPS}$ are reduced to common value $tr(QB_{11.2}^{-1}) + \mu_{11.2}^T Q\mu_{11.2}$, the risk of $\hat{\beta}_1^{RFM}$. If $\mathbf{B}_{12} \ne 0$, the results can be summarized as follows:

1. The risk of $\hat{\beta}_1^{RFM}$ remains constant while the risk of $\hat{\beta}_1^{RSM}$ is an unbounded function of Δ since $\Delta \in [0,\infty)$.
2. The risk of $\hat{\beta}_1^{RPT}$ increases as Δ moves away from zero, achieves it maximum and then decreases towards the risk of the full model estimator.
3. The risk of $\hat{\beta}_1^{RFM}$ is smaller than the risk of $\hat{\beta}_1^{RPT}$ for small values in the neighborhood of Δ and for the rest of the parameter space, $\hat{\beta}_1^{RPT}$ outperforms $\hat{\beta}_1^{RFM}$, thus, $R\big[\hat{\beta}_1^{RFM}\big] > R\big[\hat{\beta}_1^{RPT}\big]$.
4. Comparing the risks of $\hat{\beta}_1^{RSE}$ and $\hat{\beta}_1^{RFM}$, it can be seen that the estimator $\hat{\beta}_1^{RSE}$ outperforms $\hat{\beta}_1^{RFM}$ that is, $R\big[\hat{\beta}_1^{RSE}\big] \le R\big[\hat{\beta}_1^{RFM}\big]$ for all $\Delta \ge 0$.

4. Simulation Studies

In this section, we conduct a simulation study to assess the performance of the suggested estimators for finite samples. The criterion for comparing the performance of any estimator in our study is the mean square error. We simulate the response from the following LMM model

$$\mathbf{Y}_i = \mathbf{X}_i\beta + \mathbf{Z}_i\mathbf{a}_i + \epsilon_i, \tag{3}$$

where $\epsilon_i \sim \mathcal{N}(0,\sigma^2 I_{n_i})$ with $\sigma^2 = 1$. We generate random effect covariate $\mathbf{a}_i$ from a multivariate normal distribution with zero mean and covariance matrix $\mathbf{G} = 0.5 I_{2\times 2}$, where

$I_{2\times2}$ is 2×2 identity matrix. The design matrix $\mathbf{X}_i = (\mathbf{x}_{i1}, \dots, \mathbf{x}_{in_i})'$ is generated from a n_i-multivariate normal distribution with mean vector and covariance matrix Σ_x. Furthermore, we assume that the off-diagonal elements of the covariance matrix Σ_x are equal to ρ, which is the coefficient of correlation between any two predictors, with $\rho = 0.3, 0.7, 0.9$. The ratio of the largest eigenvalue to the smallest eigen-value of matrix $\mathbf{X}^T \mathbf{V}^{-1} \mathbf{X}$ is calculated as a condition number index (CNI) [24], which assesses the existence of multicollinearity in the design matrix. If the CNI is larger than 30, then the model has significant multicollinearity. Our simulations are based on the linear mixed effects model in Equation (3) with $n = 60$ and 100 subjects.

We consider a situation when the model is assumed to be sparse. In this study, our interest lies in testing the hypothesis $H_0 : \beta_2 = 0$, and our goal is to estimate the fixed effect coefficient β_1. We partition the fixed effects coefficients as $\beta = (\beta_1', \beta_2')' = (\beta_1', 0_{p_2})'$. The coefficients β_1 and β_2 are p_1 and p_2 dimensional vectors, respectively, with $p = p_1 + p_2$.

In order to investigate the behavior of the estimators, we define $\Delta^* = ||\beta - \beta_0||$, where $\beta_0 = (\beta_1^T, 0_{p_2})^T$ and $||.||$ is the euclidean norm. We considered Δ^* values between 0 and 4. If $\Delta^* = 0$, then we will have $\beta = (1, 1, 1, 1, \underbrace{0, 0, \dots, 0}_{p_2})^T$ to generate the response

under null hypothesis. On the other hand, when $\Delta^* \geq 0$, say $\Delta^* = 4$, we will have $\beta = (1, 1, 1, 1, 4, \underbrace{0, 0, \dots, 0}_{p_2 - 1})^T$ to generate the response under the local alternative hypothesis.

In our simulation study, we consider the number of fixed effect or predictor variables as $(p_1, p_2) \in \{(5, 40), (5, 500), (5, 1000)\}$. Each realization is repeated 5000 times to obtain consistent results and compute the MSE of suggested estimators with $\alpha = 0.05$.

Based on the simulated data, we calculate the mean square error (MSE) of all the estimators as $\text{MSE}(\hat{\beta}) = \frac{1}{5000} \sum_{j=1}^{5000} (\hat{\beta} - \beta)^T (\hat{\beta} - \beta)$, where $\hat{\beta}$ denotes any one of $\hat{\beta}^{\text{RSM}}, \hat{\beta}^{\text{RPT}}, \hat{\beta}^{\text{RSE}}$ and $\hat{\beta}^{\text{RPS}}$, in the jth repetition. We use the relative mean squared efficiency (RMSE), or the ratio of MSE for risk performance comparison. The RMSE of an estimator $\hat{\beta}^*$ with respect to the baseline full model ridge estimator $\hat{\beta}_1^{\text{RFM}}$ is defined as $\text{RMSE}(\hat{\beta}_1^{\text{RFM}} : \hat{\beta}_1^*) = \dfrac{\text{MSE}(\hat{\beta}_1^{\text{RFM}})}{\text{MSE}(\hat{\beta}_1^*)}$,

where β_1^* is one of the suggested estimators under consideration.

4.1. Simulation Results

In this subsection, we present the results from our simulation study. We report the results for $n = 60, 100$ and $p_1 = 5$ with different values of correlation coefficient ρ are shown in Table 1. Furthermore, we plot the RMSEs against Δ^* in Figures 1 and 2. The findings can be summarized as follows:

1. When $\Delta^* = 0$, the sub-model RSM outperforms all other estimators. As $\Delta^* = 0$ moves from zero, the RMSE of the sub-model decreases and goes to zero.
2. The pretest ridge estimator RPT outperforms shrinkage ridge and positive Stein ridge estimators in the case of $\Delta^* = 0$. However, for large number of sparse predictors p_2 while keeping p_1 and n fixed, RPT is less efficient than RPS and RSE. In the case of Δ^* being larger than zero, the RMSE of RPT decreases, and it remains below 1 for immediate values of Δ^*, after that the RMSE of RPT increases and approaches one for larger values of Δ^*.
3. RPS performs better than RSE in the entire parameter space induced by Δ^* as presented in Tables 1 and 2. Similarly, both shrinkage estimators RPS and RSE outperforms the full ridge model estimator irrespective of the corrected sub-model selected. This is consistent with the asymptotic theory presented in Section 3.
4. Δ^* which measures the degree of deviation from the Assumption 1 on the parameter space, it is clear that one cannot go wrong with the use of shrinkage estimators even if the selected sub-model is wrongly specified. As evident from Tables 1 and 2, Figures 1 and 2, if the selected sub-model is correct, that is, $\Delta^* = 0$, then the shrinkage estimators are relatively efficient compared with the ridge full model estimator. On the

other hand, if the sub-model is misspecified, the gain slowly diminishes. However, in terms of risk, the shrinkage estimators are at least as good as the full ridge model estimator. Therefore, the use of shrinkage estimators makes sense in application when a sub-model cannot be correctly specified.

5. The RMSE of the ridge-type estimators are an increasing function of the amount of multicollinearity. This indicates that the ridge-type estimators perform better than the classical estimator in the presence of multicollinearity among predictor variables.

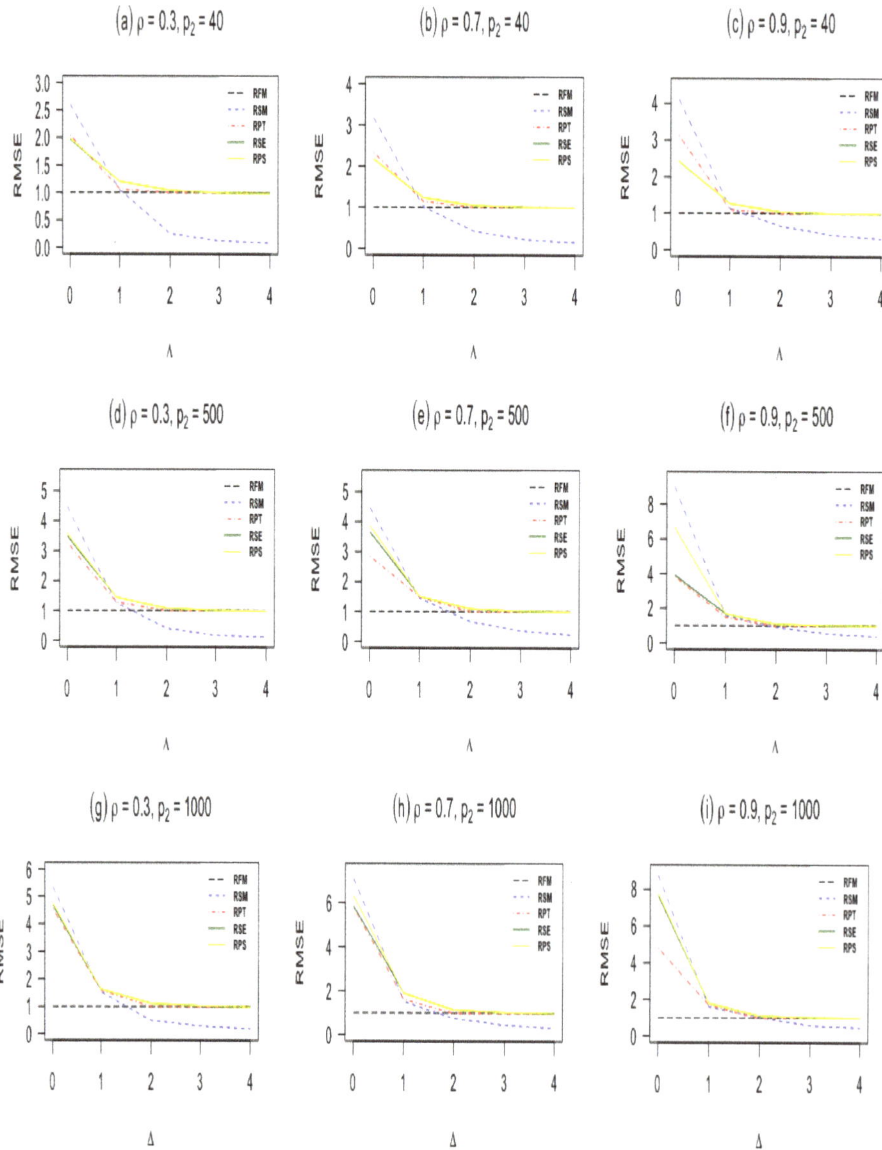

Figure 1. RMSE of estimators as a function of the non-centrality parameter Δ when $n = 60$, and $p_1 = 5$.

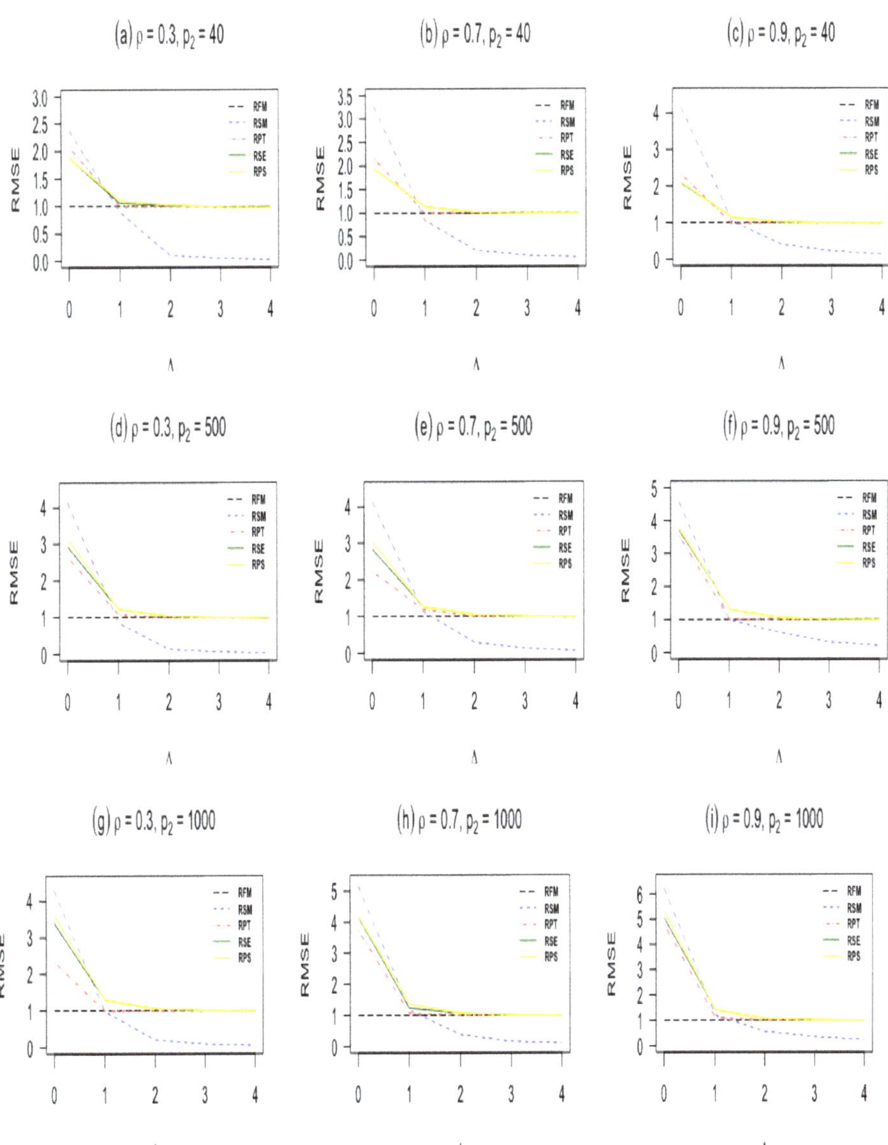

Figure 2. RMSE of estimators as a function of the non-centrality parameter Δ when $n = 100$, and $p_1 = 5$.

Table 1. RMSEs of RSM, RPT, RSE, and RPS estimators with respect to $\hat{\beta}_1^{\text{RFM}}$ when $\Delta \geq 0$ for $p_1 = 5$ and $n = 60$.

ρ	p_2	Δ	CNI	RSM	RPT	RSE	RPS
0.3	40	0	361	2.61	2.07	1.94	1.96
		1		1.05	1.07	1.20	1.25
		2		0.25	0.95	1.04	1.05
		3		0.12	0.98	0.99	1.00
		4		0.08	1.00	1.00	1.00
	500	0	613	4.48	3.29	3.48	1.96
		1		1.26	1.12	1.26	1.29
		2		0.41	0.97	1.08	1.09
		3		0.18	0.99	1.00	1.00
		4		0.13	1.00	1.00	1.00
	1000	0	693	5.36	4.53	4.67	4.71
		1		1.53	1.21	1.35	1.39
		2		0.49	1.01	1.13	1.14
		3		0.28	0.99	0.99	0.99
		4		0.10	1.00	1.00	1.00
0.7	40	0	1352	3.18	2.33	2.17	2.18
		1		1.04	1.11	1.20	1.23
		2		0.42	1.03	1.04	1.04
		3		0.23	0.98	0.99	1.00
		4		0.14	1.00	1.00	1.00
	500	0	1789	4.48	2.76	2.94	3.02
		1		1.08	1.43	1.52	1.53
		2		0.67	1.03	1.07	1.06
		3		0.35	0.98	1.00	1.00
		4		0.19	1.00	1.00	1.00
	1000	0	2134	6.82	5.24	5.30	3.02
		1		1.16	1.32	1.42	1.53
		2		0.75	1.10	1.15	1.16
		3		0.39	0.99	1.00	1.00
		4		0.11	1.00	1.00	1.00

Table 2. RMSEs of RSM, RPT, RSE, and RPS estimators with respect to $\hat{\beta}_1^{\text{RFM}}$ when $\Delta \geq 0$ for $p_1 = 5$, and $n = 100$.

ρ	p_2	Δ	CNI	RSM	RPT	RSE	RPS
0.3	40	0	150	2.38	2.09	1.88	1.90
		1		0.89	1.01	1.05	1.08
		2		0.21	0.94	1.01	1.02
		3		0.06	0.94	0.99	1.00
		4		0.02	1.00	1.00	1.00
	500	0	340	4.15	2.65	2.99	3.17
		1		0.87	1.08	1.18	1.21
		2		0.14	0.96	1.03	1.05
		3		0.06	0.99	0.99	1.00
		4		0.03	1.00	1.00	1.00
	1000	0	536	4.30	2.75	3.02	3.08
		1		0.96	1.09	1.13	1.15
		2		0.21	0.8	1.03	1.03
		3		0.09	1.00	1.00	1.00
		4		0.04	1.00	1.00	1.00
0.7	40	0	997	3.27	2.15	2.09	2.11
		1		0.85	1.02	1.09	1.10
		2		0.21	0.98	1.02	1.02
		3		0.06	0.99	0.99	0.99
		4		0.01	1.00	1.00	1.00
	500	0	1589	4.13	2.22	2.35	2.39
		1		1.04	1.19	1.21	1.20
		2		0.30	0.97	1.05	1.05
		3		0.14	1.00	1.00	1.00
		4		0.08	1.00	1.00	1.00
	1000	0	1751	5.17	3.71	4.03	4.09
		1		1.01	1.15	1.24	1.25
		2		0.39	1.04	1.07	1.06
		3		0.16	0.99	1.00	1.00
		4		0.11	1.00	1.00	1.00

4.2. Comparison with LASSO-Type Estimators

We compare our listed estimators with the LASSO and adaptive LASSO estimators. A 10-fold cross-validation is used for selecting the optimal value of the penalty parameters that minimizes the mean square errors for the LASSO-type estimators. The results for $\rho = 0.3, 0.7, 0.9$, $n = 60, 100$, $p_1 = 10$ and $p_2 = 50, 500, 1000, 2000$ are presented in Table 3. We observe the following from Table 3.

1. The performance of the sub-model estimator is the best among all estimators.
2. The pretest ridge estimator performs better than the other estimators. However, for larger values of sparse predictors p_2 the shrinkage estimators outperform the pretest estimator.
3. The performance of the LASSO and aLASSO estimators are comparable when ρ is small. The pretest and shrinkage estimators remain stable for a given value of ρ.
4. For a large number of sparse predictors p_2, the shrinkage and pretest ridge estimators outperforms the lasso-type estimators. This indicates the superiority of the shrinkage estimators over the LASSO-type estimators. Therefore shrinkage estimators are preferable when there is multicollinearity in our predictor variables.

Table 3. RMSEs of estimators with respect to $\hat{\beta}_1^{\text{RFM}}$ when $\Delta = 0$ for $p_1 = 10$.

n	ρ	p_2	CNI	RSM	RPT	RSE	RPS	LASSO	aLASSO
60	0.3	50	35.64	3.31	2.25	1.82	1.95	1.23	1.28
		500	452.76	4.13	3.71	2.61	3.01	1.47	1.52
		1000	1265.34	5.02	4.28	4.61	4.78	1.96	2.15
		2000	4567.56	7.13	5.10	6.18	6.39	2.70	3.06
	0.7	50	61.34	3.52	3.05	2.51	2.55	1.14	1.21
		500	743.17	4.49	3.65	3.41	3.50	1.36	1.58
		1000	2350.89	5.84	4.11	4.32	4.61	1.68	1.95
		2000	6908.39	8.10	5.31	6.24	6.29	1.84	2.02
	0.9	50	120.21	4.21	3.61	3.34	3.35	1.10	1.05
		500	950.98	4.82	3.3.8	3.72	3.73	1.21	1.16
		1000	5892.51	6.35	4.10	5.01	5.13	1.42	1.31
		2000	8352.73	8.51	4.63	5.24	5.38	1.61	1.35
100	0.3	50	31.21	2.91	2.54	2.12	2.23	1.32	1.36
		500	356.64	3.75	3.31	2.84	2.92	1.54	1.61
		1000	975.32	4.25	2.53	3.42	3.61	1.92	2.06
		2000	2764.84	5.61	4.25	4.91	5.08	2.31	2.46
	0.7	50	52.79	3.18	2.61	2.30	2.37	1.28	1.53
		500	578.43	4.28	3.05	3.52	3.59	1.46	2.07
		1000	1281.66	5.10	3.26	3.78	3.82	1.84	2.52
		2000	3498.30	6.12	3.01	4.26	4.33	2.27	2.41
	0.9	50	79.41	4.11	3.41	3.21	3.28	1.28	1.21
		500	681.43	4.35	3.55	3.41	3.50	1.43	1.51
		1000	1470.32	5.82	3.18	4.01	4.14	1.72	1.79
		2000	4105.90	7.04	4.57	5.22	5.32	1.87	1.96

5. Real Data Application

We consider two real data analyses using Amsterdam Growth and Health Data and a genetic and brain network connectivity edge weight data to illustrate the performance of the proposed estimators.

5.1. Amsterdam Growth and Health Data (AGHD)

The AGHD data is obtained from the Amsterdam Growth and Health Study [25]. The goal of this study is to investigate the relationship between lifestyle and health in adolescence into young adulthood. The response variable Y is the total serum cholesterol measured over six time points. There are five covariates: X_1 is the baseline fitness level measured as the maximum oxygen uptake on a treadmill, X_2 is the amount of body fat estimated by the sum of the thickness of four skinfolds, X_3 is a smoking indicator (0 = no, 1 = yes), X_4 is the gender (1 = female, 2 = male), and time measurement as X_5 and subject specific random effects.

A total of 147 subjects participated in the study where all variables were measured at $n_i = 6$ time occasions. In order to apply the proposed methods, firstly, we apply a variable selection based on AIC procedure to select the sub-model. For the AGHD data, we fit a linear mixed model with all the five covariates for both fixed and subject specific random effects by two stage selection procedure for the purpose of choosing both the random and fixed effects. The analysis found X_2 and X_5 to be significant covariates for prediction of the response variable serum cholestrol and the other variables are ignored since they are not significantly important. Based on this information, a sub-model is chosen to be X_2 and X_5 and the full model includes all the covariates. We construct the shrinkage estimators from the full-model and sub-model. In terms of null hypothesis, the restriction can be written as $\beta_2 = (\beta_1, \beta_3, \beta_4) = (0, 0, 0)$ with $p = 5$, $p_1 = 2$ and $p_2 = 3$.

To evaluate the performance of the estimators, we obtain the mean square prediction error (MSPE) using bootstrap samples. We draw 1000 bootstrap samples of the 147 subjects from the data matrix $\{(Y_{ij}, \mathbf{X}_{ij}), i = 1, 2, \ldots, 147; j = 1, 2, \ldots, 6\}$. We then calculate the relative prediction error (RPE) of β_1^* with respect to β_1^{RFM}, the full model estimator. The RPE is defined as

$$\text{RPE}(\hat{\beta}_1^{\text{RFM}} : \hat{\beta}_1^*) = \frac{\text{MSPE}(\hat{\beta}_1^*)}{\text{MSPE}(\hat{\beta}_1^{\text{RFM}})} = \frac{(\mathbf{Y} - \mathbf{X}_1\hat{\beta}_1^*)'(\mathbf{Y} - \mathbf{X}_1\hat{\beta}_1^*)}{(\mathbf{Y} - \mathbf{X}_1\hat{\beta}_1^{\text{RFM}})'(\mathbf{Y} - \mathbf{X}_1\hat{\beta}_1^{\text{RFM}})},$$

where β_1^* is one of the listed estimators. If RPE < 1, then $\hat{\beta}_1^*$ outperforms $\hat{\beta}_1^{\text{RFM}}$.

Table 4 reports the estimates, standard error of the non-sparse predictors and RPEs of the estimators with respect to the full model. As expected, the sub-model ridge estimator $\hat{\beta}_1^{\text{RSM}}$ has the minimum RPE because it is computed when the sub-model is correct, that is, $\Delta^* = 0$. It is evident by the RPE values in Table 4 that the shrinkage estimators are superior to the LASSO-type estimators. Furthermore, the positive shrinkage is more efficient than the shrinkage ridge estimator.

Table 4. Estimate, standard error for the active predictors and RPEs of estimators with respect to full-model estimator for the Amsterdam Growth and Health Study data.

	RFM	RSM	RPT	RSE	RPS	LASSO	aLASSO
Estimate(β_2)	0.381	0.395	0.392	0.389	0.390	0.624	0.611
Standard error	0.104	0.102	0.100	0.009	0.008	0.081	0.079
Estimate (β_5)	0.137	0.125	0.131	0.130	0.133	0.101	0.105
Standard error	0.012	0.010	0.009	0.011	0.010	0.013	0.012
RPE	1.000	0.723	0.841	0.838	0.831	0.986	0.973

5.2. Resting-State Effective Brain Connectivity and Genetic Data

This data comprises longitudinal resting-state functional magnetic resonance imaging (rs-fMRI) effective brain connectivity network and genetic study [26] data obtained from a sample of 111 subjects with a total of 319 rs-fMRI scans from the Alzheimer's Disease Neuroimaging Initiative (ADNI) database. The 111 subjects comprise 36 cognitively normal (CN), 63 mild cognitive impairment (MCI) and 12 Alzheimer's Disease (AD) subjects. The response is a network connection between regions of interest estimated from an rs-fMRI scan within the Default Mode Network (DMN), and we observe a longitudinal sequence of such connections for each subject with the number of repeated measurements. The DMN consists of a set of brain regions that tend to be active in resting-state, when a subject is mind wandering with no intended task. For this data analysis, we consider the network edge weight from the left intraparietal cortex to posterior cingulate cortex (LIPC $\rightarrow$ PCC) as our response. The genetic data are single nucleotide polymorphism (SNPs) from non-sex chromosomes, i.e., chromosome 1 to chromosome 22. SNPs with minor allele frequency less than 5% are removed as are SNPs with a Hardy–Weinberg equilibrium p-value lower than 10^{-6} or a missing rate greater than 5%. After preprocessing we are left with 1,220,955 SNPs and the longitudinal rs-fMRI effective connectivity network using the 111 subjects with rs-fMRI data. The response is network edge weight. There are SNPs which are the fixed effects and subject specific random effects.

In order to apply the proposed methods, we use a genome- wide association study (GWAS) for screening the genetic data to 100 SNPs. We implement a second screening by applying multinomial logistic regression to identify a smaller subset of the 100 SNPs that are potentially associated with disease (CN/MCI/AD). This yields a subset of top 10 SNPs. This showed the top 10 SNPs are the most important predictors and the other 90 SNPs are ignored as not significant. We now have two models, which are the full model with all 100 SNPs and sub-model with 10 SNPs selected. Finally, we construct the pretest and shrinkage estimators from the full-model and sub-model.

We draw 1000 bootstrap samples with replacements from the corresponding data matrix $\{(Y_{ij}, \mathbf{X}_{ij}), i = 1, \ldots, 111; j = 1, \ldots, n_i\}$. We report the RPE of the estimators based on the bootstrap simulation with respect to the full model ridge estimator in Table 5. We observe that the RPE of the sub-model, pretest, shrinkage and positive shrinkage ridge estimators outperforms the full model estimator. Clearly, the sub-model ridge estimator has the smallest RPE since it's computed when the candidate sub-model is correct, i.e., $\Delta = 0$. Both shrinkage ridge estimators outperform the pretest ridge estimator. Particularly, the positive shrinkage performed better than the shrinkage estimator. The performance of both shrinkage and pretest ridge estimators are better than the LASSO-type estimators. Thus, the data analysis is in line with our simulation and theoretical findings.

Table 5. RPEs of estimators.

	RFM	RSM	RPT	RSE	RPS	LASSO	aLASSO
RPE	1.000	0.802	0.947	0.932	0.928	1.051	1.190

6. Conclusions

In this paper, we present efficient estimation strategies for the linear mixed effect model when there exists multicollinearity among predictor variables for high-dimensional data application. We considered the estimation of fixed effects parameters in the linear mixed model when some of the predictors may have a very weak influence on the response of interest. We introduced pretest and shrinkage estimation in our model using the ridge estimation as the reference estimator. In addition, we established the asymptotic properties of the pretest and shrinkage ridge estimators. Our theoretical findings demonstrate that the shrinkage ridge estimators outperform the full model ridge estimator and perform relatively better than the sub-model estimator in a wide range of the parameter space.

Additionally, a Monte Carlo simulation was conducted to investigate and assess the finite sample behavior of proposed estimators when the model is sparse (restrictions on parameters hold). As expected, the sub-model ridge estimator outshines all other estimators when the restrictions hold. However, when this assumption is violated, the shrinkage and pretest ridge estimators outperform the sub-model estimator. Furthermore, when the number of sparse predictors are extremely large relative to the sample size, the shrinkage estimators outperform the pretest ridge estimator. These numerical results are consistent with our asymptotic result. We also assess the relative performance of the LASSO-type estimators with our ridge-type estimators. We observe that the performance of pretest and shrinkage ridge estimators are superior to the LASSO-type estimators when predictors are highly correlated. For our real data application, the shrinkage ridge estimators are superior with the smallest relative prediction error compared to the LASSO-type estimators.

In summary, the results of the data analyses strongly confirm the findings of the simulation study and suggest the use of the shrinkage ridge estimation strategy when no prior information about the parameter subspace is available. The results of our simulation study and real data application are consistent with available results in [27–29].

In our future work, we will focus on other penalty estimators like the Elastic-Net, the minimax concave penalty (MCP), and the smoothly clipped absolute deviation method (SCAD) as estimation strategy in LMM for high-dimensional data. These estimators will be assessed and compared with the proposed ridge-type estimators. Another interesting extension will be integrating two sub-models by incorporating ridge-type estimation strategies in the linear mixed effect models. The goal is to improve the estimation accuracy of the non-sparse set of the fixed effects parameters by combining an over-fitted model estimator with an under-fitted one [27,29]. This approach will include combining two sub-models produced by two different variable selection techniques from the LMM [28].

Author Contributions: Conceptualization, E.A.O. and S.E.A.; methodology, E.A.O. and F.S.N.; formal analysis, E.A.O.; writing—original draft preparation, E.A.O.; writing—review and editing, E.A.O., S.E.A. and F.S.N.; supervision, F.S.N. and S.E.A.; funding acquisition, F.S.N. and S.E.A. All authors have read and agreed to the published version of the manuscript.

Funding: This research was funded by Natural Sciences and Engineering Research Council of Canada (NSERC).

Institutional Review Board Statement: Not applicable.

Informed Consent Statement: Not applicable.

Data Availability Statement: Publicly available datasets were analyzed in this study. This data can be found here https://pubmed.ncbi.nlm.nih.gov/22434862/ (accessed on 20 April 2021).

Acknowledgments: Research is supported by the Visual and Automated Disease Analytics (VADA) graduate training program.

Conflicts of Interest: The authors declare no conflict of interest.

Appendix A

Proof of Proposition 1. The asymptotic relationship between the sub-model and full model estimators of β_1, we use the argument and equation: $\hat{\mathbf{Y}} = \mathbf{Y} - \mathbf{X}_2\hat{\beta}_2^{\mathrm{RFM}}$, where

$$
\begin{aligned}
\hat{\beta}_1^{\mathrm{RFM}} &= \arg\min_{\beta_1} \left\{ (\hat{\mathbf{Y}} - \mathbf{X}_1\beta_1)^{\mathrm{T}}\mathbf{V}^{-1}(\hat{\mathbf{Y}} - \mathbf{X}_1\beta_1) + \lambda||\beta_1||^2 \right\} \\
&= \left[\mathbf{X}_1^{\mathrm{T}}\mathbf{V}^{-1}\mathbf{X}_1 + \lambda\mathrm{I}_{p_1}\right]^{-1}\mathbf{X}_1^{\mathrm{T}}\mathbf{V}^{-1}\hat{\mathbf{Y}} \\
&= \left[\mathbf{X}_1^{\mathrm{T}}\mathbf{V}^{-1}\mathbf{X}_1 + \lambda\mathrm{I}_{p_1}\right]^{-1}\mathbf{X}_1^{\mathrm{T}}\mathbf{V}^{-1}\mathbf{Y} - \left[\mathbf{X}_1^{\mathrm{T}}\mathbf{V}^{-1}\mathbf{X}_1 + \lambda\mathrm{I}_{p_1}\right]^{-1}\mathbf{X}_1^{\mathrm{T}}\mathbf{V}^{-1}\mathbf{X}_2\hat{\beta}_2^{\mathrm{RFM}} \\
&= \hat{\beta}_1^{\mathrm{RSM}} - \left[\mathbf{X}_1\mathbf{V}^{-1}\mathbf{X}_1 + \lambda\mathrm{I}_{p_1}\right]^{-1}\mathbf{X}_1^{\mathrm{T}}\mathbf{V}^{-1}\mathbf{X}_2\hat{\beta}_2^{\mathrm{RFM}} \\
&= \hat{\beta}_1^{\mathrm{RSM}} - \mathbf{B}_{11}^{-1}\mathbf{B}_{12}\hat{\beta}_2^{\mathrm{RFM}}
\end{aligned}
$$

From Theorem 1, we partition $\sqrt{n}(\hat{\beta}^{\mathrm{RFM}} - \beta)$ as $\sqrt{n}(\hat{\beta}^{\mathrm{RFM}} - \beta) = (\sqrt{n}(\hat{\beta}_1^{\mathrm{RFM}} - \beta_1), \sqrt{n}(\hat{\beta}_2^{\mathrm{RFM}} - \beta_2))$. We obtain $\sqrt{n}(\hat{\beta}_1^{\mathrm{RFM}} - \beta_1) \xrightarrow{D} \mathcal{N}_{p_1}(-\mu_{11.2}, \mathbf{B}_{11.2}^{-1})$, where $\mathbf{B}_{11.2}^{-1} = \mathbf{B}_{11} - \mathbf{B}_{12}\mathbf{B}_{22}^{-1}\mathbf{B}_{21}$. We have shown that $\hat{\beta}_1^{\mathrm{RSM}} = \hat{\beta}_1^{\mathrm{RFM}} + \mathbf{B}_{11}^{-1}\mathbf{B}_{12}\hat{\beta}_2^{\mathrm{RFM}}$. Using this expression and under the local alternative $\{K_n\}$, we obtain the following expressions

$$
\begin{aligned}
\varphi_2 &= \sqrt{n}(\hat{\beta}_1^{\mathrm{RSM}} - \beta_1) \\
&= \sqrt{n}(\hat{\beta}_1^{\mathrm{RFM}} + \mathbf{B}_{11}^{-1}\mathbf{B}_{12}\hat{\beta}_2^{\mathrm{RFM}} - \beta_1) \\
&= \varphi_1 + \mathbf{B}_{11}^{-1}\mathbf{B}_{12}\sqrt{n}\hat{\beta}_2^{\mathrm{RFM}}, \\
\varphi_3 &= \sqrt{n}(\hat{\beta}_1^{\mathrm{RFM}} - \hat{\beta}_1^{\mathrm{RSM}}) \\
&= \sqrt{n}(\hat{\beta}_1^{\mathrm{RFM}} - \beta_1) - \sqrt{n}(\hat{\beta}_1^{\mathrm{RSM}} - \beta_1) \\
&= \varphi_1 - \varphi_2.
\end{aligned}
$$

Since φ_2 and φ_3 are linear functions of φ_1, as $n \to \infty$, they are also asymptotically normally distributed. Their mean vectors and covariance matrices are as follows:

$$E(\varphi_1) = E\left(\sqrt{n}(\hat{\beta}_1^{\mathrm{RFM}} - \beta_1)\right) = -\mu_{11.2}$$

$$E(\varphi_2) = E\left(\varphi_1 + \mathbf{B}_{11}^{-1}\mathbf{B}_{12}\sqrt{n}\hat{\beta}_2^{\mathrm{RFM}}\right)$$

$$= E(\varphi_1) + \mathbf{B}_{11}^{-1}\mathbf{B}_{12}\sqrt{n}E(\hat{\beta}_2^{\mathrm{RFM}})$$

$$= -\mu_{11.2} + \mathbf{B}_{11}^{-1}\mathbf{B}_{12}\kappa = -(\mu_{11.2} - \delta) = -\gamma$$

$$E(\varphi_3) = E(\varphi_1 - \varphi_2) = -\mu_{11.2} - (-(\mu_{11.2} - \delta)) = \delta$$

$$Var(\varphi_1) = \mathbf{B}_{22.1}^{-1}$$

$$Var(\varphi_2) = Var\left(\varphi_1 + \mathbf{B}_{11}^{-1}\mathbf{B}_{12}\sqrt{n}\hat{\beta}_2^{\mathrm{RFM}}\right)$$

$$= Var(\varphi_1) + \mathbf{B}_{11}^{-1}\mathbf{B}_{12}\mathbf{B}_{22.1}^{-1}\mathbf{B}_{21}\mathbf{B}_{11}^{-1}$$

$$+ 2Cov\left[\sqrt{n}(\hat{\beta}_1^{\mathrm{RFM}} - \beta_1), \sqrt{n}(\hat{\beta}_2^{\mathrm{RFM}} - \beta_2)\right](\mathbf{B}_{11}^{-1}\mathbf{B}_{12})^{\mathrm{T}}$$

$$= \mathbf{B}_{22.1}^{-1} - \mathbf{B}_{11}^{-1}\mathbf{B}_{12}\mathbf{B}_{22.1}^{-1}\mathbf{B}_{21}\mathbf{B}_{11}^{-1} = \mathbf{B}_{11}^{-1}$$

$$Var(\varphi_3) = Var\left(\sqrt{n}(\hat{\beta}_1^{\mathrm{RFM}} - \hat{\beta}_1^{\mathrm{RSM}})\right)$$

$$= Var\left(\sqrt{n}(\hat{\beta}_1^{\mathrm{RFM}} - \hat{\beta}_1^{\mathrm{RFM}} - \mathbf{B}_{11}^{-1}\mathbf{B}_{12}\hat{\beta}_2^{\mathrm{RFM}})\right)$$

$$= \mathbf{B}_{11}^{-1}\mathbf{B}_{12}Var\left[\sqrt{n}\hat{\beta}_2^{\mathrm{RFM}}\right](\mathbf{B}_{11}^{-1}\mathbf{B}_{12})^{\mathrm{T}}$$

$$= \mathbf{B}_{11}^{-1}\mathbf{B}_{12}\mathbf{B}_{22.1}^{-1}\mathbf{B}_{21}\mathbf{B}_{11}^{-1} = \boldsymbol{\Phi}$$

$$Cov(\varphi_1, \varphi_3) = Cov\left[\sqrt{n}(\hat{\beta}_1^{\mathrm{RFM}} - \beta_1), \sqrt{n}(\hat{\beta}_1^{\mathrm{RFM}} - \hat{\beta}_1^{\mathrm{RSM}})\right]$$

$$= Var\left(\sqrt{n}(\hat{\beta}_1^{\mathrm{RFM}} - \beta_1)\right) - Cov\left[\sqrt{n}(\hat{\beta}_1^{\mathrm{RFM}} - \beta_1), \sqrt{n}(\hat{\beta}_1^{\mathrm{RSM}} - \beta_1)\right]$$

$$= Var(\varphi_1) - Cov\left[\sqrt{n}(\hat{\beta}_1^{\mathrm{RFM}} - \beta_1), \sqrt{n}(\hat{\beta}_1^{\mathrm{RFM}} - \beta_1) + \sqrt{n}\mathbf{B}_{11}^{-1}\mathbf{B}_{12}\hat{\beta}_2^{\mathrm{RFM}}\right]$$

$$= \mathbf{B}_{11}^{-1}\mathbf{B}_{12}\mathbf{B}_{22.1}^{-1}\mathbf{B}_{21}\mathbf{B}_{11}^{-1} = \boldsymbol{\Phi}$$

$$Cov(\varphi_2, \varphi_3) = Cov\left[\sqrt{n}(\hat{\beta}_1^{\mathrm{RSM}} - \beta_1), \sqrt{n}(\hat{\beta}_1^{\mathrm{RFM}} - \hat{\beta}_1^{\mathrm{RSM}})\right]$$

$$= Cov\left[\sqrt{n}(\hat{\beta}_1^{\mathrm{RSM}} - \beta_1), \sqrt{n}(\hat{\beta}_1^{\mathrm{RFM}} - \beta_1)\right] - Var\left(\sqrt{n}(\hat{\beta}_1^{\mathrm{RSM}} - \beta_1)\right)$$

$$= \mathbf{B}_{11.2}^{-1} - \mathbf{B}_{11}^{-1}\mathbf{B}_{12}\mathbf{B}_{22.1}^{-1}\mathbf{B}_{21}\mathbf{B}_{11}^{-1} - \mathbf{B}_{11}^{-1}$$

$$= \mathbf{B}_{11.2}^{-1} - (\mathbf{B}_{11.2}^{-1} - \mathbf{B}_{11}^{-1}) - \mathbf{B}_{11}^{-1} = 0$$

Therefore, the asymptotic distributions of the vectors φ_2 and φ_3 are obtained as follows:

$$\varphi_2 = \sqrt{n}(\hat{\beta}_1^{\mathrm{RSM}} - \beta_1) \xrightarrow{D} \mathcal{N}_{p_1}(-\gamma, \mathbf{B}_{11}^{-1})$$

$$\varphi_3 = \sqrt{n}(\hat{\beta}_1^{\mathrm{RFM}} - \hat{\beta}_1^{\mathrm{RSM}}) \xrightarrow{D} \mathcal{N}_{p_1}(\delta, \boldsymbol{\Phi})$$

$\square$

Appendix B

We next introduce the lemmas given in [30] to aid with the proof of the bias and covariance of the estimators.

Lemma A1. *Let* $V = (V_1, V_2, \ldots V_p)^T$ *be a p-dimensional normal vector distributed as* $\mathcal{N}_p(\boldsymbol{\mu}_v, \boldsymbol{\Sigma}_p)$, *then for a measurable function* Ψ, *we have*

$$E\left[V\Psi(V^T V)\right] = \boldsymbol{\mu}_v E\left[\Psi \chi^2_{p+2}(\Delta)\right]$$

$$E\left[VV^T\Psi(V^T V)\right] = \boldsymbol{\Sigma}_p E\left[\Psi \chi^2_{p+2}(\Delta)\right] + \boldsymbol{\mu}_v \boldsymbol{\mu}_v^T E\left[\Psi \chi^2_{p+4}(\Delta)\right]$$

where $\chi^2_k(\Delta)$ *is a non-central chi-square distribution with k degrees of freedom and non-centrality parameter* Δ.

Appendix B.1

Proof of Theorem 2.

$$\text{ADB}(\hat{\boldsymbol{\beta}}_1^{\text{RFM}}) = E\{ \lim_{n\to\infty} \sqrt{n}(\hat{\boldsymbol{\beta}}_1^{\text{RFM}} - \boldsymbol{\beta}_1)\}$$

$$= -\boldsymbol{\mu}_{11.2}.$$

$$\text{ADB}(\hat{\boldsymbol{\beta}}_1^{\text{RSM}}) = E\{ \lim_{n\to\infty} \sqrt{n}(\hat{\boldsymbol{\beta}}_1^{\text{RSM}} - \boldsymbol{\beta}_1)\}$$

$$= E\{ \lim_{n\to\infty} \sqrt{n}(\hat{\boldsymbol{\beta}}_1^{\text{RFM}} - \mathbf{B}_{11}^{-1}\mathbf{B}_{12}\hat{\boldsymbol{\beta}}_2^{\text{RFM}} - \boldsymbol{\beta}_1)\}$$

$$= E\{ \lim_{n\to\infty} \sqrt{n}(\hat{\boldsymbol{\beta}}_1^{\text{RFM}} - \boldsymbol{\beta}_1)\} - E\{ \lim_{n\to\infty} \sqrt{n}(\mathbf{B}_{11}^{-1}\mathbf{B}_{12}\hat{\boldsymbol{\beta}}_2^{\text{RFM}})\}$$

$$= -\boldsymbol{\mu}_{11.2} - E\{ \lim_{n\to\infty} \sqrt{n}(\mathbf{B}_{11}^{-1}\mathbf{B}_{12}\hat{\boldsymbol{\beta}}_2^{\text{RFM}})\}$$

$$= -\boldsymbol{\mu}_{11.2} - \mathbf{B}_{11}^{-1}\mathbf{B}_{12}\boldsymbol{\kappa} = -(\boldsymbol{\mu}_{11.2} + \boldsymbol{\delta}) = -\boldsymbol{\gamma}.$$

Using Lemma 1,

$$\text{ADB}(\hat{\boldsymbol{\beta}}_1^{\text{RPT}}) = E\{ \lim_{n\to\infty} \sqrt{n}(\hat{\boldsymbol{\beta}}_1^{\text{RPT}} - \boldsymbol{\beta}_1)\}$$

$$= E\{ \lim_{n\to\infty} \sqrt{n}(\hat{\boldsymbol{\beta}}_1^{\text{RFM}} - (\hat{\boldsymbol{\beta}}_1^{\text{RFM}} - \hat{\boldsymbol{\beta}}_1^{\text{RSM}})I(L_n \leq d_{n,\alpha}) - \boldsymbol{\beta}_1)\}$$

$$= E\{ \lim_{n\to\infty} \sqrt{n}(\hat{\boldsymbol{\beta}}_1^{\text{RFM}} - \boldsymbol{\beta}_1)\} - E\{ \lim_{n\to\infty} \sqrt{n}(\hat{\boldsymbol{\beta}}_1^{\text{RFM}} - \hat{\boldsymbol{\beta}}_1^{\text{RSM}})I(L_n \leq d_{n,\alpha})\}$$

$$= -\boldsymbol{\mu}_{11.2} - E\{ \lim_{n\to\infty} \sqrt{n}(\hat{\boldsymbol{\beta}}_1^{\text{RFM}} - \hat{\boldsymbol{\beta}}_1^{\text{RSM}})I(L_n \leq d_{n,\alpha}))\}$$

$$= -\boldsymbol{\mu}_{11.2} - \boldsymbol{\delta}\mathbf{H}_{p_2+2}(\chi^2_{p_2}; \Delta).$$

$$\text{ADB}(\hat{\boldsymbol{\beta}}_1^{\text{RSE}}) = E\{ \lim_{n\to\infty} \sqrt{n}(\hat{\boldsymbol{\beta}}_1^{\text{RSE}} - \boldsymbol{\beta}_1)\}$$

$$= E\{ \lim_{n\to\infty} \sqrt{n}(\hat{\boldsymbol{\beta}}_1^{\text{RFM}} - (\hat{\boldsymbol{\beta}}_1^{\text{RFM}} - \hat{\boldsymbol{\beta}}_1^{\text{RSM}})(p_2 - 2)L_n^{-1} - \boldsymbol{\beta}_1)\}$$

$$= E\{ \lim_{n\to\infty} \sqrt{n}(\hat{\boldsymbol{\beta}}_1^{\text{RFM}} - \boldsymbol{\beta}_1)\} - E\{ \lim_{n\to\infty} \sqrt{n}(\hat{\boldsymbol{\beta}}_1^{\text{RFM}} - \hat{\boldsymbol{\beta}}_1^{\text{RSM}})(p_2 - 2)L_n^{-1}\}$$

$$= -\boldsymbol{\mu}_{11.2} - E\{ \lim_{n\to\infty} \sqrt{n}(\hat{\boldsymbol{\beta}}_1^{\text{RFM}} - \hat{\boldsymbol{\beta}}_1^{\text{RSM}})(p_2 - 2)L_n^{-1}\}$$

$$= -\boldsymbol{\mu}_{11.2} - (p_2 - 2)\boldsymbol{\delta}E(\chi^{-2}_{p_2+2}(\Delta)).$$

$$
\begin{aligned}
\mathrm{ADB}(\hat{\beta}_1^{\mathrm{RPS}}) &= E\big\{ \lim_{n\to\infty} \sqrt{n}(\hat{\beta}_1^{\mathrm{RPS}} - \beta_1) \big\} \\
&= E\big\{ \lim_{n\to\infty} \sqrt{n}(\hat{\beta}_1^{\mathrm{RSM}} + (\hat{\beta}_1^{\mathrm{RFM}} - \hat{\beta}_1^{\mathrm{RSM}})(1 - (p_2 - 2)\mathrm{L}_n^{-1})\mathrm{I}(\mathrm{L}_n > p_2 - 2) - \beta_1) \big\} \\
&= E\big\{ \sqrt{n}\big[\hat{\beta}_1^{\mathrm{RSM}} + (\hat{\beta}_1^{\mathrm{RFM}} - \hat{\beta}_1^{\mathrm{RSM}})(1 - \mathrm{I}(\mathrm{L}_n \le p_2 - 2)) \\
&\quad - (\hat{\beta}_1^{\mathrm{RFM}} - \hat{\beta}_1^{\mathrm{RSM}})(p_2 - 2)\mathrm{L}_n^{-1}\mathrm{I}(\mathrm{L}_n > p_2 - 2) - \beta_1\big] \big\} \\
&= E\big\{ \lim_{n\to\infty} \sqrt{n}(\hat{\beta}_1^{\mathrm{RFM}} - \beta_1) \big\} - E\big\{ \lim_{n\to\infty} \sqrt{n}(\hat{\beta}_1^{\mathrm{RFM}} - \hat{\beta}_1^{\mathrm{RSM}})(p_2 - 2)\mathrm{I}(\mathrm{L}_n \le p_2 - 2) \big\} \\
&\quad - E\big\{ \lim_{n\to\infty} \sqrt{n}(\hat{\beta}_1^{\mathrm{RFM}} - \hat{\beta}_1^{\mathrm{RSM}})(p_2 - 2)\mathrm{L}_n^{-1}\mathrm{I}(\mathrm{L}_n > p_2 - 2) \\
&= -\mu_{11.2} - \delta \mathbf{H}_{p_2+2}(\chi^2_{p_2-2}; \Delta) \big\} - (p_2 - 2)\delta E\big\{ \chi^{-2}_{p_2+2}(\Delta)\mathrm{I}(\chi^{-2}_{p_2+2} > p_2 - 2) \big\}.
\end{aligned}
$$

$\square$

Appendix B.2

In order to compute the risk functions, we first compute the asymptotic covariance of the estimators. The asymptotic covariance of an estimator $\hat{\beta}_1^*$ is expressed as

$$
\mathrm{Cov}(\hat{\beta}_1^*) = \lim_{n\to\infty} E\big\{ n(\hat{\beta}_1^* - \beta_1)(\hat{\beta}_1^* - \beta_1)^{\mathrm{T}} \big\}.
$$

Proof of Theorem 3. We first start by computing the asymptotic covariance of the estimator $\hat{\beta}_1^{\mathrm{RFM}}$ as:

$$
\begin{aligned}
\mathrm{Cov}(\hat{\beta}_1^{\mathrm{RFM}}) &= E\big\{ \lim_{n\to\infty} \sqrt{n}(\hat{\beta}_1^{\mathrm{RFM}} - \beta_1)\sqrt{n}(\hat{\beta}_1^{\mathrm{RFM}} - \beta_1)^{\mathrm{T}} \big\} \\
&= E(\varphi_1 \varphi_1^{\mathrm{T}}) = \mathrm{Cov}(\varphi_1 \varphi_1^{\mathrm{T}}) + E(\varphi_1)E(\varphi_1^{\mathrm{T}}) \\
&= \mathbf{B}_{11.2}^{-1} + \mu_{11.2}\mu_{11.2}^{\mathrm{T}}.
\end{aligned}
$$

Furthermore, similarly, the asymptotic covariance of the estimator $\hat{\beta}_1^{\mathrm{RSM}}$ is obtained as:

$$
\begin{aligned}
\mathrm{Cov}(\hat{\beta}_1^{\mathrm{RSM}}) &= E\big\{ \lim_{n\to\infty} \sqrt{n}(\hat{\beta}_1^{\mathrm{RSM}} - \beta_1)\sqrt{n}(\hat{\beta}_1^{\mathrm{RSM}} - \beta_1)^{\mathrm{T}} \big\} \\
&= E(\varphi_2 \varphi_2^{\mathrm{T}}) = Cov(\varphi_2 \varphi_2^{\mathrm{T}}) + E(\varphi_2)E(\varphi_2^{\mathrm{T}}) \\
&= \mathbf{B}_{11}^{-1} + \gamma\gamma^{\mathrm{T}}.
\end{aligned}
$$

The asymptotic covariance of the estimator $\hat{\beta}_1^{\mathrm{RPT}}$ is obtained as:

$$
\begin{aligned}
\mathrm{Cov}(\hat{\beta}_1^{\mathrm{RPT}}) &= E\big\{ \lim_{n\to\infty} \sqrt{n}(\hat{\beta}_1^{\mathrm{RPT}} - \beta_1)\sqrt{n}(\hat{\beta}_1^{\mathrm{RPT}} - \beta_1)^{\mathrm{T}} \big\} \\
&= E\big\{ \lim_{n\to\infty} n\big[(\hat{\beta}_1^{\mathrm{RFM}} - \beta_1) - (\hat{\beta}_1^{\mathrm{RFM}} - \hat{\beta}_1^{\mathrm{RSM}})I(\mathrm{L}_n \le d_{n,\alpha})\big] \\
&\quad \big[(\hat{\beta}_1^{\mathrm{RFM}} - \beta_1) - (\hat{\beta}_1^{\mathrm{RFM}} - \hat{\beta}_1^{\mathrm{RSM}})I(\mathrm{L}_n \le d_{n,\alpha})\big]^{\mathrm{T}} \big\} \\
&= E\Big\{ [\varphi_1 - \varphi_3 I(\mathrm{L}_n \le d_{n,\alpha})][\varphi_1 - \varphi_3 I(\mathrm{L}_n \le d_{n,\alpha})]^{\mathrm{T}} \Big\} \\
&= E\Big\{ \varphi_1 \varphi_1^{\mathrm{T}} - 2\varphi_3 \varphi_1^{\mathrm{T}} I(\mathrm{L}_n \le d_{n,\alpha}) + \varphi_3 \varphi_3^{\mathrm{T}} I(\mathrm{L}_n \le d_{n,\alpha}) \Big\}
\end{aligned}
$$

Thus, we need to find $E\{\varphi_1 \varphi_1^{\mathrm{T}}\}$, $E\{\varphi_3 \varphi_1^{\mathrm{T}} I(\mathrm{L}_n \le d_{n,\alpha})\}$ and $E\{\varphi_3 \varphi_3^{\mathrm{T}} I(\mathrm{L}_n \le d_{n,\alpha})\}$. The first term is $E\{\varphi_1 \varphi_1^{\mathrm{T}}\} = \mathbf{B}_{11.2}^{-1} + \mu_{11.2}\mu_{11.2}^{\mathrm{T}}$. Using Lemma 1, the third term is computed as:

$$
E\{\varphi_3 \varphi_3^{\mathrm{T}} I(\mathrm{L}_n \le d_{n,\alpha})\} = \mathbf{\Phi}\mathbf{H}_{p_2+2}(\chi^2_{p_2}; \Delta) + \delta\delta^{\mathrm{T}}\mathbf{H}_{p_2+4}(\chi^2_{p_2}; \Delta).
$$

The second term $E\{\varphi_3 \varphi_1^{\mathrm{T}} I(\mathrm{L}_n \le d_{n,\alpha})\}$ can be computed from normal theory as

$$E\left\{\varphi_3\varphi_1^{\mathrm{T}}I(L_n \leq d_{n,\alpha})\right\} = E\left\{E(\varphi_3\varphi_1^{\mathrm{T}}I(L_n \leq d_{n,\alpha})|\varphi_3)\right\} = E\left\{\varphi_3 E(\varphi_1^{\mathrm{T}}I(L_n \leq d_{n,\alpha})|\varphi_3)\right\}$$

$$= E\left\{\varphi_3[-\mu_{11.2} + (\varphi_3 - \delta)]^{\mathrm{T}}I(L_n \leq d_{n,\alpha})\right\}$$

$$= -E\left\{\varphi_3\mu_{11.2}I(L_n \leq d_{n,\alpha})\right\} + E\left\{\varphi_3(\varphi_3 - \delta)^{\mathrm{T}}I(L_n \leq d_{n,\alpha})\right\}$$

$$= -\mu_{11.2}^{\mathrm{T}}E\left\{\varphi_3 I(L_n \leq d_{n,\alpha})\right\} + E\left\{\varphi_3\varphi_3^{\mathrm{T}}I(L_n \leq d_{n,\alpha})\right\}$$

$$- E\left\{\varphi_3\delta^{\mathrm{T}}I(L_n \leq d_{n,\alpha})\right\}$$

$$= -\mu_{11.2}^{\mathrm{T}}\delta\mathbf{H}_{p_2+2}(\chi^2_{p_2};\Delta) + \left\{Cov(\varphi_3\varphi_3^{\mathrm{T}})\mathbf{H}_{p_2+2}(\chi^2_{p_2};\Delta)\right.$$

$$\left. + E(\varphi_3)E(\varphi_3^{\mathrm{T}})\mathbf{H}_{p_2+4}(\chi^2_{p_2};\Delta) - \delta\delta^{\mathrm{T}}\mathbf{H}_{p_2+2}(\chi^2_{p_2};\Delta)\right\}$$

$$= -\mu_{11.2}^{\mathrm{T}}\delta\mathbf{H}_{p_2+2}(\chi^2_{p_2};\Delta) + \boldsymbol{\Phi}\mathbf{H}_{p_2+2}(\chi^2_{p_2};\Delta) + \delta\delta^{\mathrm{T}}\mathbf{H}_{p_2+4}(\chi^2_{p_2};\Delta)$$

$$- \delta\delta^{\mathrm{T}}\mathbf{H}_{p_2+2}(\chi^2_{p_2};\Delta)$$

Putting all the terms together and simplifying, we obtain

$$\mathrm{Cov}(\hat{\boldsymbol{\beta}}_1^{\mathrm{RPT}})$$

$$=\mu_{11.2}\mu_{11.2}^{\mathrm{T}} + 2\mu_{11.2}^{\mathrm{T}}\delta\mathbf{H}_{p_2+2}(\chi^2_{p_2};\Delta) + \mathbf{B}_{11.2}^{-1} - \boldsymbol{\Phi}\mathbf{H}_{p_2+2}(\chi^2_{p_2};\Delta) - \delta\delta^{\mathrm{T}}\mathbf{H}_{p_2+4}(\chi^2_{p_2};\Delta)$$

$$+2\delta\delta^{\mathrm{T}}\mathbf{H}_{p_2+2}(\chi^2_{p_2};\Delta)$$

$$=\mathbf{B}_{11.2}^{-1} + \mu_{11.2}\mu_{11.2}^{\mathrm{T}} + 2\mu_{11.2}^{\mathrm{T}}\delta\mathbf{H}_{p_2+2}(\chi^2_{p_2};\Delta) - \boldsymbol{\Phi}\mathbf{H}_{p_2+2}(\chi^2_{p_2};\Delta)$$

$$+ \delta\delta^{\mathrm{T}}[2\mathbf{H}_{p_2+2}(\chi^2_{p_2};\Delta) - \mathbf{H}_{p_2+4}(\chi^2_{p_2};\Delta)].$$

The asymptotic covariance of the estimator $\hat{\boldsymbol{\beta}}_1^{\mathrm{RSE}}$ can be obtained as

$$\mathrm{Cov}(\hat{\boldsymbol{\beta}}_1^{\mathrm{RSE}}) = E\{\lim_{n\to\infty}\sqrt{n}(\hat{\boldsymbol{\beta}}_1^{\mathrm{RSE}} - \boldsymbol{\beta}_1)\sqrt{n}(\hat{\boldsymbol{\beta}}_1^{\mathrm{RSE}} - \boldsymbol{\beta}_1)^{\mathrm{T}}\}$$

$$= E\{\lim_{n\to\infty}n[(\hat{\boldsymbol{\beta}}_1^{\mathrm{RFM}} - \boldsymbol{\beta}_1) - (\hat{\boldsymbol{\beta}}_1^{\mathrm{RFM}} - \hat{\boldsymbol{\beta}}_1^{\mathrm{RSM}})(p_2 - 2)L_n^{-1}]$$

$$[(\hat{\boldsymbol{\beta}}_1^{\mathrm{RFM}} - \boldsymbol{\beta}_1) - (\hat{\boldsymbol{\beta}}_1^{\mathrm{RFM}} - \hat{\boldsymbol{\beta}}_1^{\mathrm{RSM}})(p_2 - 2)L_n^{-1}]^{\mathrm{T}}\}$$

$$= E\left\{[\varphi_1 - \varphi_3(p_2 - 2)L_n^{-1}][\varphi_1 - \varphi_3(p_2 - 2)L_n^{-1}]^{\mathrm{T}}\right\}$$

$$= E\left\{\varphi_1\varphi_1^{\mathrm{T}} - 2(p_2 - 2)\varphi_3\varphi_1^{\mathrm{T}}L_n^{-1} + (p_2 - 2)^2\varphi_3\varphi_3^{\mathrm{T}}L_n^{-2}\right\}$$

We need to compute $E\{\varphi_3\varphi_3^{\mathrm{T}}L_n^{-2}\}$ and $E\{\varphi_3\varphi_1^{\mathrm{T}}L_n^{-1}\}$. By using Lemma 1, the first term is obtained as follows:

$$E\{\varphi_3\varphi_3^{\mathrm{T}}L_n^{-2}\} = \boldsymbol{\Phi}E(\chi^{-4}_{p_2+2}(\Delta)) + \delta\delta^{\mathrm{T}}E(\chi^{-4}_{p_2+4}(\Delta)).$$

The second term is computed from normal theory

$$E\left\{\varphi_3\varphi_1^{\mathrm{T}}L_n^{-1}\right\} = E\left\{E(\varphi_3\varphi_1^{\mathrm{T}}L_n^{-1}|\varphi_3)\right\} = E\left\{\varphi_3 E(\varphi_1^{\mathrm{T}}L_n^{-1}|\varphi_3)\right\}$$

$$= E\left\{\varphi_3[-\mu_{11.2} + (\varphi_3 - \delta)]^{\mathrm{T}}L_n^{-1}\right\}$$

$$= -E\left\{\varphi_3\mu_{11.2}L_n^{-1}\right\} + E\left\{\varphi_3(\varphi_3 - \delta)^{\mathrm{T}}L_n^{-1}\right\}$$

$$= -\mu_{11.2}^{\mathrm{T}}E\left\{\varphi_3 L_n^{-1}\right\} + E\left\{\varphi_3\varphi_3^{\mathrm{T}}L_n^{-1}\right\} - E\left\{\varphi_3\delta^{\mathrm{T}}L_n^{-1}\right\}$$

From above, we can find $E\{\varphi_3\delta^{\mathrm{T}}L_n^{-1}\} = \delta\delta^{\mathrm{T}}E\big(\chi^{-2}_{p_2+2}(\Delta)\big)$ and $E\{\varphi_3 L_n^{-1}\} = \delta E\big(\chi^{-2}_{p_2+2}(\Delta)\big)$. Putting these terms together and simplifying, we obtain

$$\mathrm{Cov}(\hat{\beta}_1^{\mathrm{RSE}}) = \mathbf{B}_{11.2}^{-1} + \mu_{11.2}\mu_{11.2}^{\mathrm{T}} + 2(p_2 - 2)\mu_{11.2}^{\mathrm{T}}\delta E\big(\chi^{-2}_{p_2+2}(\Delta)\big)$$
$$- (p_2 - 2)\Phi\big\{2E\big(\chi^{-2}_{p_2+2}(\Delta)\big) - (p_2 - 2)E\big(\chi^{-4}_{p_2+2}(\Delta)\big)\big\}$$
$$+ (p_2 - 2)\delta\delta^{\mathrm{T}}\big\{ -2E\big(\chi^{-2}_{p_2+4}(\Delta)\big) + 2E\big(\chi^{-2}_{p_2+2}(\Delta)\big) + (p_2 - 2)E\big(\chi^{-4}_{p_2+4}(\Delta)\big)\big\}.$$

Since $\hat{\beta}_1^{\mathrm{RPS}} = \hat{\beta}_1^{\mathrm{RSE}} - (\hat{\beta}_1^{\mathrm{RFM}} - \hat{\beta}_1^{\mathrm{RSM}})\{1 - (p_2 - 2)L_n^{-1}\}I(L_n \leq p_2 - 2).$

We derive the covariance of the estimator $\hat{\beta}_1^{\mathrm{RPS}}$ as follows.

$$\mathrm{Cov}(\hat{\beta}_1^{\mathrm{RPS}}) = E\Big\{ \lim_{n\to\infty} \sqrt{n}(\hat{\beta}_1^{\mathrm{RPS}} - \beta_1)\sqrt{n}(\hat{\beta}_1^{\mathrm{RPS}} - \beta_1)^{\mathrm{T}}\Big\}$$

$$= E\Big\{ \lim_{n\to\infty} \sqrt{n}(\hat{\beta}_1^{\mathrm{RSE}} - \beta_1) - \sqrt{n}(\hat{\beta}_1^{\mathrm{RFM}} - \hat{\beta}_1^{\mathrm{RSM}})\{1 - (p_2 - 2)L_n^{-1}\}I(L_n \leq p_2 - 2)$$

$$\times \Big[\sqrt{n}(\hat{\beta}_1^{\mathrm{RSE}} - \beta_1) - \sqrt{n}(\hat{\beta}_1^{\mathrm{RFM}} - \hat{\beta}_1^{\mathrm{RSM}})\{1 - (p_2 - 2)L_n^{-1}\}I(L_n \leq p_2 - 2)\Big]^{T}\Big\}$$

$$= E\Big\{ \lim_{n\to\infty} \sqrt{n}(\hat{\beta}_1^{\mathrm{RSE}} - \beta_1)\sqrt{n}(\hat{\beta}_1^{\mathrm{RSE}} - \beta_1)^{\mathrm{T}} - 2\varphi_3\sqrt{n}(\hat{\beta}_1^{\mathrm{RSE}} - \beta_1)^{\mathrm{T}}\{1 - (p_2 - 2)L_n^{-1}\}I(L_n \leq p_2 - 2)$$

$$+ \varphi_3\varphi_3^{\mathrm{T}}\{1 - (p_2 - 2)L_n^{-1}\}^2 I(L_n \leq p_2 - 2)\Big\}$$

$$= \mathrm{Cov}(\hat{\beta}_1^{\mathrm{RSE}}) - 2E\Big\{ \lim_{n\to\infty} \varphi_3\sqrt{n}(\hat{\beta}_1^{\mathrm{RSE}} - \beta_1)^{\mathrm{T}}\{1 - (p_2 - 2)L_n^{-1}\}^2 I(L_n \leq p_2 - 2)\Big\}$$

$$+ E\Big\{ \lim_{n\to\infty} \varphi_3\varphi_3^{\mathrm{T}}\{1 - (p_2 - 2)L_n^{-1}\}^2 I(L_n \leq p_2 - 2)\Big\}$$

$$= \mathrm{Cov}(\hat{\beta}_1^{\mathrm{RSE}}) - 2E\Big\{ \lim_{n\to\infty} \varphi_3\varphi_1^{\mathrm{T}}\{1 - (p_2 - 2)L_n^{-1}\}I(L_n \leq p_2 - 2)\Big\}$$

$$+ 2E\Big\{ \lim_{n\to\infty} \varphi_3\varphi_3^{\mathrm{T}}(p_2 - 2)L_n^{-1}\{1 - (p_2 - 2)L_n^{-1}\}I(L_n \leq p_2 - 2)\Big\}$$

$$+ E\Big\{ \lim_{n\to\infty} \varphi_3\varphi_3^{\mathrm{T}}\{1 - (p_2 - 2)L_n^{-1}\}^2 I(L_n \leq p_2 - 2)\Big\}$$

$$= \mathrm{Cov}(\hat{\beta}_1^{\mathrm{RSE}}) - 2E\Big\{ \lim_{n\to\infty} \varphi_3\varphi_1^{\mathrm{T}}\{1 - (p_2 - 2)L_n^{-1}\}I(L_n \leq p_2 - 2)\Big\}$$

$$- E\Big\{ \lim_{n\to\infty} \varphi_3\varphi_3^{\mathrm{T}}(p_2 - 2)^2 L_n^{-2}I(L_n \leq p_2 - 2)\Big\} + E\Big\{ \lim_{n\to\infty} \varphi_3\varphi_3^{\mathrm{T}}I(L_n \leq p_2 - 2)\Big\}$$

We first compute the last term in the equation above $E\big\{\varphi_3\varphi_3^{\mathrm{T}}I(L_n \leq p_2 - 2)\big\}$ as $E\big\{\varphi_3\varphi_3^{\mathrm{T}}I(L_n \leq p_2 - 2)\big\} = \Phi H_{p_2+2}(p_2 - 2; \Delta) + \delta\delta^{\mathrm{T}}H_{p_2+4}(p_2 - 2; \Delta)$. Using Lemma 1 and from the normal theory, we find,

$$E\left\{\boldsymbol{\varphi}_3\boldsymbol{\varphi}_1^{\mathrm{T}}\{1-(p_2-2)\mathrm{L}_n^{-1}\}\mathrm{I}(\mathrm{L}_n \le p_2-2)\right\}$$

$$= E\left\{E(\boldsymbol{\varphi}_3\boldsymbol{\varphi}_1^{\mathrm{T}}\{1-(p_2-2)\mathrm{L}_n^{-1}\}\mathrm{I}(\mathrm{L}_n \le p_2-2)|\boldsymbol{\varphi}_3)\right\}$$

$$= E\left\{\boldsymbol{\varphi}_3 E(\boldsymbol{\varphi}_1^{\mathrm{T}}\{1-(p_2-2)\mathrm{L}_n^{-1}\}\mathrm{I}(\mathrm{L}_n \le p_2-2)|\boldsymbol{\varphi}_3)\right\}$$

$$= E\left\{\boldsymbol{\varphi}_3[\boldsymbol{\mu}_{11.2}+(\boldsymbol{\varphi}_3-\boldsymbol{\delta})]^{\mathrm{T}}\{1-(p_2-2)\mathrm{L}_n^{-1}\}\mathrm{I}(\mathrm{L}_n \le p_2-2)\right\}$$

$$= -\boldsymbol{\mu}_{11.2}E\left(\boldsymbol{\varphi}_3\{1-(p_2-2)\mathrm{L}_n^{-1}\}\mathrm{I}(\mathrm{L}_n \le p_2-2)\right)$$

$$+ E\left(\boldsymbol{\varphi}_3\boldsymbol{\varphi}_3^{\mathrm{T}}\{1-(p_2-2)\mathrm{L}_n^{-1}\}\mathrm{I}(\mathrm{L}_n \le p_2-2)\right)$$

$$- E\left(\boldsymbol{\varphi}_3\boldsymbol{\delta}^{\mathrm{T}}\{1-(p_2-2)\mathrm{L}_n^{-1}\}\mathrm{I}(\mathrm{L}_n \le p_2-2)\right)$$

$$= -\boldsymbol{\delta}\boldsymbol{\mu}_{11.2}^{\mathrm{T}}E\left(\{1-(p_2-2)\chi_{p_2+2}^{-2}(\Delta)\}\mathrm{I}\left(\chi_{p_2+2}^{-2}(\Delta) \le p_2-2\right)\right)$$

$$+ \boldsymbol{\Phi}E\left(\{1-(p_2-2)\chi_{p_2+2}^{-2}(\Delta)\}\mathrm{I}\left(\chi_{p_2+2}^{-2}(\Delta) \le p_2-2\right)\right)$$

$$+ \boldsymbol{\delta}\boldsymbol{\delta}^{\mathrm{T}}E\left(\{1-(p_2-2)\chi_{p_2+4}^{-2}(\Delta)\}\mathrm{I}\left(\chi_{p_2+4}^{-2}(\Delta) \le p_2-2\right)\right)$$

$$- \boldsymbol{\delta}\boldsymbol{\delta}^{\mathrm{T}}E\left(\{1-(p_2-2)\chi_{p_2+4}^{-2}(\Delta)\}\mathrm{I}\left(\chi_{p_2+4}^{-2}(\Delta) \le p_2-2\right)\right).$$

$$E\left\{\boldsymbol{\varphi}_3\boldsymbol{\varphi}_3^{\mathrm{T}}(p_2-2)^2\mathrm{L}_n^{-2}\mathrm{I}(\mathrm{L}_n \le p_2-2)\right\} = (p_2-2)^2\boldsymbol{\Phi}E\left(\chi_{p_2+2}^{-4}(\Delta)\mathrm{I}\left(\chi_{p_2+2}^{2}(\Delta) \le p_2-2\right)\right)$$

$$+ (p_2-2)^2\boldsymbol{\delta}\boldsymbol{\delta}^{\mathrm{T}}E\left(\chi_{p_2+2}^{-4}(\Delta)\mathrm{I}\left(\chi_{p_2+2}^{2}(\Delta) \le p_2-2\right)\right)$$

Putting all the terms together, we obtain

$$\mathrm{Cov}(\hat{\boldsymbol{\beta}}_1^{\mathrm{RPS}}) = \mathrm{Cov}(\hat{\boldsymbol{\beta}}_1^{\mathrm{RSE}}) + 2\boldsymbol{\delta}\boldsymbol{\mu}_{11.2}^{\mathrm{T}}E\left(\{1-(p_2-2)\chi_{p_2+2}^{-2}(\Delta)\}\mathrm{I}\left(\chi_{p_2+2}^{2}(\Delta) \le p_2-2\right)\right)$$

$$- 2\boldsymbol{\Phi}E\left(\{1-(p_2-2)\chi_{p_2+2}^{-2}(\Delta)\}\mathrm{I}\left(\chi_{p_2+2}^{2}(\Delta) \le p_2-2\right)\right)$$

$$- 2\boldsymbol{\delta}\boldsymbol{\delta}^{\mathrm{T}}E(\{1-(p_2-2)\chi_{p_2+4}^{-2}(\Delta)\}\mathrm{I}(\chi_{p_2+4}^{2}(\Delta) \le p_2-2))$$

$$+ 2\boldsymbol{\delta}\boldsymbol{\delta}^{\mathrm{T}}E\left(\{1-(p_2-2)\chi_{p_2+2}^{-2}(\Delta)\}\mathrm{I}\left(\chi_{p_2+2}^{2}(\Delta) \le p_2-2\right)\right)$$

$$- (p_2-2)^2\boldsymbol{\Phi}E\left(\chi_{p_2+2}^{-4}(\Delta)\mathrm{I}\left(\chi_{p_2+2,\alpha}^{2}(\Delta) \le p_2-2\right)\right)$$

$$- (p_2-2)^2\boldsymbol{\delta}\boldsymbol{\delta}^{\mathrm{T}}E\left(\chi_{p_2+2}^{-4}(\Delta)\mathrm{I}\left(\chi_{p_2+2}^{2}(\Delta) \le p_2-2\right)\right)$$

$$+ \boldsymbol{\Phi}\mathrm{H}_{p_2+2}(p_2-2;\Delta) + \boldsymbol{\delta}\boldsymbol{\delta}^{\mathrm{T}}\mathrm{H}_{p_2+4}(p_2-2;\Delta).$$

$\square$

References

1. Laird, N.M.; Ware, J.H. Random-effects models for longitudinal data. *Biometrics* **1982**, *38*, 963–974. [CrossRef]
2. Longford, N. Regression analysis of multilevel data with measurement error. *Br. J. Math. Stat. Psychol.* **1993**, *46*, 301–311. [CrossRef]
3. Tibshirani, R. Regression shrinkage and selection via the lasso. *J. R. Stat. Soc. Ser. B (Methodol.)* **1996**, *58*, 267–288. [CrossRef]

4. Zou, H. The adaptive lasso and its oracle properties. *J. Am. Stat. Assoc.* **2006**, *101*, 1418–1429. [CrossRef]
5. Tran, M.N. The loss rank criterion for variable selection in linear regression analysis. *Scand. J. Stat.* **2011**, *38*, 466–479. [CrossRef]
6. Huang, J.; Ma, S.; Zhang, C.H. Adaptive Lasso for sparse high-dimensional regression models. *Stat. Sin.* **2008**, *18*, 1603–1618.
7. Kim, Y.; Choi, H.; Oh, H.S. Smoothly clipped absolute deviation on high dimensions. *J. Am. Stat. Assoc.* **2008**, *103*, 1665–1673. [CrossRef]
8. Wang, H.; Leng, C. Unified LASSO estimation by least squares approximation. *J. Am. Stat. Assoc.* **2007**, *102*, 1039–1048. [CrossRef]
9. Yuan, M.; Lin, Y. Model selection and estimation in regression with grouped variables. *J. R. Stat. Soc. Ser. B (Stat. Methodol.* **2006**, *68*, 49–67. [CrossRef]
10. Leng, C.; Lin, Y.; Wahba, G. A note on the lasso and related procedures in model selection. *Stat. Sin.* **2006**, *16*, 1273–1284.
11. Park, T.; Casella, G. The bayesian lasso. *J. Am. Stat. Assoc.* **2008**, *103*, 681–686. [CrossRef]
12. Greenlaw, K.; Szefer, E.; Graham, J.; Lesperance, M.; Nathoo, F.S.; Initiative, A.D.N. A Bayesian group sparse multi-task regression model for imaging genetics. *Bioinformatics* **2017**, *33*, 2513–2522. [CrossRef] [PubMed]
13. Ahmed, S.E.; Nicol, C.J. An application of shrinkage estimation to the nonlinear regression model. *Comput. Stat. Data Anal.* **2012**, *56*, 3309–3321. [CrossRef]
14. Ahmed, S.E.; Raheem, S.E. Shrinkage and absolute penalty estimation in linear regression models. *Wiley Interdiscip. Rev. Comput. Stat.* **2012**, *4*, 541–553. [CrossRef]
15. Lisawadi, S.; Kashif Ali Shah, M.; Ejaz Ahmed, S. Model selection and post estimation based on a pretest for logistic regression models. *J. Stat. Comput. Simul.* **2016**, *86*, 3495–3511. [CrossRef]
16. Ahmed, S.E.; Opoku, E.A. Submodel selection and post-estimation of the linear mixed models. In *Proceedings of the Tenth International Conference on Management Science and Engineering Management*; Springer: Berlin/Heidelberg, Germany, 2017; pp. 633–646.
17. Raheem, S.E.; Ahmed, S.E.; Doksum, K.A. Absolute penalty and shrinkage estimation in partially linear models. *Comput. Stat. Data Anal.* **2012**, *56*, 874–891. [CrossRef]
18. Geladi, P.; Kowalski, B.R. Partial least-squares regression: A tutorial. *Anal. Chim. Acta* **1986**, *185*, 1–17. [CrossRef]
19. Liu, K. Using Liu-type estimator to combat collinearity. *Commun. Stat.-Theory Methods* **2003**, *32*, 1009–1020. [CrossRef]
20. Hoerl, A.E.; Kennard, R.W. Ridge regression: Biased estimation for nonorthogonal problems. *Technometrics* **1970**, *12*, 55–67. [CrossRef]
21. Yüzbaşı, B.; Ejaz Ahmed, S. Shrinkage and penalized estimation in semi-parametric models with multicollinear data. *J. Stat. Comput. Simul.* **2016**, *86*, 3543–3561. [CrossRef]
22. Yüzbası, B.; Ahmed, S.E.; Güngör, M. Improved penalty strategies in linear regression models. *REVSTAT J.* **2017**, *15*, 251–276.
23. Knight, K.; Fu, W. Asymptotics for lasso-type estimators. *Ann. Stat.* **2000**, *28*, 1356–1378.
24. Belsley, D.A. *Conditioning Diagnostics: Collinearity and Weak Data in Regression*; Number 519.536 B452; Wiley: Hoboken, NJ, USA, 1991.
25. Twisk, J.; Kemper, H.; Mellenbergh, G. Longitudinal development of lipoprotein levels in males and females aged 12–28 years: The Amsterdam Growth and Health Study. *Int. J. Epidemiol.* **1995**, *24*, 69–77. [CrossRef] [PubMed]
26. Nie, Y.; Opoku, E.; Yasmin, L.; Song, Y.; Wang, J.; Wu, S.; Scarapicchia, V.; Gawryluk, J.; Wang, L.; Cao, J.; et al. Spectral dynamic causal modelling of resting-state fMRI: An exploratory study relating effective brain connectivity in the default mode network to genetics. *Stat. Appl. Genet. Mol. Biol.* **2020**, *19*. [CrossRef]
27. Ahmed, S.E.; Kim, H.; Yıldırım, G.; Yüzbaşı, B. High-Dimensional Regression Under Correlated Design: An Extensive Simulation Study. In *International Workshop on Matrices and Statistics*; Springer: Berlin/Heidelberg, Germany, 2016; pp. 145–175.
28. Ejaz Ahmed, S.; Yüzbaşı, B. Big data analytics: Integrating penalty strategies. *Int. J. Manag. Sci. Eng. Manag.* **2016**, *11*, 105–115. [CrossRef]
29. Ahmed, S.E.; Yüzbaşı, B. High dimensional data analysis: Integrating submodels. In *Big and Complex Data Analysis*; Springer: Berlin/Heidelberg, Germany, 2017; pp. 285–304.
30. Judge, G.G.; Bock, M.E. *The Statistical Implication of Pre-Test and Steinrule Estimators in Econometrics*; Elsevier: Amsterdam, The Netherlands, 1978.

Article

Edge-Preserving Denoising of Image Sequences

Fan Yi * and Peihua Qiu

Department of Biostatistics, University of Florida, Gainesville, FL 32603, USA; pqiu@ufl.edu
* Correspondence: yifan@ufl.edu; Tel.: +1-352-745-4977

Abstract: To monitor the Earth's surface, the satellite of the NASA Landsat program provides us image sequences of any region on the Earth constantly over time. These image sequences give us a unique resource to study the Earth's surface, changes of the Earth resource over time, and their implications in agriculture, geology, forestry, and more. Besides natural sciences, image sequences are also commonly used in functional magnetic resonance imaging (fMRI) of medical studies for understanding the functioning of brains and other organs. In practice, observed images almost always contain noise and other contaminations. For a reliable subsequent image analysis, it is important to remove such contaminations in advance. This paper focuses on image sequence denoising, which has not been well-discussed in the literature yet. To this end, an edge-preserving image denoising procedure is suggested. The suggested method is based on a jump-preserving local smoothing procedure, in which the bandwidths are chosen such that the possible spatio-temporal correlations in the observed image intensities are accommodated properly. Both theoretical arguments and numerical studies show that this method works well in the various cases considered.

Keywords: bandwidth selection; correlation; edge-preserving image denoising; image sequence; jump regression analysis; local smoothing; nonparametric regression; spatio-temporal data

Citation: Yi, F.; Qiu, P. Edge-Preserving Denoising of Image Sequences. *Entropy* **2021**, *23*, 1332. https://doi.org/10.3390/e23101332

Academic Editors: Amelia Carolina Sparavigna, Farouk Nathoo and S. Ejaz Ahmed

Received: 2 September 2021
Accepted: 7 October 2021
Published: 12 October 2021

Publisher's Note: MDPI stays neutral with regard to jurisdictional claims in published maps and institutional affiliations.

1. Introduction

The Landsat project, led by the US Geological Survey (USGS) and NASA, has launched eight satellites since 1972 to continuously provide scientifically valuable images of the Earth's surface. These images can be freely accessed by researchers around the world (cf., Zanter [1]). This rich archive of Landsat images has become a major resource for scientific research about the Earth's surface and its resources in different scientific disciplines, including forest science, climate science, agriculture, ecology, fire science, and many more. As an example, Figure 1 shows two images of the Las Vegas area in Nevada taken in 1984 and 2007, respectively. These two images clearly show the increasing urban sprawl of Las Vegas during the 23-year period, and consequently, the environment in that region has changed dramatically. The current satellite (i.e., the Landsat 8) can deliver an image of a given region roughly every 16 days. So, we have a sequence of images of that region collected sequentially over time, stored in the Landsat database, which is increasing all the time. Image sequences are commonly used in many other applications, including functional magnetic resonance imaging (fMRI) in neuroscience and quality control in manufacturing industries (Qiu [2]). In practice, observed images usually contain noise and other contaminations (Gonzalez and Woods [3]). For reliable subsequent image analyses, such contaminations should be removed in advance. In the image processing literature, the removal of noise from an observed image is referred to as image denoising. This paper focuses on image denoising for analyzing observed image sequences.

In the literature, there has been extensive discussion on image denoising (Qiu [4]). Many early methods in the computer science literature are based on the Markov random field (MRF) framework, in which observed image intensities of an image are assumed to have the Markov property that the observed intensity at a given pixel depends only on the observed intensities in a neighborhood of the given pixel (Geman and Geman [5]).

Then, if the true image is assumed to have a prior distribution which is also an MRF, its posterior distribution would be an MRF too, and consequently, the true image can be estimated by the maximum a posteriori (MAP) estimator (e.g., Geman and Geman [5], Besag [6], Fessler et al. [7]). Other popular image denoising methods include those based on diffusion equations (e.g., Perona and Malik [8], Weickert [9]), total variation (Beck and Teboulle [10], Rudin et al. [11], Yuan et al. [12]), wavelet transformations (e.g., Chang et al. [13], Mrázek [14]), jump regression analysis (e.g., Gijbels et al. [15], Qiu [16], Qiu [17], Qiu and Mukherjee [18]), adaptive weights smoothing (e.g., Polzehl and Spokoiny [19]), spatial adaption (e.g., Kervrann and Boulanger [20]) and more. Besides noise removal, edge-preserving is important for image denoising because edges are important structures of the images. Some of the methods mentioned above can preserve edges well, such as the ones based on jump regression analysis, total variation, and wavelet transformations. Thorough surveys of popular edge-preserving image denoising methods can be found in Jain and Tyagi [21] and Qiu [4].

Figure 1. Two Landsat images of the Las Vegas area taken in 1984 (**left panel**) and 2007 (**right panel**).

Although there are already some existing methods for edge-preserveing image denoising, almost all of them handle observed images taken at a single time point. So far, we have not found much discussion about denoising image sequences, which is the focus of the current paper. A given image sequence often describes a gradual change in appearance over time, subject to the underlying process. For instance, the sequence of images of the Las Vegas area acquired by the Landsat satellite (cf., Figure 1) describes the gradual change of the Earth's surface in that area over time. As mentioned above, two consecutive images in the sequence acquired by the current Landsat satellite are only about 16 days apart. So, their difference should be very small. However, the images could be substantially different after a long period of time, as shown in Figure 1. In such applications, it should be reasonable to assume that edge locations in different images either do not change or change gradually over time. To handle such image sequences, the neighboring images should be useful when denoising the image at a given time point, or information in neighboring images should be shared during image denoising. By noticing such features of image sequences, we propose an edge-preserving image denoising procedure for analyzing image sequences in this paper. Our proposed method is based on the jump regression analysis (JRA) used for regression modeling when the underlying regression function has jumps or other singularities (Qiu [22]). It is a local smoothing procedure, and the possible spatio-temporal correlation in the observed image data has been accommodated properly in its construction. Both theoretical arguments and numerical studies show that this method works well in various different cases.

The remaining parts of the article are organized as follows. The proposed method is described in detail in Section 2. Its statistical properties and the numerical studies about its performance in different finite-sample cases are presented in Section 3. Several concluding remarks are provided in Section 4. Some technical details are given in Appendix A.

2. Materials and Methods

This section describes our proposed method in two parts. A JRA model for describing an image sequence and the model estimation are discussed in Section 2.1. Selection of several parameters used in model estimation is discussed in Section 2.2.

2.1. JRA Model and Its Estimation

To describe an image sequence, let us consider the following JRA model:

$$Z_{ijk} = f(x_i, y_j; t_k) + \varepsilon_{ijk}, \quad i = 1, 2, \ldots, n_x, j = 1, 2, \ldots, n_y, k = 1, 2, \ldots, n_t, \tag{1}$$

where Z_{ijk} is the observed image intensity level at the (i, j)-th pixel (x_i, y_j) and at the k-th time point t_k, $f(x_i, y_j; t_k)$ is the true image intensity level, and ε_{ijk} is the pointwise random noise with mean 0 and variance σ^2. In model (1), spatio-temporal data correlation is allowed, namely, $\{\varepsilon_{ijk}\}$ could be correlated over i, j and k. For image data, the pixel locations are usually regularly spaced. Without loss of generality, it is assumed that they are equally spaced in the design space $\Omega = [0, 1] \times [0, 1]$, namely, $(x_i, y_j) = (i/n_x, j/n_y)$, for all i and j, where n_x and n_y are the numbers of rows and columns, respectively. The observation times $\{t_k, k = 1, 2, \ldots, n_t\}$ are also assumed to be equally spaced in the time interval $[0, 1]$. The true image intensity function $f(x, y; t)$, for $(x, y) \in \Omega$, is continuous in the design space Ω at each $t \in [0, 1]$, except on the edges where it has jumps.

To estimate the unknown image intensity function $f(x, y; t)$ in model (1), we consider using a local smoothing method, instead of a global smoothing method (e.g., smoothing spline method), because of a large amount of data involved in the current problem. Likewise, it has been well-discussed in the JRA literature that conventional smoothing methods (e.g., conventional local kernel smoothing methods) would not work well for estimating models like (1) where the true image intensity function $f(x, y; t)$ has jumps at the edges, because the jumps would be blurred by such conventional methods (cf., Qiu [22]). In this paper, we suggest a jump-preserving local smoothing method for estimating (1), described in detail below. For a given point $(x, y; t) \in \Omega \times [0, 1]$, define a local neighborhood

$$O(x, y; t) = \{\left(x', y'; t'\right) : \left(x', y'; t'\right) \in \Omega \times [0, 1],$$

$$\sqrt{\frac{(x' - x)^2}{h_x^2} + \frac{(y' - y)^2}{h_y^2}} \leq 1, |t' - t|/h_t \leq 1\},$$

where h_x, h_y and h_t are the bandwidths in the $x-$, $y-$, and $t-$axis, respectively. In $O(x, y; t)$, we first consider the following local linear kernel (LLK) smoothing procedure (Fan and Gijbels [23]):

$$\min_{a,b,c,d} \sum_{i=1}^{n_x} \sum_{j=1}^{n_y} \sum_{k=1}^{n_t} \left\{Z_{ijk} - \left[a + b(x_i - x) + c(y_j - y) + d(t_k - t)\right]\right\}^2$$

$$K\left(\frac{x_i - x}{h_x}, \frac{y_j - y}{h_y}\right) K\left(\frac{t_k - t}{h_t}\right), \tag{2}$$

where $K(v)$ is a density kernel function with the support $\{v : |v| \leq 1\}$. The solutions to (a, b, c, d) of the minimization problem (2) are denoted as $\widehat{a}(x, y; t)$, $\widehat{b}(x, y; t)$, $\widehat{c}(x, y; t)$, and $\widehat{d}(x, y; t)$, respectively. It can be checked that they have the following expressions:

$$\begin{bmatrix} \widehat{a}(x, y; t) \\ \widehat{b}(x, y; t) \\ \widehat{c}(x, y; t) \\ \widehat{d}(x, y; t) \end{bmatrix} = \begin{bmatrix} m_{000} & m_{100} & m_{010} & m_{001} \\ m_{100} & m_{200} & m_{110} & m_{101} \\ m_{010} & m_{110} & m_{020} & m_{011} \\ m_{001} & m_{101} & m_{011} & m_{002} \end{bmatrix}^{-1} \begin{bmatrix} \sum_{ijk} Z_{ijk} K_{ijk} \\ \sum_{ijk} (x_i - x) Z_{ijk} K_{ijk} \\ \sum_{ijk} (y_j - y) Z_{ijk} K_{ijk} \\ \sum_{ijk} (t_k - t) Z_{ijk} K_{ijk} \end{bmatrix}, \tag{3}$$

where $\sum_{ijk}$ denotes $\sum_{i=1}^{n_x} \sum_{j=1}^{n_y} \sum_{k=1}^{n_t}$, K_{ijk} denotes $K(\frac{x_i-x}{h_x}, \frac{y_j-y}{h_y})K(\frac{t_k-t}{h_t})$, and $m_{rsl} = \sum_{ijk}(x_i - x)^r(y_j - y)^s(t_k - t)^l K_{ijk}$, for $r, s, l = 0, 1, 2$. The LLK estimator of $f(x,y;t)$ is defined to be $\hat{a}(x,y;t)$. The estimated gradient direction of $f(x,y;t)$ at $(x,y;t)$ is $\hat{G}(x,y;t) = (\hat{b}(x,y;t), \hat{c}(x,y;t), \hat{d}(x,y;t))'$ which indicates the direction in which the estimated plane in $O(x,y;t)$ by the LLK procedure (2) increases the fastest. If there is an edge surface in $O(x,y;t)$, then $\hat{G}(x,y;t)$ would be (approximately) orthogonal to that surface.

In cases when there are no edges in the neighborhood $O(x,y;t)$, $\hat{a}(x,y;t)$ would be a good estimate of $f(x,y;t)$. Otherwise, it cannot be a good estimate because $\hat{a}(x,y;t)$ is a weighted average of all observed image intensities in $O(x,y;t)$, the jumps in the image intensity surface would be smoothed out in the weighted average, and the estimate $\hat{a}(x,y;t)$ would be biased for estimating $f(x,y;t)$. To overcome that limitation, we consider the following one-sided smoothing idea. Let $O(x,y;t)$ be divided into two parts $O^{(1)}(x,y;t)$ and $O^{(2)}(x,y;t)$ by a plane that passes $(x,y;t)$ and is perpendicular to $\hat{G}(x,y;t)$. See Figure 2 for an example.

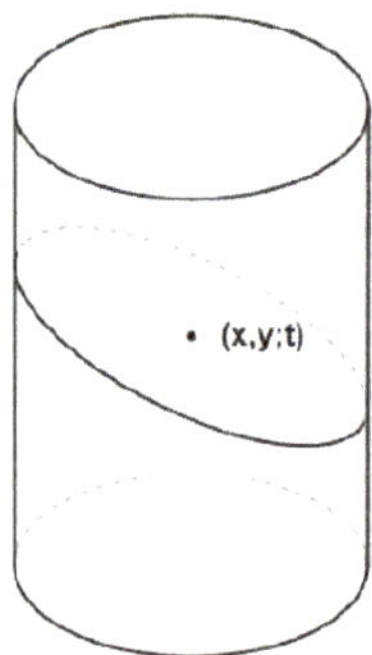

Figure 2. The neighborhood $O(x,y;t)$ is divided into two parts by a plane that passes $(x,y;t)$ and is perpendicular to the estimated gradient direction $\hat{G}(x,y;t)$.

Then, in cases when there is an edge surface in $O(x,y;t)$, that plane would be (approximately) parallel to the edge surface. Consequently, at least one of $O^{(1)}(x,y;t)$ and $O^{(2)}(x,y;t)$ would be (mostly) located on a single side of the edge surface in such cases. Now, let us consider the following one-sided LLK smoothing procedure: for $l = 1, 2$,

$$
\min_{a,b,c,d} \sum_{(x_i,y_j;t_k)\in O^{(l)}(x,y;t)} \left\{ Z_{ijk} - [a + b(x_i - x) + c(y_j - y) + d(t_k - t)] \right\}^2
$$
$$
K\left(\frac{x_i - x}{h_x}, \frac{y_j - y}{h_y}\right) K\left(\frac{t_k - t}{h_t}\right).
\tag{4}
$$

The solutions of (4) to (a, b, c, d) are denoted as $(\hat{a}^{(l)}(x,y;t), \hat{b}^{(l)}(x,y;t), \hat{c}^{(l)}(x,y;t), \hat{d}^{(l)}(x,y;t))$, for $l = 1, 2$. Intuitively, when there are no edges in $O(x,y;t)$, $\hat{a}(x,y;t)$, $\hat{a}^{(1)}(x,y;t)$ and $\hat{a}^{(2)}(x,y;t)$ are all consistent estimates of $f(x,y;t)$ under some regular conditions. In such cases, $\hat{a}(x,y;t)$ would be preferred since it averages more observations and consequently it would have a smaller variance. When there are edges in $O(x,y;t)$, $\hat{a}(x,y;t)$ would not be a good estimate of $f(x,y;t)$ as explained above, but one of $\hat{a}^{(1)}(x,y;t)$ and $\hat{a}^{(2)}(x,y;t)$ should estimate $f(x,y;t)$ well. Therefore, in all cases, at least one of the three estimators $\hat{a}(x,y;t)$, $\hat{a}^{(1)}(x,y;t)$ and $\hat{a}^{(2)}(x,y;t)$ should estimate $f(x,y;t)$ well.

Next, we need to choose a good estimator from $\hat{a}(x,y;t)$, $\hat{a}^{(1)}(x,y;t)$ and $\hat{a}^{(2)}(x,y;t)$ based on the observed data, which is not straightforward, partly because we do not know in advance whether there are edges in the neighborhood $O(x,y;t)$ and whether the edges are mostly contained in $O^{(1)}(x,y;t)$ or $O^{(2)}(x,y;t)$ if the answer to the first question is positive. To overcome this difficulty, let us consider the following weighted residual mean squares (WRMS) of the fitted local plane by the LLK procedure (2):

$$e(x,y;t) = \left\{ \sum_{ijk} [Z_{ijk} - \widehat{a}(x,y;t) - \widehat{b}(x,y;t)(x_i - x) - \widehat{c}(x,y;t)(y_j - y) - \widehat{d}(x,y;t)(t_k - t)]^2 K_{ijk} \right\} / \sum_{ijk} K_{ijk}. \tag{5}$$

The above WRMS measures how well the fitted local plane describes the observed data in $O(x,y;t)$. If there are edges in $O(x,y;t)$, this quantity would be relatively large, due mainly to the jumps in the image intensity surface. Otherwise, it would be relatively small. So, the quantity $e(x,y;t)$ contains useful information about the existence of edges in $O(x,y;t)$. Similarly, we can define WRMS values for the two one-sided local planes fitted in $O^{(1)}(x,y;t)$ and $O^{(2)}(x,y;t)$. They are denoted as $e^{(1)}(x,y;t)$ and $e^{(2)}(x,y;t)$. Based on these WRMS values, we define our edge-preserving estimator of $f(x,y;t)$ to be

$$\begin{aligned}
\widehat{f}(x,y;t) = {} & \widehat{a}(x,y;t)I(D(x,y;t) \leq u) \\
& + \widehat{a}^{(1)}(x,y;t)I(D(x,y;t) > u)I(e^{(1)}(x,y;t) < e^{(2)}(x,y;t)) \\
& + \widehat{a}^{(2)}(x,y;t)I(D(x,y;t) > u)I(e^{(1)}(x,y;t) > e^{(2)}(x,y;t)) \\
& + \frac{\widehat{a}^{(1)}(x,y;t) + \widehat{a}^{(2)}(x,y;t)}{2} I(D(x,y;t) > u)I(e^{(1)}(x,y;t) = e^{(2)}(x,y;t)),
\end{aligned} \tag{6}$$

where $D(x,y;t) = \max(e(x,y;t) - e^{(1)}(x,y;t), e(x,y;t) - e^{(2)}(x,y;t))$, $I(\cdot)$ is the indicator function, and $u > 0$ is a threshold parameter. By (6), it is obvious that $\widehat{f}(x,y;t)$ is defined to be one of $\widehat{a}(x,y;t)$, $\widehat{a}^{(1)}(x,y;t)$ and $\widehat{a}^{(2)}(x,y;t)$. The quantity $\widehat{a}(x,y;t)$, which is obtained from the entire neighborhood $O(x,y;t)$, is chosen if the observed data indicate no edges in $O(x,y;t)$, supported by the event $D(x,y;t) \leq u$. Otherwise, one of the two one-sided quantities, $\widehat{a}^{(1)}(x,y;t)$ and $\widehat{a}^{(2)}(x,y;t)$, with a smaller WRMS value is chosen. Although, theoretically, the event $(e^{(1)}(x,y;t) = e^{(2)}(x,y;t))$ would have probability zero of happening, the last term on the right-hand-side of (6) is still included for completeness of the definition of $\widehat{f}(x,y;t)$ and for the consideration that $e^{(1)}(x,y;t)$ and $e^{(2)}(x,y;t)$ could be considered the same in certain algorithms when their values are close.

2.2. Parameter Selection

In our proposed method described in Section 2.1, there are four parameters; h_x, h_y, h_t and u, that need to be chosen properly in advance. For that purpose, it is natural to consider the cross validation (CV) procedure, especially in the current research problem where the observed data are quite large in size. However, it has been well-demonstrated in the literature that the conventional CV procedure would not work well in cases when the observed data are autocorrelated, because it cannot effectively distinguish the data correlation structure from the mean structure (cf., Altman [24], Opsomer et al. [25]). In the current problem, spatio-temperal data correlation is possible in almost all applications. Thus, the conventional CV procedure is not feasible in such cases. In the univariate regression setup, Brabanter et al. [26] suggested a modified CV procedure for choosing smoothing parameters in cases with correlated data. This procedure is generalized here for choosing the parameters h_x, h_y, h_t and u used in the proposed method, which is described below. Let the modified CV score for choosing h_x, h_y, h_t and u be defined as

$$CV(h_x, h_y, h_t, u) = \frac{1}{n_x n_y n_t} \sum_{ijk} \left[\widehat{f}_{-(ijk)}(x_i, y_j; t_k) - Z(x_i, y_j; t_k) \right]^2, \tag{7}$$

where $\hat{f}_{-(ijk)}(x_i, y_j; t_k)$ is the leave-one-out estimate of $f(x_i, y_j; t_k)$ by (2)–(6) after the observation Z_{ijk} is removed from the estimation process and after the kernel function is replaced by the so-called ϵ-optimal bimodal kernel function $K_\epsilon(v)$ defined to be

$$K_\epsilon(v) = \frac{4}{4 - 3\epsilon - \epsilon^3} \times \begin{cases} \frac{3}{4}(1 - v^2)I(|v| \le 1), & \text{if } |v| \ge \epsilon, \\ \frac{3(1-\epsilon^2)}{4\epsilon}|v|, & \text{if } |v| < \epsilon, \end{cases} \tag{8}$$

where $0 < \epsilon < 1$ is a parameter. Based on a large simulation study, Brabanter et al. [26] suggested choosing ϵ to be 0.1, which is adopted in this paper. Then, by the above modified CV procedure, (7) and (8), the parameters h_x, h_y, h_t and u can be chosen by minimizing the modified CV score $CV(h_x, h_y, h_t, u)$.

3. Results

3.1. Statistical Properties

In this part, we discuss some statistical properties of the proposed edge-preserving image sequence denoising method (2)–(6). First, we have the following proposition.

Proposition 1. *Assume that i) the kernel function $K(v)$ used in (2) is a Lipschitz-1 continuous density function, and ii) the noise terms $\{\varepsilon_{ijk}, i = 1, 2, \ldots, n_x, j = 1, 2, \ldots, n_y, k = 1, 2, \ldots, n_t\}$ in model (1) form a strong mixing stochastic process with the following strong mixing coefficients:*

$$\alpha(d) = \sup_{(ijk),(i'j'k')} \sup_{A,B} \Big\{ |P(A \cap B) - P(A)P(B)|, A \in \sigma(\varepsilon_{ijk}), B \in \sigma(\varepsilon_{i'j'k'}),$$

$$\max\{|i - i'|, |j - j'|, |k - k'|\} > d \Big\},$$

which have the property that $\alpha(d) \le c_1 \sigma^2 \rho^{c_2 d}$, where $c_1, c_2 > 0$ and $0 < \rho < 1$ are constants, and iii) $E(\varepsilon_{111}^6) < \infty$. Let $N = n_x n_y n_t$, $H = h_x h_y h_t$, $n_{min} = \min(n_x, n_y, n_t)$, and $h_{min} = \min(h_x, h_y, h_t)$. Then, for any $(x, y; t) \in \Omega_h = [h_x, 1 - h_x] \times [h_y, 1 - h_y] \times [h_t, 1 - h_t]$, we have

$$\left| \frac{1}{NH} \sum_{ijk} K\left(\frac{x_i - x}{h_x}, \frac{y_i - y}{h_y}\right) K\left(\frac{t_i - t}{h_t}\right) - 1 \right| = O\left(\frac{1}{n_{min} h_{min}}\right),$$

$$E\left[\left| \frac{1}{NH} \sum_{ijk} \varepsilon_{ijk} K\left(\frac{x_i - x}{h_x}, \frac{y_i - y}{h_y}\right) K\left(\frac{t_i - t}{h_t}\right) \right|^2 \right] = O\left(\frac{1}{NH}\right),$$

$$E\left[\left| \frac{1}{NH} \sum_{ijk} (\varepsilon_{ijk}^2 - \sigma^2) K\left(\frac{x_i - x}{h_x}, \frac{y_i - y}{h_y}\right) K\left(\frac{t_i - t}{h_t}\right) \right|^2 \right] = O\left(\frac{1}{NH}\right).$$

Based on the results in Proposition 1, we can derive the following properties of the LLK estimates defined in (3).

Theorem 1. *Besides the conditions in Proposition 1, we further assume that the true image intensity function $f(x, y; t)$ has continuous first-order partial derivatives with respect to x, y and t in the design space Ω except at the edge curves. Then, for any $(x, y; t) \in \Omega_h \setminus J_h$, we have*

$$\begin{bmatrix} \hat{a}(x, y; t) \\ \hat{b}(x, y; t) \\ \hat{c}(x, y; t) \\ \hat{d}(x, y; t) \end{bmatrix} = \begin{bmatrix} f(x, y; t) \\ f_x'(x, y; t) \\ f_y'(x, y; t) \\ f_t'(x, y; t) \end{bmatrix} + \begin{bmatrix} O(h_x^2 + h_y^2 + h_t^2) \\ O(\frac{h_x^2 + h_y^2 + h_t^2}{h_x}) \\ O(\frac{h_x^2 + h_y^2 + h_t^2}{h_y}) \\ O(\frac{h_x^2 + h_y^2 + h_t^2}{h_t}) \end{bmatrix} + \begin{bmatrix} O_p(\frac{1}{\sqrt{NH}}) \\ O_p(\frac{1}{h_x \sqrt{NH}}) \\ O_p(\frac{1}{h_y \sqrt{NH}}) \\ O_p(\frac{1}{h_t \sqrt{NH}}) \end{bmatrix}.$$

for any $(x, y, t) \in J_h \setminus S_h$, we have

$$
\begin{bmatrix} \hat{a}(x,y;t) \\ \hat{b}(x,y;t) \\ \hat{c}(x,y;t) \\ \hat{d}(x,y;t) \end{bmatrix} = \begin{bmatrix} f_-(x_\tau, y_\tau; t_\tau) + d_\tau \xi^{(2)}_{000} \\ \frac{d_\tau}{\xi_{200}h_x}\zeta^{(2)}_{100} \\ \frac{d_\tau}{\xi_{020}h_y}\zeta^{(2)}_{010} \\ \frac{d_\tau}{\xi_{002}h_t}\zeta^{(2)}_{001} \end{bmatrix} + \begin{bmatrix} O(\sqrt{h_x^2 + h_y^2 + h_t^2}) \\ O(\frac{\sqrt{h_x^2+h_y^2+h_t^2}}{h_x}) \\ O(\frac{\sqrt{h_x^2+h_y^2+h_t^2}}{h_y}) \\ O(\frac{\sqrt{h_x^2+h_y^2+h_t^2}}{h_t}) \end{bmatrix} + \begin{bmatrix} O_p(\frac{1}{\sqrt{NH}}) \\ O_p(\frac{1}{h_x\sqrt{NH}}) \\ O_p(\frac{1}{h_y\sqrt{NH}}) \\ O_p(\frac{1}{h_t\sqrt{NH}}) \end{bmatrix}, \quad (9)
$$

where $\xi_{rsl} = \int_{\Omega \times [0,1]} u^r v^s w^l K(u,v)K(w)\, dudvdw$, $\xi^{(2)}_{rsl} = \int_{Q^{(2)}} u^r v^s w^l K(u,v)K(w)\, dudvdw$, for $r, s, l = 0, 1, 2$, J is the closure of the set of all jump points of $f(x,y;t)$, $J_h = \{(x,y;t) : (x,y;t) \in \Omega_h, \sqrt{(x-x^*)^2/h_x^2 + (y-y^*)^2/h_y^2} \le 1, |t - t^*|/h_t \le 1, \text{ for any } (x^*, y^*, t^*) \in J\}$, S is the set of singular points in J, including the crossing points of two or more edges, points on an edge surface at which the edge surface does not have a unique tangent surface, and points in J at which the jump sizes in $f(x,y;t)$ are zero, $S_h = \{(x,y;t) : (x,y;t) \in \Omega_h, \sqrt{(x-x^*)^2/h_x^2 + (y-y^*)^2/h_y^2} \le 1, |t - t^*|/h_t \le 1, \text{ for any } (x^*, y^*, t^*) \in S\}$, $(x_\tau, y_\tau; t_\tau) \in J \setminus S$ is the projection of $(x,y;t)$ to J with the Euclidean distance between the two points being $c\sqrt{h_x^2 + h_y^2 + h_t^2}$, for a constant $0 < c < 1$, and $f_-(x_\tau, y_\tau; t_\tau)$ is the smaller one of the two one-sided limits of $f(x,y;t)$ at $(x_\tau, y_\tau; t_\tau)$. In cases when $O(x,y;t)$ contains jumps, without loss of generality, it is assumed that $O(x,y;t)$ is divided by the edge surface into two parts I_1 and I_2 with a positive jump size d_τ from I_1 to I_2 at $(x_\tau, y_\tau; t_\tau)$, and $Q^{(1)}$ and $Q^{(2)}$ are the two corresponding parts in the support of $K(u,v)K(w)$.

The next two theorems establish the consistency of the proposed edge-preserving image denoising procedure (2)–(6). First, we have the following theorem about the WRMS values defined in (5).

Theorem 2. *Assume that the conditions in Theorem 1 are satisfied, $h_x^2 + h_y^2 + h_t^2 = o(1)$, $(h_x^2 + h_y^2 + h_t^2)/h_{min} = o(1)$, $1/(NH) = o(1)$ and $1/(NHh_{min}^2) = o(1)$. Then, we have the following results: for any $(x,y;t) \in \Omega_h \setminus J_h$,*

$$
\begin{aligned}
e(x,y;t) &= \sigma^2 + o_p(1), \\
e^{(l)}(x,y;t) &= \sigma^2 + o_p(1), \quad \text{for } l = 1, 2;
\end{aligned}
\quad (10)
$$

for any $(x,y;t) \in J_h \setminus S_h$,

$$
\begin{aligned}
e(x,y;t) &= \sigma^2 + d_\tau C_\tau^2 + o_p(1), \\
e^{(l)}(x,y;t) &= \sigma^2 + d_\tau \left[C_\tau^{(l)}\right]^2 + o_p(1), \quad \text{for } l = 1, 2,
\end{aligned}
\quad (11)
$$

where

$$
\begin{aligned}
C_\tau = \Bigg(& \int\!\!\int\!\!\int_{Q^{(1)}} \left[\xi^{(2)}_{000} + \frac{\xi^{(2)}_{100}}{\xi_{200}}u + \frac{\xi^{(2)}_{010}}{\xi_{020}}v + \frac{\xi^{(2)}_{001}}{\xi_{002}}w \right]^2 K(u,v)K(w)dudvdw + \\
& \int\!\!\int\!\!\int_{Q^{(2)}} \left[1 - \xi^{(2)}_{000} - \frac{\xi^{(2)}_{100}}{\xi_{200}}u - \frac{\xi^{(2)}_{010}}{\xi_{020}}v - \frac{\xi^{(2)}_{001}}{\xi_{002}}w \right]^2 K(u,v)K(w)dudvdw \Bigg)^{1/2}.
\end{aligned}
$$

and

$$C_{\tau}^{(l)} = \left(2 \int \int \int_{Q^{(1l)}} \left[B_{0l} + \frac{B_{1l}}{\zeta_{200}} u + \frac{B_{2l}}{\zeta_{020}} v + \frac{B_{3l}}{\zeta_{002}} w \right]^2 K(u,v)K(w)dudvdw + \right.$$

$$\left. 2 \int \int \int_{Q^{(2l)}} \left[1 - B_{0l} - \frac{B_{1l}}{\zeta_{200}} u - \frac{B_{2l}}{\zeta_{020}} v - \frac{B_{3l}}{\zeta_{002}} w \right]^2 K(u,v)K(w)dudvdw \right)^{1/2}.$$

*with the quantities $Q^{(1l)}$, $Q^{(2l)}$, B_{0l}, B_{1l}, B_{2l} and B_{3l} defined as follows. Let $\overrightarrow{g} = (\frac{d_\tau}{\zeta_{200}h_x}\zeta_{100}^{(2)},$
$\frac{d_\tau}{\zeta_{020}h_y}\zeta_{010}^{(2)}, \frac{d_\tau}{\zeta_{002}h_t}\zeta_{001}^{(2)})$. Then, from (9), $\overrightarrow{g}$ is actually the asymptotic direction of the gradient vector $\widehat{G}(x,y;t)$. Let $\widetilde{O}^{(l)}(x,y;t)$, for $l = 1,2$, be two halves of the neighborhood $O(x,y;t)$ separated by a plane passing the point $(x,y;t)$ in the direction perpendicular to $\overrightarrow{g}$ and $\widetilde{Q}^{(l)}$ be the two corresponding parts in the support of $K(u,v)K(w)$. Then, $Q^{(1l)} = Q^{(1)} \cap \widetilde{Q}^{(l)}$, $Q^{(2l)} = Q^{(2)} \cap \widetilde{Q}^{(l)}$, $B_{0l} = \int \int \int_{Q^{(2l)}} K(u,v)K(w)dudvdw$, $B_{1l} = \int \int \int_{Q^{(2l)}} uK(u,v)K(w)dudvdw$, $B_{2l} = \int \int \int_{Q^{(2l)}} vK(u,v)K(w)dudvdw$, and $B_{3l} = \int \int \int_{Q^{(2l)}} wK(u,v)K(w)dudvdw$, for $l = 1,2$.*

Theorem 3. *Under the conditions in Theorem 2 and the extra assumption that threshold parameter $u = u_N \to 0$ as $N \to \infty$, we have, for any $(x,y;t) \in \Omega_h$,*

$$\widehat{f}(x,y;t) = f(x,y;t) + o_p(1).$$

The proofs of these theoretical results are given in Appendix A.

3.2. Numerical Studies

In this part, we study the numerical performance of our proposed method for denoising an image sequence. First, we consider a simulation example in which the true image intensity function in model (1) has the following expression:

$$f(x,y;t) = \begin{cases} -2(x-0.5)^2 - 2(y-0.5)^2 - 0.1\sin(2\pi t) + 1, & \text{if } r(x,y;t) \le 0.25^2, \\ -2(x-0.5)^2 - 2(y-0.5)^2 - 0.1\sin(2\pi t), & \text{otherwise,} \end{cases}$$

where $r(x,y;t) = (x-0.5)^2 + (y-0.5)^2 + 0.01\sin(2\pi t)$, $(x,y) \in \Omega = [0,1] \times [0,1]$, and $t \in [0,1]$. At a given value of t, $f(x,y;t)$ has a circular edge curve $r(x,y;t) = 0.25^2$ with a constant jump size 1 in $f(x,y;t)$ at the edges. The radius of the circular edge curve, $\sqrt{0.25^2 - 0.01\sin(2\pi t)}$, changes periodically over $t \in [0,1]$. The image intensity function $f(x,y;t)$ at $t = 0.01$ and 0.25 and its temporal profile $f(0.25, 0.25; t)$ are shown in Figure 3. It can be seen that both the image intensity level at a given pixel and the edge curve change gradually when t changes in $[0,1]$.

(a)   (b)

Figure 3. (a) The true image intensity function $f(x,y;t)$ at $t = 0.01$ (left) and $t = 0.25$ (right). (b) The temporal profile $f(0.25, 0.25; t)$ when t changes in $[0,1]$.

In model (1), the random errors $\{\varepsilon_{ijk}, i = 1,2,\ldots,n_x, j = 1,2,\ldots,n_y, k = 1,2,\ldots,n_t\}$ are generated by the function spatialnoise() in the R-package neuRosim (cf., Welvaert et al. [27]). In that R function, there are two parameters ρ and σ to specify in advance, where

ρ controls the data autocorrelation in all three dimensions and σ is the common standard deviation of the random errors. In all our examples, σ is fixed at 0.1, 0.2 or 0.3, and ρ is fixed at 0.1, 0.3 or 0.5, to study the possible impact of data noise level and data correlation on the performance of the proposed method. Without loss of generality, we set $n_x = n_y$ in all examples. In the model estimation procedure (2)–(6), we set $h_x = h_y$, and the kernel function $K(v)$ is chosen to be the following truncated Gaussian density function:

$$K(v) = \begin{cases} \frac{\exp(-v^2/2)-\exp(-0.5)}{2\pi-3\pi\exp(-0.5)}, & \text{if } |v| \leq 1, \\ 0, & \text{otherwise.} \end{cases}$$

In cases when $\sigma = 0.1, 0.2$ or 0.3, $n_x = 64$ or 128, $n_t = 50$ or 100, $\rho = 0.1, 0.3$ or 0.5, the MSE values of the estimator $\widehat{f}(x,y;t)$ defined in (6) are presented in Table 1, along with the corresponding parameters h_x, h_t and u selected by the modified CV procedure (7) and (8). In each case considered, the MSE value is computed based on 10 replicated simulations. For comparison purposes, the optimal MSE value of the estimator $\widehat{f}(x,y;t)$, when its parameters $(h_x, h_t$ and $u)$ are chosen such that the MSE value reaches the minimum in each case considered, is also presented in the table, along with the corresponding parameter values. From the table, we can draw the following conclusions. (i) The MSE values are smaller when either n_x or n_t is larger, which confirms the consistency results discussed in Section 3.1. (ii) When ρ is larger (i.e., the spatio-temporal data correlation is stronger), the MSE values are larger. So, data correlation does have an impact on the performance of the proposed method, which is intuitively reasonable. (iii) By comparing the MSE and the optimal MSE values, we can see that the MSE values are usually larger than their optimal values, but their differences are not that big in almost all cases considered. This conclusion indicates that the modified CV procedure (7) and (8) for determining the values of the parameters (h_x, h_t, u) is quite effective. (iv) The parameter values chosen by the modified CV procedure (7) and (8) are quite close to the optimal parameter values in most cases considered.

Table 1. In each entry, MSE of $\widehat{f}(x,y;t)$ in (6) is presented in the first line with its standard error (in parenthesis); the corresponding values of (h_x, h_t, u) chosen by the modified CV procedure (7) and (8) is presented in the second line; the optimal MSE is presented in the third line with its standard error (in parenthesis); the optimal values of (h_{xy}, h_t, u) are presented in the fourth line. MSE in the table has been multiplied by 10^3 and standard error has been multiplied by 10^5.

		$n_t = 50$		$n_t = 100$	
σ	ρ	$n_x = 64$	$n_x = 128$	$n_x = 64$	$n_x = 128$
0.1	0.1	0.65(0.80)	0.30(0.25)	0.48(0.43)	0.26(0.10)
		(0.03, 0.10, 0.05)	(0.03, 0.08, 0.025)	(0.03, 0.10, 0.05)	(0.02, 0.07, 0.05)
		0.32(0.46)	0.20(0.14)	0.37(0.36)	0.19(0.08)
		(0.04, 0.07, 0.025)	(0.03, 0.05, 0.025)	(0.03, 0.08, 0.025)	(0.02, 0.05, 0.025)
	0.3	0.60(0.45)	0.33(0.16)	0.59(0.39)	0.33(0.15)
		(0.04, 0.10, 0.05)	(0.03, 0.07, 0.025)	(0.03, 0.10, 0.05)	(0.02, 0.07, 0.025)
		0.49(0.35)	0.30(0.16)	0.50(0.37)	0.29(0.22)
		(0.04, 0.08, 0.025)	(0.03, 0.06, 0.025)	(0.03, 0.08, 0.025)	(0.03, 0.04, 0.025)
	0.5	1.25(1.24)	0.80(0.22)	0.81(0.55)	0.64(0.21)
		(0.03, 0.10, 0.05)	(0.02, 0.07, 0.025)	(0.03, 0.10, 0.05)	(0.02, 0.04, 0.025)
		0.77(0.65)	0.49(0.24)	0.74(0.46)	0.45(0.25)
		(0.04, 0.09, 0.025)	(0.03, 0.06, 0.025)	(0.03, 0.09, 0.025)	(0.03, 0.04, 0.025)

Table 1. *Cont.*

σ	ρ	$n_t = 50$		$n_t = 100$	
		$n_x = 64$	$n_x = 128$	$n_x = 64$	$n_x = 128$
0.2	0.1	1.14(1.13)	0.68(0.38)	1.02(0.74)	0.56(0.26)
		(0.04, 0.10, 0.025)	(0.03, 0.08, 0.025)	(0.04, 0.10, 0.025)	(0.03, 0.07, 0.025)
		1.11(0.86)	0.66(0.33)	0.93(0.71)	0.54(0.31)
		(0.04, 0.09, 0.025)	(0.03, 0.07, 0.025)	(0.04, 0.08, 0.025)	(0.03, 0.05, 0.025)
	0.3	1.69(0.91)	1.03(0.54)	1.32(1.08)	0.78(0.41)
		(0.04, 0.10, 0.025)	(0.03, 0.08, 0.025)	(0.04, 0.10, 0.025)	(0.03, 0.07, 0.025)
		1.69(1.24)	1.03(0.54)	1.29(1.12)	0.78(0.41)
		(0.04, 0.11, 0.025)	(0.03, 0.08, 0.025)	(0.04, 0.09, 0.025)	(0.03, 0.07, 0.025)
	0.5	3.25(1.74)	2.88(0.78)	1.95(1.85)	2.61(0.58)
		(0.04, 0.07, 0.025)	(0.02, 0.07, 0.025)	(0.04, 0.09, 0.025)	(0.02, 0.04, 0.025)
		2.59(2.23)	1.54(1.32)	1.91(1.78)	1.21(0.43)
		(0.05, 0.10, 0.025)	(0.04, 0.09, 0.025)	(0.04, 0.11, 0.025)	(0.03, 0.08, 0.025)
0.3	0.1	2.32(1.91)	1.26(1.03)	1.59(0.81)	0.92(0.34)
		(0.05, 0.13, 0.025)	(0.04, 0.09, 0.025)	(0.04, 0.11, 0.025)	(0.03, 0.08, 0.025)
		2.28(2.58)	1.26(1.03)	1.59(0.65)	0.92(0.34)
		(0.05, 0.11, 0.025)	(0.04, 0.09, 0.025)	(0.04, 0.10, 0.025)	(0.03, 0.08, 0.025)
	0.3	3.15(2.28)	1.72(1.37)	2.26(1.53)	1.36(0.50)
		(0.05, 0.13, 0.025)	(0.04, 0.09, 0.025)	(0.04, 0.11, 0.025)	(0.03, 0.08, 0.025)
		3.14(2.45)	1.71(1.52)	2.21(1.31)	1.33(0.41)
		(0.05, 0.14, 0.025)	(0.04, 0.10, 0.025)	(0.04, 0.13, 0.025)	(0.04, 0.09, 0.025)
	0.5	6.78(3.46)	6.81(2.00)	4.18(2.72)	6.33(1.43)
		(0.04, 0.09, 0.05)	(0.02, 0.07, 0.05)	(0.04, 0.10, 0.025)	(0.02, 0.04, 0.05)
		4.46(4.94)	2.48(2.38)	3.18(3.42)	1.88(0.56)
		(0.06, 0.16, 0.025)	(0.05, 0.11, 0.025)	(0.05, 0.14, 0.025)	(0.04, 0.10, 0.025)

Next, we compare our proposed method, denoted as NEW, with some alternative methods described below. The first alternative method is the conventional LLK procedure (2), by which $f(x, y; t)$ is estimated by $\hat{a}(x, y; t)$ defined in (3). Its bandwidths are chosen by the conventional CV procedure, without considering any possible spatio-temporal data correlation. As explained in Section 2.1, this estimator would blur edges while removing noise. The second alternative method is to use $\hat{a}(x, y; t)$ for estimating $f(x, y; t)$, but its bandwidths are chosen by the modified CV procedure (7) and (8). The above two alternative methods are denoted as LLK-C and LLK, respectively, where LLK-C denotes the first conventional LLK procedure that does not accommodate data correlation. The third alternative method is the one by Gijbels et al. [15] which is used for edge-preserving image denoising of a single image. To apply this method to the current problem, individual images collected at different time points can be denoised by it separately. This method assumes that the observed image intensities at different pixels are independent of each other, and thus their bandwidths can be chosen by the conventional CV procedure. This method is denoted as GLQ. The fourth alternative method is to use $\widehat{f}(x, y; t)$ in (6) to estimate $f(x, y; t)$, but the parameters (h_x, h_t, u) are chosen by the conventional CV procedure. This method is denoted as NEW-C. By considering all these four alternative methods (i.e., LLK-C, LLK, GLQ and NEW-C), we can check whether the current problem to denoise an image sequence can be handled properly by the conventional LLK procedure with or without using the modified CV procedure, by an existing edge-preserving image denoising method designed for denoising a single image, or by the proposed method without considering the possible spatio-temporal data correlation. To evaluate their performance, in addition to the regular MSE criterion, we also consider the following edge-preservation (EP) criterion originally discussed in Hall and Qiu [28]:

$$EP(\widehat{f}) = |JS(\widehat{f}) - JS(f)| / JS(f),$$

where

$$JS(f) = \frac{1}{(n_x - 2)(n_y - 2)(n_t - 2)} \sum_{i=2}^{n_x-1} \sum_{j=2}^{n_y-1} \sum_{k=2}^{n_t-1} \left([f(x_{i+1}, y_j; t_k) - f(x_{i-1}, y_j; t_k)]^2 + \right.$$

$$\left. [f(x_i, y_{j+1}; t_k) - f(x_i, y_{j-1}; t_k)]^2 + [f(x_i, y_j; t_{k+1}) - f(x_i, y_j; t_{k-1})]^2 \right)^{1/2},$$

and $JS(\widehat{f})$ is defined similarly. According to Hall and Qiu [28], $JS(f)$ is a reasonable measure of the cumulative jump magnitude of f at the edge locations. So, $EP(\widehat{f})$ provides a measure of the percentage of the cumulative jump magnitude of f that has been lost during data smoothing by using the estimator $\widehat{f}$. By this explanation, the smaller its value, the better. In cases when $\sigma = 0.1, 0.2$ or 0.3, $n_x = 128$, $n_t = 100$, and $\rho = 0.1, 0.3$ or 0.5, the MSE and EP values of the related methods are presented in Table 2. From the table, it can be seen that the proposed method NEW has the smallest MSE values with quite large margins among all five methods in all cases considered, except the case when $\sigma = 0.1$ and $\rho = 0.1$ where NEW-C has a lightly smaller MSE value than that of NEW due to the weak data correlation in that case. Likewise, NEW has much smaller EP values in all cases considered, compared to the four competing methods. This example confirms that it is necessary to consider edge-preserving procedures when denoising image sequences and the possible spatio-temporal data correlation should be taken into account during the denoising process. It also confirms the benefit to share useful information among neighboring images when denoising an image sequence.

Table 2. In each entry, the first line is the MSE value with its standard error (in parenthesis), and the second line is the EP value. MSE values in the table are in the unit of 10^3 and the standard error values are in the unit of 10^5.

σ	ρ	LLK-C	LLK	GLQ	NEW-C	NEW
0.1	0.1	2.06(0.08)	2.10(0.06)	0.60(0.18)	0.24(0.11)	0.26(0.10)
		73.68%	18.43%	28.24%	12.32%	7.48%
	0.3	3.04(0.14)	2.28(0.09)	0.95(0.18)	2.93(0.40)	0.33(0.15)
		124.48%	34.40%	43.69%	131.28%	10.58%
	0.5	3.89(0.24)	3.23(0.21)	1.42(0.42)	3.77(0.48)	0.64(0.21)
		141.47%	95.86%	57.40%	148.17%	28.86%
0.2	0.1	4.16(0.25)	2.93(0.15)	1.51(0.38)	0.86(0.25)	0.56(0.26)
		142.65%	51.78%	54.40%	39.01%	9.14%
	0.3	9.39(0.52)	3.67(0.25)	2.87(0.51)	9.60(0.78)	0.78(0.41)
		291.31%	82.84%	94.59%	295.72%	15.08%
	0.5	12.80(0.94)	11.21(0.86)	7.75(1.32)	13.12(1.16)	2.61(0.58)
		326.38%	289.71%	203.86%	334.62%	84.24%
0.3	0.1	7.88(0.57)	3.94(0.26)	3.17(0.86)	1.01(0.37)	0.92(0.34)
		235.43%	82.24%	73.18%	23.36%	15.41%
	0.3	19.97(1.15)	5.56(0.50)	12.36(0.63)	19.97(1.16)	1.36(0.50)
		461.12%	133.33%	261.31%	461.13%	25.78%
	0.5	27.64(2.09)	23.75(1.92)	15.75(1.71)	28.04(2.29)	6.33(1.43)
		514.22%	458.82%	292.50%	518.16%	144.58%

In the cases when $\sigma = 0.2$ and $\rho = 0.1, 0.3$ or 0.5, Figure 4 shows the observed images at $t = 0.5$ in the first column, and the denoised images by the methods LLK-C, LLK, GLQ, NEW-C and NEW in columns 2–6. From the figure, it can be seen that the denoised images by NEW are the best in removing noise and preserving edges. As a comparison, the denoised images by LLK-C, and NEW-C are quite noisy because their selected bandwidths by the conventional CV procedure are relatively small due to the fact the conventional CV procedure cannot distinguish the data correlation from the mean structure, as discussed in Section 2.2. The denoised images by LLK are quite blurry because the method does not take the edges into account when denoising the images. The denoised images by GLQ are quite blurry as well since GLQ denoises individual images at different time points separately and the serial data correlation is ignored in this method.

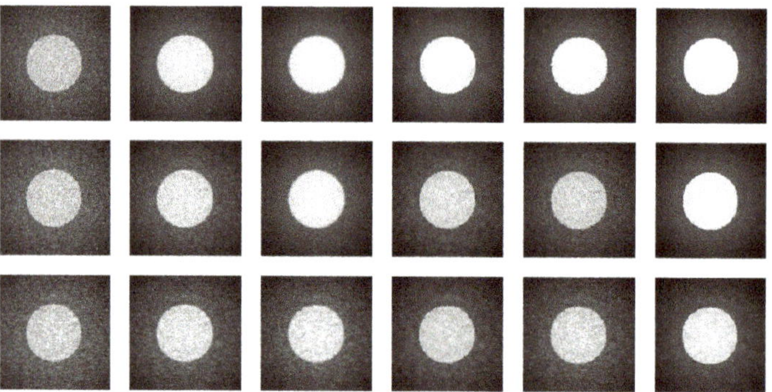

Figure 4. The first column shows the observed images at $t = 0.5$ when $\sigma = 0.2$ and $\rho = 0.1$ (1st row), 0.3 (2nd row), and 0.5 (3rd row). Second to sixth columns show the denoised images by LLK-C, LLK, GLQ, NEW-C and NEW, respectively.

Next, we apply the proposed method NEW and the four alternative methods LLK-C, LLK, GLQ and NEW-C to a sequence of cell images that records the vasculogenesis process. The sequence has 100 images, and each image has 128×128 pixels. A detailed description of the data can be found in Svoboda et al. [29]. The 1st, 50th and 100th images of the sequence are shown in Figure 5.

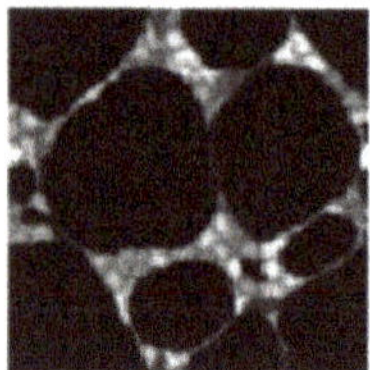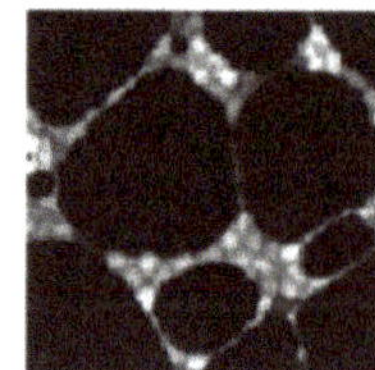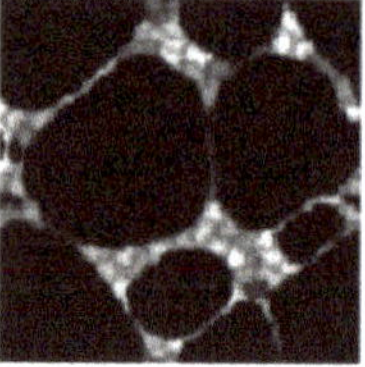

Figure 5. The 1st, 50th and 100th cell images of the image sequence for describing a vasculogenesis process.

In the image denoising literature, to test the noise removal ability of a image denoising method, it is a common practice to add random noise at a certain level to the test images and then apply the image denoising method to the noisy test images (cf., Gijbels et al. [15]). To follow this convention, spatio-temporally correlated noise is first generated using the R-package neuRosim and then added to the sequence of 100 cell images described above. When generating the noise, σ is chosen to be 0.1, 0.2 or 0.3 and ρ is chosen to be 0.1, 0.3 or 0.5, as in the simulation examples presented above. The MSE and EP values of the five image denoising methods based on 10 replicated simulations are presented in Table 3. From the table, it can be seen that NEW still has smaller MSE and EP values in this example, compared to the four competing methods, except in a small number of cases when σ and ρ are relatively small.

Table 3. Results for denoising a sequence of 100 cell images. In each entry, the first line is the MSE value and its standard error (in parenthesis), and the second line is the EP value. MSE values in the table are in the unit of 10^3 and the standard errors are in the unit of 10^5.

σ	ρ	LLK-C	LLK	GLQ	NEW-C	NEW
0.1	0.1	1.69(0.11)	0.97(0.08)	1.67(0.12)	1.69(0.12)	1.35(0.12)
		63.30%	5.53%	18.88%	63.31%	18.52%
	0.3	2.36(0.16)	1.43(0.14)	1.94(0.18)	2.36(0.16)	1.51(0.19)
		77.54%	31.64%	25.72%	77.55%	7.28%
	0.5	3.21(0.25)	2.82(0.24)	2.28(0.29)	3.21(0.25)	1.92(0.31)
		88.68%	75.95%	30.68%	88.68%	10.11%
0.2	0.1	3.22(17.00)	1.47(5.54)	3.93(0.29)	3.22(17.00)	1.67(0.25)
		85.64%	13.57%	76.53%	85.64%	16.28%
	0.3	8.71(0.56)	2.34(0.35)	5.00(0.43)	8.71(0.56)	2.17(0.45)
		189.74%	42.07%	91.44%	189.75%	4.88%
	0.5	12.12(0.94)	10.35(0.88)	6.41(0.86)	12.14(0.96)	4.48(0.90)
		213.90%	187.93%	102.68%	214.07%	59.86%
0.3	0.1	3.16(0.50)	2.01(0.28)	5.47(0.53)	3.16(0.50)	1.93(0.40)
		47.15%	22.46%	54.20%	47.15%	10.91%
	0.3	19.30(1.23)	4.29(0.71)	10.11(0.85)	19.30(1.23)	2.82(0.77)
		308.32%	79.75%	161.91%	308.32%	14.37%
	0.5	26.96(2.09)	22.88(1.95)	13.36(1.82)	27.00(2.13)	8.75(1.85)
		345.91%	306.28%	180.35%	346.14%	113.48%

The 50th observed test image after the spatio-temporally correlated noise with $\rho = 0.1$, 0.3 or 0.5 being added is shown in the first column of Figure 6. The denoised images by the five methods LLK-C, LLK, GLQ, NEW-C and NEW are shown in columns 2–6 of the figure. It can be seen that similar conclusions to those from Figure 4 can be made here, and the denoised images by NEW look reasonably well, as the algorithm work well in removing noise and preserving edges.

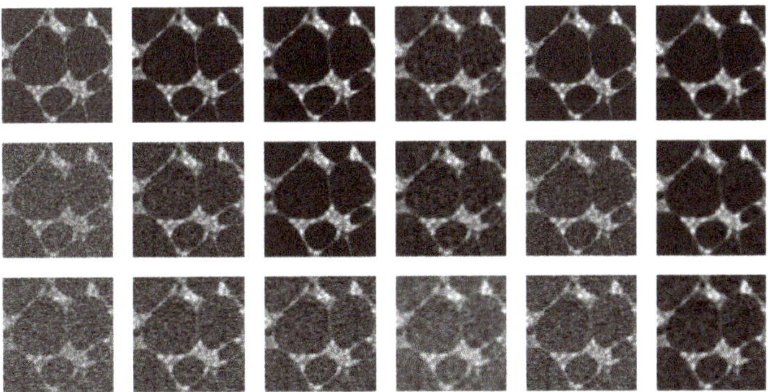

Figure 6. First column shows the 50th observed cell image after the spatio-temporally correlated noise with $\rho = 0.1$ (1st row), 0.3 (2nd row) or 0.5 (3rd row) being added. The second to sixth columns show the denoised images by LLK-C, LLK, GLQ, NEW-C and NEW, respectively.

Finally, we apply the five methods considered in the above examples to a sequence of Landsat images of the Salton Sea region. The Salton Sea is the largest inland lake located at the southern border of California, US, and has a great impact on the local ecosystem (Shuford et al. [30]). The Landsat images used here were taken during the time period of 27 May 2000 and 24 December 2001. There are a total of 20 images collected at roughly equally-spaced time points, and each image has 100×100 pixels. In this example, we consider the case when $\sigma = 0.3$ and $\rho = 0.3$. The MSE values of the five methods LLK-C, LLK, GLQ, NEW-C, and NEW calculated in the same way as before are 9.70, 4.78, 12.03, 9.77, and 4.82, respectively. Their EP values are respectively 85.54%, 20.18%, 109.91%, 86.15%, and 19.14%. So, we can see that NEW method has the best edge-preserving performance among the five methods in this example, and NEW and LLK have the best overall noise removal performance. The 10th noisy observed test image taken on 28 April 2001 and its denoised versions by the five methods are shown in Figure 7. It can be seen from the figure that the denoised images by the methods LLK-C, GLQ, and NEW-C are still quite noisy, and the noise in the images generated by NEW and LLK is mostly removed while the edges are preserved reasonably well.

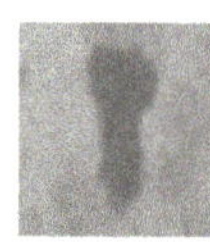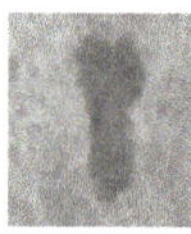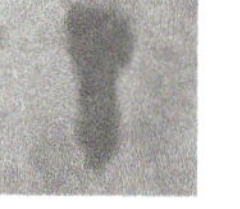

Figure 7. The first image is the observed landsat image of the Salton Sea region taken on 28 April 2001 after the spatio-temporally correlated noise with $\sigma = 0.3$ and $\rho = 0.3$ being added. Second to sixth images are its denoised versions by LLK-C, LLK, GLQ, NEW-C, and NEW, respectively.

4. Conclusions

In this paper, we have described our proposed edge-preserving image denoising method for handling image sequences. Some major features of the proposed method include (i) helpful information in neighboring images is shared during image denoising, (ii) edge structures in the observed images can be preserved when removing noise, and (iii) possible sptio-temporal data correlation can be accommodated in the related local smoothing procedure. Theoretical arguments given in Section 3.1 and numerical studies presented in Section 3.2 show that the proposed method works well in various cases considered. There are still some issues about the proposed method for future research. For instance, in the proposed local smoothing procedure (2)–(6), each of the bandwidths (h_x, h_y, h_t) is chosen by the modified CV procedure (7) and (8) to be the same in the entire design space $\Omega \times [0, 1]$. Intuitively, relatively small bandwidths are preferred at places where the image intensity surface $f(x, y; t)$ has large curvature and relatively large bandwidths are preferred at places where the curvature of $f(x, y; t)$ is small. Thus, in some applications where the curvature of $f(x, y; t)$ could change quite dramatically in the design space, variable bandwidths might be helpful. Such issues will be studied carefully in our future research.

Author Contributions: Methodology, P.Q.; Formal analysis, F.Y.; Writing—original draft preparation, F.Y.; Writing—review and editing, P.Q.; Funding acquisition, P.Q.; Supervision, P.Q. All authors have read and agreed to the published version of the manuscript.

Funding: This research was funded by the National Science Foundation grant DMS-1914639.

Data Availability Statement: Publicly available datasets were analyzed in this study. They can be found from the links: https://cbia.fi.muni.cz/datasets/ and https://earthexplorer.usgs.gov.

Acknowledgments: We thank the four referees for many constructive comments and suggestions about the paper which greatly improved its quality. This research is supported in part by the National Science Foundation grant DMS-1914639.

Conflicts of Interest: The authors declare no conflicts of interest.

Appendix A

Appendix A.1. Proof of Proposition 1

Define $B_h(x,y,t) = \{(x',y';t') : \sqrt{(|x'-x|/h_x)^2 + (|y'-y|/h_y)^2} \leq 1, |t-t'| \leq h_t, (x',y';t') \in [0,1] \times [0,1] \times [0,1]\}$, $\Delta_{ijk} = [x_{i-1},x_i] \times [y_{j-1},y_j] \times [t_{k-1},t_k]$, $x_0 = y_0 = t_0 = 0$. Then it can be seen that

$$\left| \frac{1}{NH} \sum_{ijk} K\left(\frac{x_i-x}{h_x}, \frac{y_i-y}{h_y}\right) K\left(\frac{t_k-t}{h_t}\right) - 1 \right|$$

$$= \left| \frac{1}{H} \sum_{ijk} \int\int\int_{\Delta_{ijk}} K\left(\frac{x_i-x}{h_x}, \frac{y_i-y}{h_y}\right) K\left(\frac{t_k-t}{h_t}\right) dudvdw - 1 \right|$$

$$= \left| \frac{1}{H} \sum_{ijk} \int\int\int_{\Delta_{ijk}} K\left(\frac{x_i-x}{h_x}, \frac{y_i-y}{h_y}\right) K\left(\frac{t_k-t}{h_t}\right) dudvdw - \right.$$
$$\left. \frac{1}{H} \int\int\int_{B_h(x,y,t)} K\left(\frac{u-x}{h_x}, \frac{v-y}{h_y}\right) K\left(\frac{w-t}{h_t}\right) dudvdw \right|$$

$$= \left| \frac{1}{H} \sum_{ijk} \int\int\int_{\Delta_{ijk}} K\left(\frac{x_i-x}{h_x}, \frac{y_i-y}{h_y}\right) K\left(\frac{t_k-t}{h_t}\right) dudvdw - \right.$$
$$\left. \frac{1}{H} \sum_{ijk} \int\int\int_{B_h(x,y,t) \cap \Delta_{ijk}} K\left(\frac{u-x}{h_x}, \frac{v-y}{h_y}\right) K\left(\frac{w-t}{h_t}\right) dudvdw \right|$$

$$= \left| \frac{1}{H} \sum_{ijk} \int\int\int_{B_h(x,y,t)^c \cap \Delta_{ijk}} K\left(\frac{x_i-x}{h_x}, \frac{y_i-y}{h_y}\right) K\left(\frac{t_k-t}{h_t}\right) dudvdw + \right.$$
$$\frac{1}{H} \sum_{ijk} \int\int\int_{B_h(x,y,t) \cap \Delta_{ijk}} K\left(\frac{x_i-x}{h_x}, \frac{y_i-y}{h_y}\right) K\left(\frac{t_k-t}{h_t}\right) dudvdw -$$
$$\left. \frac{1}{H} \sum_{ijk} \int\int\int_{B_h(x,y,t) \cap \Delta_{ijk}} K\left(\frac{u-x}{h_x}, \frac{v-y}{h_y}\right) K\left(\frac{w-t}{h_t}\right) dudvdw \right|$$

$$\leq O\left(\frac{1}{n_{min}h_{min}}\right) + \frac{1}{H} \sum_{ijk} \int\int\int_{B_h(x,y,t) \cap \Delta_{ijk}} \left| K\left(\frac{x_i-x}{h_x}, \frac{y_j-y}{h_y}\right) K\left(\frac{t_k-t}{h_t}\right) - \right.$$
$$\left. K\left(\frac{u-x}{h_x}, \frac{v-y}{h_y}\right) K\left(\frac{w-t}{h_t}\right) \right| dudvdw$$

$$\leq O\left(\frac{1}{n_{min}h_{min}}\right) + \frac{1}{H} \sum_{ijk} \int\int\int_{B_h(x,y,t) \cap \Delta_{ijk}} \frac{(1+\sqrt{2})C}{n_{min}h_{min}} dudvdw$$

$$= O\left(\frac{1}{n_{min}h_{min}}\right) + \frac{1}{H} \frac{(1+\sqrt{2})C}{n_{min}h_{min}} \int\int\int_{B_h(x,y,t)} 1 \, dudvdw$$

$$= O\left(\frac{1}{n_{min}h_{min}}\right),$$

where $C \geq 0$ is the Lipschitz constant that satisfies the condition $|K(u) - K(u')| \leq C|u-u'|$. So, the first result in Proposition 1 is valid.

To prove the second result, it can be checked that

$$
E\left|\frac{1}{NH}\sum_{ijk}\varepsilon_{ijk}K\left(\frac{x_i-x}{h_x},\frac{y_i-y}{h_y}\right)K\left(\frac{t_i-x}{h_t}\right)\right|^2
$$

$$
= Var\left(\frac{1}{NH}\sum_{ijk}\varepsilon_{ijk}K\left(\frac{x_i-x}{h_x},\frac{y_i-y}{h_y}\right)K\left(\frac{t_k-x}{h_t}\right)\right)
$$

$$
= \frac{1}{N^2H^2}\sum_{ijk}\sum_{i'j'k'}K\left(\frac{x_i-x}{h_x},\frac{y_i-y}{h_y}\right)K\left(\frac{t_k-x}{h_t}\right)
$$
$$
K\left(\frac{x_{i'}-x}{h_x},\frac{y_{i'}-y}{h_y}\right)K\left(\frac{t_{k'}-x}{h_t}\right)Cov(\varepsilon_{ijk},\varepsilon_{i'j'k'})
$$

$$
\leq \frac{1}{N^2H^2}\sum_{ijk}\sum_{i'j'k'}K\left(\frac{x_i-x}{h_x},\frac{y_i-y}{h_y}\right)K\left(\frac{t_k-x}{h_t}\right)
$$
$$
K\left(\frac{x_{i'}-x}{h_x},\frac{y_{i'}-y}{h_y}\right)K\left(\frac{t_{k'}-x}{h_t}\right)c_1\sigma^2\rho^{c_2\max\{|i-i'|,|j-j'|,|k-k'|\}}
$$

$$
\leq \frac{1}{N^2H^2}\sum_{ijk}K\left(\frac{x_i-x}{h_x},\frac{y_i-y}{h_y}\right)K\left(\frac{t_k-x}{h_t}\right)c_1\sigma^2 24\int_0^\infty \tau^2\rho^\tau d\tau
$$

$$
= O\left(\frac{1}{NH}\right).
$$

Similarly, it can be checked that

$$
E\left|\frac{1}{NH}\sum_{ijk}(\varepsilon_{ijk}^2-\sigma^2)K\left(\frac{x_i-x}{h_x},\frac{y_i-y}{h_y}\right)K\left(\frac{t_i-x}{h_t}\right)\right|^2
$$

$$
= Var\left(\frac{1}{NH}\sum_{ijk}\varepsilon_{ijk}^2K\left(\frac{x_i-x}{h_x},\frac{y_i-y}{h_y}\right)K\left(\frac{t_k-x}{h_t}\right)\right)
$$

$$
= \frac{1}{N^2H^2}\sum_{ijk}\sum_{i'j'k'}K\left(\frac{x_i-x}{h_x},\frac{y_i-y}{h_y}\right)K\left(\frac{t_k-x}{h_t}\right)
$$
$$
K\left(\frac{x_{i'}-x}{h_x},\frac{y_{i'}-y}{h_y}\right)K\left(\frac{t_{k'}-x}{h_t}\right)Cov(\varepsilon_{ijk}^2,\varepsilon_{i'j'k'}^2)
$$

$$
\leq \frac{1}{N^2H^2}\sum_{ijk}\sum_{i'j'k'}K\left(\frac{x_i-x}{h_x},\frac{y_i-y}{h_y}\right)K\left(\frac{t_k-x}{h_t}\right)
$$
$$
K\left(\frac{x_{i'}-x}{h_x},\frac{y_{i'}-y}{h_y}\right)K\left(\frac{t_{k'}-x}{h_t}\right)12(c_1\sigma^2\rho^{c_2\max\{|i-i'|,|j-j'|,|k-k'|\}})^{1/4}E(\varepsilon_{111}^4)
$$

$$
\leq \frac{1}{N^2H^2}\sum_{ijk}K\left(\frac{x_i-x}{h_x},\frac{y_i-y}{h_y}\right)K\left(\frac{t_k-x}{h_t}\right)12(c_1\sigma^2 24\int_0^\infty \tau^2\rho^\tau d\tau)^{1/3}(E(\varepsilon_{111}^6))^{2/3}
$$

$$
= O\left(\frac{1}{NH}\right).
$$

The first inequality in the above expression is based on the result in Davydov [31]. So, the third result is valid.

Appendix A.2. Proof of Theorem 1

We first consider the case when $(x, y; t) \in \Omega_h \setminus J_h$. By Taylor expansion, we have

$$
\begin{aligned}
Z_{ijk} =& f(x_i, y_j; t_k) + \epsilon_{ijk} \\
=& f(x, y; t) + (x_i - x)f'_x(x, y; t) + (y_j - y)f'_y(x, y; t) + (t_k - t)f'_t(x, y; t) + \\
& O(h_x^2 + h_y^2 + h_t^2) + \epsilon_{ijk}.
\end{aligned}
$$

So, it can be checked that

$$
\begin{bmatrix}
\sum_{ijk} Z_{ijk} K_{ijk} \\
\sum_{ijk} (x_i - x) Z_{ijk} K_{ijk} \\
\sum_{ijk} (y_j - y) Z_{ijk} K_{ijk} \\
\sum_{ijk} (t_k - t) Z_{ijk} K_{ijk}
\end{bmatrix}
= M
\begin{bmatrix}
f(x, y; t) \\
f'_x(x, y; t) \\
f'_y(x, y; t) \\
f'_t(x, y; t)
\end{bmatrix}
+
\begin{bmatrix}
\sum_{ijk} O(h_x^2 + h_y^2 + h_t^2) K_{ijk} \\
\sum_{ijk} (x_i - x) O(h_x^2 + h_y^2 + h_t^2) K_{ijk} \\
\sum_{ijk} (y_j - y) O(h_x^2 + h_y^2 + h_t^2) K_{ijk} \\
\sum_{ijk} (t_k - t) O(h_x^2 + h_y^2 + h_t^2) K_{ijk}
\end{bmatrix}
+
\begin{bmatrix}
\sum_{ijk} \epsilon_{ijk} K_{ijk} \\
\sum_{ijk} (x_i - x) \epsilon_{ijk} K_{ijk} \\
\sum_{ijk} (y_j - y) \epsilon_{ijk} K_{ijk} \\
\sum_{ijk} (t_k - t) \epsilon_{ijk} K_{ijk}
\end{bmatrix},
$$

where

$$
M =
\begin{bmatrix}
m_{000} & m_{100} & m_{010} & m_{001} \\
m_{100} & m_{200} & m_{110} & m_{101} \\
m_{010} & m_{110} & m_{020} & m_{011} \\
m_{001} & m_{101} & m_{011} & m_{002}
\end{bmatrix}.
$$

From Expression (3), we have

$$
\begin{bmatrix}
\widehat{a}(x, y; t) \\
\widehat{b}(x, y; t) \\
\widehat{c}(x, y; t) \\
\widehat{d}(x, y; t)
\end{bmatrix}
=
\begin{bmatrix}
f(x, y; t) \\
f'_x(x, y; t) \\
f'_y(x, y; t) \\
f'_t(x, y; t)
\end{bmatrix}
+ M^{-1}
\begin{bmatrix}
\sum_{ijk} O(h_x^2 + h_y^2 + h_t^2) K_{ijk} \\
\sum_{ijk} (x_i - x) O(h_x^2 + h_y^2 + h_t^2) K_{ijk} \\
\sum_{ijk} (y_j - y) O(h_x^2 + h_y^2 + h_t^2) K_{ijk} \\
\sum_{ijk} (t_k - t) O(h_x^2 + h_y^2 + h_t^2) K_{ijk}
\end{bmatrix}
+
M^{-1}
\begin{bmatrix}
\sum_{ijk} \epsilon_{ijk} K_{ijk} \\
\sum_{ijk} (x_i - x) \epsilon_{ijk} K_{ijk} \\
\sum_{ijk} (y_j - y) \epsilon_{ijk} K_{ijk} \\
\sum_{ijk} (t_k - t) \epsilon_{ijk} K_{ijk}
\end{bmatrix}.
$$

By some simple algebraic manipulations, we have

$$
M^{-1} =
\begin{bmatrix}
O(\frac{1}{NH}) & O(\frac{1}{NH \cdot h_x}) & O(\frac{1}{NH \cdot h_y}) & O(\frac{1}{NH \cdot h_t}) \\
O(\frac{1}{NH \cdot h_x}) & O(\frac{1}{NH \cdot h_x^2}) & O(\frac{1}{NH \cdot h_x h_y}) & O(\frac{1}{NH \cdot h_x h_t}) \\
O(\frac{1}{NH \cdot h_y}) & O(\frac{1}{NH \cdot h_x h_y}) & O(\frac{1}{NH \cdot h_y^2}) & O(\frac{1}{NH \cdot h_y h_t}) \\
O(\frac{1}{NH \cdot h_t}) & O(\frac{1}{NH \cdot h_x h_t}) & O(\frac{1}{NH \cdot h_y h_t}) & O(\frac{1}{NH \cdot h_t^2})
\end{bmatrix}.
$$

Then,

$$
\begin{bmatrix}
\widehat{a}(x, y; t) \\
\widehat{b}(x, y; t) \\
\widehat{c}(x, y; t) \\
\widehat{d}(x, y; t)
\end{bmatrix}
=
\begin{bmatrix}
f(x, y; t) \\
f'_x(x, y; t) \\
f'_y(x, y; t) \\
f'_t(x, y; t)
\end{bmatrix}
+
\begin{bmatrix}
O(h_x^2 + h_y^2 + h_t^2) \\
O(\frac{h_x^2 + h_y^2 + h_t^2}{h_x}) \\
O(\frac{h_x^2 + h_y^2 + h_t^2}{h_y}) \\
O(\frac{h_x^2 + h_y^2 + h_t^2}{h_t})
\end{bmatrix}
+
\begin{bmatrix}
O_p(\frac{1}{\sqrt{NH}}) \\
O_p(\frac{1}{h_x \sqrt{NH}}) \\
O_p(\frac{1}{h_y \sqrt{NH}}) \\
O_p(\frac{1}{h_t \sqrt{NH}})
\end{bmatrix}.
$$

Now, we consider the case when $(x, y; t) \in J_h \setminus S_h$. If $(x_i, y_j; t_k) \in I_1$, then we have

$$
\begin{aligned}
Z_{ijk} &= f(x_i, y_j; t_k) + \varepsilon_{ijk} \\
&= f_-(x_\tau, y_\tau; t_\tau) + O\left(\sqrt{h_x^2 + h_y^2 + h_t^2}\right) + \varepsilon_{ijk},
\end{aligned}
$$

and if $(x_i, y_j; t_k) \in I_2$, we have

$$
\begin{aligned}
Z_{ijk} &= f(x_i, y_j; t_k) + \varepsilon_{ijk} \\
&= f_-(x_\tau, y_\tau; t_\tau) + d_\tau + O\left(\sqrt{h_x^2 + h_y^2 + h_t^2}\right) + \varepsilon_{ijk}.
\end{aligned}
$$

By some similar arguments to those in the case considered above, we have

$$
\begin{bmatrix} \widehat{a}(x,y;t) \\ \widehat{b}(x,y;t) \\ \widehat{c}(x,y;t) \\ \widehat{d}(x,y;t) \end{bmatrix}
=
\begin{bmatrix}
f_-(x_\tau, y_\tau; t_\tau) + d_\tau \dfrac{\sum_{(x_i,y_j;t_k)\in I_2} K_{ijk}}{\sum_{ijk} K_{ijk}} \\[2ex]
\dfrac{d_\tau}{h_x} \dfrac{\sum_{(x_i,y_j;t_k)\in I_2}[(x_i-x)/h_x]K_{ijk}}{\sum_{ijk}[(x_i-x)/h_x]^2 K_{ijk}} \\[2ex]
\dfrac{d_\tau}{h_y} \dfrac{\sum_{(x_i,y_j;t_k)\in I_2}[(y_j-y)/h_y]K_{ijk}}{\sum_{ijk}[(y_j-y)/h_y]^2 K_{ijk}} \\[2ex]
\dfrac{d_\tau}{h_t} \dfrac{\sum_{(x_i,y_j;t_k)\in I_2}[(t_k-t)/h_t]K_{ijk}}{\sum_{ijk}[(t_k-t)/h_t]^2 K_{ijk}}
\end{bmatrix}
+
$$

$$
\begin{bmatrix}
O\left(\sqrt{h_x^2+h_y^2+h_t^2}\right) \\[1ex]
O\left(\dfrac{\sqrt{h_x^2+h_y^2+h_t^2}}{h_x}\right) \\[1ex]
O\left(\dfrac{\sqrt{h_x^2+h_y^2+h_t^2}}{h_y}\right) \\[1ex]
O\left(\dfrac{\sqrt{h_x^2+h_y^2+h_t^2}}{h_t}\right)
\end{bmatrix}
+
\begin{bmatrix}
O_p\left(\dfrac{1}{\sqrt{NH}}\right) \\[1ex]
O_p\left(\dfrac{1}{h_x\sqrt{NH}}\right) \\[1ex]
O_p\left(\dfrac{1}{h_y\sqrt{NH}}\right) \\[1ex]
O_p\left(\dfrac{1}{h_t\sqrt{NH}}\right)
\end{bmatrix}
$$

$$
=
\begin{bmatrix}
f_-(x_\tau, y_\tau; t_\tau) + d_\tau \zeta_{000}^{(2)} \\[1ex]
\dfrac{d_\tau}{\zeta_{200}h_x}\zeta_{100}^{(2)} \\[1ex]
\dfrac{d_\tau}{\zeta_{020}h_y}\zeta_{010}^{(2)} \\[1ex]
\dfrac{d_\tau}{\zeta_{002}h_t}\zeta_{001}^{(2)}
\end{bmatrix}
+
\begin{bmatrix}
O\left(\sqrt{h_x^2+h_y^2+h_t^2}\right) \\[1ex]
O\left(\dfrac{\sqrt{h_x^2+h_y^2+h_t^2}}{h_x}\right) \\[1ex]
O\left(\dfrac{\sqrt{h_x^2+h_y^2+h_t^2}}{h_y}\right) \\[1ex]
O\left(\dfrac{\sqrt{h_x^2+h_y^2+h_t^2}}{h_t}\right)
\end{bmatrix}
+
\begin{bmatrix}
O_p\left(\dfrac{1}{\sqrt{NH}}\right) \\[1ex]
O_p\left(\dfrac{1}{h_x\sqrt{NH}}\right) \\[1ex]
O_p\left(\dfrac{1}{h_y\sqrt{NH}}\right) \\[1ex]
O_p\left(\dfrac{1}{h_t\sqrt{NH}}\right)
\end{bmatrix}
$$

Appendix A.3. Proof of Theorem 2

We prove the second equations in (10) and (11) here. The first equations can be proved similarly. For simplicity, we write $\widehat{a}^{(l)}(x,y;t)$, $\widehat{b}^{(l)}(x,y;t)$, $\widehat{c}^{(l)}(x,y;t)$, $\widehat{d}^{(l)}(x,y;t)$, $O^{(l)}(x,y;t)$ and $\widetilde{O}^{(l)}(x,y;t)$ as $\widehat{a}^{(l)}$, $\widehat{b}^{(l)}$, $\widehat{c}^{(l)}$, $\widehat{d}^{(l)}$, $O^{(l)}$ and $\widetilde{O}^{(l)}$, respectively from now on. First, by Proposition 1, it is easy to show that

$$
\frac{\sum_{ijk} \varepsilon_{ijk} K\left(\frac{x_i-x}{h_x}, \frac{y_j-y}{h_y}\right) K\left(\frac{t_j-x}{h_t}\right)}{\sum_{ijk} K\left(\frac{x_i-x}{h_x}, \frac{y_j-y}{h_y}\right) K\left(\frac{t_j-x}{h_t}\right)} = O_p\left(\frac{1}{\sqrt{NH}}\right),
\tag{A1}
$$

$$
\frac{\sum_{ijk} (\varepsilon_{ijk}^2 - \sigma^2) K\left(\frac{x_i-x}{h_x}, \frac{y_j-y}{h_y}\right) K\left(\frac{t_j-x}{h_t}\right)}{\sum_{ijk} K\left(\frac{x_i-x}{h_x}, \frac{y_j-y}{h_y}\right) K\left(\frac{t_j-x}{h_t}\right)} = o_p(1).
\tag{A2}
$$

Let us first consider the case when $(x, y; t) \in \Omega_h \setminus J_h$. In such a case, it can be checked that

$$
\begin{aligned}
e^{(l)}(x, y; t) &= \left\{ \sum_{(x_i, y_j; t_k) \in O^{(l)}} [\varepsilon_{ijk} + f(x_i, y_j; t_k) - \widehat{a}^{(l)} - \widehat{b}^{(l)}(x_i - x) - \right. \\
&\quad \left. \widehat{c}^{(l)}(y_j - y) - \widehat{d}^{(l)}(t_k - t)]^2 K_{ijk} \right\} / \sum_{(x_i, y_j; t_k) \in O^{(l)}} K_{ijk} \\
&= \left\{ \sum_{(x_i, y_j; t_k) \in O^{(l)}} \varepsilon_{ijk}^2 K_{ijk} \right\} / \sum_{(x_i, y_j; t_k) \in O^{(l)}} K_{ijk} + \\
&\quad \left\{ 2 \sum_{(x_i, y_j; t_k) \in O^{(l)}} \varepsilon_{ijk}[f(x_i, y_j; t_k) - \widehat{a}^{(l)} - \widehat{b}^{(l)}(x_i - x) - \right. \\
&\quad \left. \widehat{c}^{(l)}(y_j - y) - \widehat{d}^{(l)}(t_k - t)]K_{ijk} \right\} / \sum_{(x_i, y_j; t_k) \in O^{(l)}} K_{ijk} + \\
&\quad \left\{ \sum_{(x_i, y_j; t_k) \in O^{(l)}} [f(x_i, y_j; t_k) - \widehat{a}^{(l)} - \widehat{b}^{(l)}(x_i - x) - \right. \\
&\quad \left. \widehat{c}^{(l)}(y_j - y) - \widehat{d}^{(l)}(t_k - t)]^2 K_{ijk} \right\} / \sum_{(x_i, y_j; t_k) \in O^{(l)}} K_{ijk} \\
&=: \ A_1^{(l)}(x, y; t) + A_2^{(l)}(x, y; t) + A_3^{(l)}(x, y; t).
\end{aligned}
$$

Similar to (A2), we have

$$
A_1^{(l)}(x, y; t) = \sigma^2 + o_p(1). \tag{A3}
$$

Taylor expansion of $f(x_i, y_j; t_k)$ at point $(x, y; t)$, results in Theorem 1, and by similar arguments for (A1), we have

$$
\begin{aligned}
A_2^{(l)}(x, y; t) &\leq \ 2|f(x, y; t) - \widehat{a}^{(l)}| \left| \frac{\sum_{(x_i, y_j; t_k) \in O^{(l)}} \varepsilon_{ijk} K_{ijk}}{\sum_{(x_i, y_j; t_k) \in O^{(l)}} K_{ijk}} \right| + \\
&\quad 2h_x |f_x'(x, y; t) - \widehat{b}^{(l)}| \left| \frac{\sum_{(x_i, y_j; t_k) \in O^{(l)}} \varepsilon_{ijk} \frac{x_i - x}{h_x} K_{ijk}}{\sum_{(x_i, y_j; t_k) \in O^{(l)}} K_{ijk}} \right| + \\
&\quad 2h_y |f_y'(x, y; t) - \widehat{c}^{(l)}| \left| \frac{\sum_{(x_i, y_j; t_k) \in O^{(l)}(x, y; t)} \varepsilon_{ijk} \frac{y_j - y}{h_y} K_{ijk}}{\sum_{(x_i, y_j; t_k) \in O^{(l)}} K_{ijk}} \right| + \\
&\quad 2h_t |f_t'(x, y; t) - \widehat{d}^{(l)}| \left| \frac{\sum_{(x_i, y_j; t_k) \in O^{(l)}} \varepsilon_{ijk} \frac{t_k - t}{h_t} K_{ijk}}{\sum_{(x_i, y_j; t_k) \in O^{(l)}} K_{ijk}} \right| \\
&= \ o_p(1).
\end{aligned} \tag{A4}
$$

Similarly, we have

$$
A_3^{(l)}(x, y; t) = o_p(1). \tag{A5}
$$

By combining (A3)–(A5), we have

$$
e^{(l)}(x, y; t) = \sigma^2 + o_p(1).
$$

Now, let us consider the case when $(x, y; t) \in J_h \setminus S_h$. Similar to the above case, let us write

$$e^{(l)}(x, y; t) = A_1^{(l)}(x, y; t) + A_2^{(l)}(x, y; t) + A_3^{(l)}(x, y; t).$$

Here, we still have

$$A_1^{(l)}(x, y; t) = \sigma^2 + o_p(1). \tag{A6}$$

For $A_2^{(l)}(x, y; t)$, we have

$$
\begin{aligned}
A_2^{(l)}(x, y; t) =\ & \left\{ 2 \sum_{(x_i, y_j; t_k) \in I^1 \cap O^{(l)}} \varepsilon_{ijk} [f(x_i, y_j; t_k) - \widehat{a}^{(l)} - \widehat{b}^{(l)}(x_i - x) - \right. \\
& \left. \widehat{c}^{(l)}(y_j - y) - \widehat{d}^{(l)}(t_k - t)] K_{ijk} \right\} / \sum_{(x_i, y_j; t_k) \in O^{(l)}} K_{ijk} + \\
& \left\{ 2 \sum_{(x_i, y_j; t_k) \in I^2 \cap O^{(l)}} \varepsilon_{ijk} [f(x_i, y_j; t_k) - \widehat{a}^{(l)} - \widehat{b}^{(l)}(x_i - x) - \right. \\
& \left. \widehat{c}^{(l)}(y_j - y) - \widehat{d}^{(l)}(t_k - t)] K_{ijk} \right\} / \sum_{(x_i, y_j; t_k) \in O^{(l)}} K_{ijk} \\
=:\ & A_{21}^{(l)}(x, y; t) + A_{22}^{(l)}(x, y; t).
\end{aligned}
$$

By the results in Theorem 1, we have

$$
\begin{aligned}
A_{21}^{(l)}(x, y; t) =\ & \frac{2 \sum_{(x_i, y_j; t_k) \in I^1 \cap O^{(l)}} \varepsilon_{ijk} \left[f(x_i, y_j; t_k) - f_-(x_\tau, y_\tau; t_\tau) \right] K_{ijk}}{\sum_{(x_i, y_j; t_k) \in O^{(l)}} K_{ijk}} - \\
& \frac{(D_1 + o_p(1)) \sum_{(x_i, y_j; t_k) \in I^1 \cap O^{(l)}} \varepsilon_{ijk} K_{ijk}}{\sum_{(x_i, y_j; t_k) \in O^{(l)}} K_{ijk}} - \\
& \frac{(D_2 + o_p(1)) \sum_{(x_i, y_j; t_k) \in I^1 \cap O^{(l)}} \varepsilon_{ijk} \frac{x_i - x}{h_x} K_{ijk}}{\sum_{(x_i, y_j; t_k) \in O^{(l)}} K_{ijk}} - \\
& \frac{(D_3 + o_p(1)) \sum_{(x_i, y_j; t_k) \in I^1 \cap O^{(l)}} \varepsilon_{ijk} \frac{y_j - y}{h_y} K_{ijk}}{\sum_{(x_i, y_j; t_k) \in O^{(l)}} K_{ijk}} - \\
& \frac{(D_4 + o_p(1)) \sum_{(x_i, y_j; t_k) \in I^1 \cap O^{(l)}} \varepsilon_{ijk} \frac{t_k - t}{h_t} K_{ijk}}{\sum_{(x_i, y_j; t_k) \in O^{(l)}} K_{ijk}},
\end{aligned}
$$

where D_1, D_2, D_3 and D_4 are constants. By similar arguments for (A1), we can conclude that

$$A_{21}^{(l)} = o_p(1).$$

Similarly, we have

$$A_{22}^{(l)} = o_p(1).$$

So,

$$A_2^{(l)} = o_p(1). \tag{A7}$$

By similar arguments to those about Proposition 1, we have

$$\left| \frac{1}{NH} \sum_{(x_i, y_j; t_k) \in O^{(l)}} K_{ijk} - \frac{1}{2} \right| = o(1).$$

For a function $\phi(x, y; t)$ satisfying the condition that $\sup_{x^2+y^2+t^2\leq 1} |\phi(x, y; t)| \leq b_\phi < \infty$, we can have

$$\left| \frac{1}{NH} \sum_{(x_i,y_j;t_k)\in I^1\cap O^{(l)}} \phi(\frac{x_i - x}{h_x}, \frac{y_j - y}{h_y}; \frac{t_k - t}{h_t}) K_{ijk} - \right.$$

$$\left. \frac{1}{NH} \sum_{(x_i,y_j;t_k)\in I^1\cap \widetilde{O}^{(l)}} \phi(\frac{x_i - x}{h_x}, \frac{y_j - y}{h_y}; \frac{t_k - t}{h_t}) K_{ijk} \right|$$

$$\leq \quad b_\phi ||K|| \frac{1}{NH} \sum_{(x_i,y_j;t_k)\in O^{(l)}\Delta\widetilde{O}^{(l)}} 1$$

$$= \quad o(1),$$

where $O^{(l)}\Delta\widetilde{O}^{(l)} = (O^{(l)} \cup \widetilde{O}^{(l)}) \setminus (O^{(l)} \cap \widetilde{O}^{(l)})$. The last equation above is a direct conclusion of (9). By the above results, we have

$$A_3^{(l)}(x, y; t) = \frac{2}{NH} \sum_{(x_i,y_j;t_k)\in O^{(l)}} \left[f(x_i, y_j; t_k) - \widehat{a}^{(l)} - \widehat{b}^{(l)}(x_i - x) - \right. \tag{A8}$$

$$\left. \widehat{c}^{(l)}(y_j - y) - \widehat{d}^{(l)}(t_k - t) \right]^2 K_{ijk}$$

$$= \frac{2}{NH} \sum_{(x_i,y_j;t_k)\in O^{(l)}} \left[f(x_i, y_j; t_k) - f_-(x_\tau, y_\tau; t_\tau) - d_\tau B_{0l} - \frac{d_\tau B_{1l}}{\xi_{200}} \frac{x_i - x}{h_x} - \right.$$

$$\left. \frac{d_\tau B_{2l}}{\xi_{020}} \frac{y_j - y}{h_y} - \frac{d_\tau B_{3l}}{\xi_{002}} \frac{t_k - t}{h_t} \right]^2 K_{ijk} + o_p(1)$$

$$= \frac{2}{NH} \left(\sum_{(x_i,y_j;t_k)\in I^1\cap O^{(l)}} + \sum_{(x_i,y_j;t_k)\in I^2\cap O^{(l)}} \right)$$

$$\left[f(x_i, y_j; t_k) - f_-(x_\tau, y_\tau; t_\tau) - d_\tau B_{0l} - \frac{d_\tau B_{1l}}{\xi_{200}} \frac{x_i - x}{h_x} - \right.$$

$$\left. \frac{d_\tau B_{2l}}{\xi_{020}} \frac{y_j - y}{h_y} - \frac{d_\tau B_{3l}}{\xi_{002}} \frac{t_k - t}{h_t} \right]^2 K_{ijk} + o_p(1)$$

$$= \frac{2}{NH} \left(\sum_{(x_i,y_j;t_k)\in I^1\cap \widetilde{O}^{(l)}} + \sum_{(x_i,y_j;t_k)\in I^2\cap \widetilde{O}^{(l)}} \right)$$

$$\left[f(x_i, y_j; t_k) - f_-(x_\tau, y_\tau; t_\tau) - d_\tau B_{0l} - \frac{d_\tau B_{1l}}{\xi_{200}} \frac{x_i - x}{h_x} - \right.$$

$$\left. \frac{d_\tau B_{2l}}{\xi_{020}} \frac{y_j - y}{h_y} - \frac{d_\tau B_{3l}}{\xi_{002}} \frac{t_k - t}{h_t} \right]^2 K_{ijk} + o_p(1)$$

$$= \frac{2}{NH} \sum_{(x_i,y_j;t_k)\in I^1\cap \widetilde{O}^{(l)}} \left[- d_\tau B_{0l} - \frac{d_\tau B_{1l}}{\xi_{200}} \frac{x_i - x}{h_x} - \right.$$

$$\left. \frac{d_\tau B_{2l}}{\xi_{020}} \frac{y_j - y}{h_y} - \frac{d_\tau B_{3l}}{\xi_{002}} \frac{t_k - t}{h_t} \right]^2 K_{ijk} +$$

$$\frac{2}{NH} \sum_{(x_i,y_j;t_k)\in I^2\cap \widetilde{O}^{(l)}} \left[d_\tau - d_\tau B_{0l} - \frac{d_\tau B_{1l}}{\xi_{200}} \frac{x_i - x}{h_x} - \right.$$

$$\left. \frac{d_\tau B_{2l}}{\xi_{020}} \frac{y_j - y}{h_y} - \frac{d_\tau B_{3l}}{\xi_{002}} \frac{t_k - t}{h_t} \right]^2 K_{ijk} + o_p(1)$$

$$
= 2d_\tau^2 \int\int\int_{Q^{(1l)}} \left[B_{0l} + \frac{B_{1l}}{\varsigma_{200}} u + \frac{B_{2l}}{\varsigma_{020}} v + \frac{B_{3l}}{\varsigma_{002}} w \right]^2 K(u,v)K(w)\,dudvdw \;+
$$

$$
2d_\tau^2 \int\int\int_{Q^{(2l)}} \left[1 - B_{0l} - \frac{B_{1l}}{\varsigma_{200}} u - \frac{B_{2l}}{\varsigma_{020}} v - \frac{B_{3l}}{\varsigma_{002}} w \right]^2 K(u,v)K(w)\,dudvdw
$$

$$
+o_p(1)
$$

$$
= d_\tau^2 (C_\tau^{(l)})^2 + o_p(1),
$$

where

$$
C_\tau^{(l)} = \left(2\int\int\int_{Q^{(1l)}} \left[B_{0l} + \frac{B_{1l}}{\varsigma_{200}} u + \frac{B_{2l}}{\varsigma_{020}} v + \frac{B_{3l}}{\varsigma_{002}} w \right]^2 K(u,v)K(w)\,dudvdw \;+ \right.
$$

$$
\left. 2\int\int\int_{Q^{(2l)}} \left[1 - B_{0l} - \frac{B_{1l}}{\varsigma_{200}} u - \frac{B_{2l}}{\varsigma_{020}} v - \frac{B_{3l}}{\varsigma_{002}} w \right]^2 K(u,v)K(w)\,dudvdw \right)^{1/2}.
$$

Then by equation (A6)–(A8), we have

$$
e^{(l)}(x,y;t) = \sigma^2 + d_\tau^2 (C_\tau^{(l)})^2 + o_p(1).
$$

Similarly, we can prove that

$$
e(x,y;t) = \sigma^2 + d_\tau^2 (C_\tau)^2 + o_p(1),
$$

where

$$
C_\tau = \left(\int\int\int_{Q^{(1)}} \left[\varsigma_{000}^{(2)} + \frac{\varsigma_{100}^{(2)}}{\varsigma_{200}} u + \frac{\varsigma_{010}^{(2)}}{\varsigma_{020}} v + \frac{\varsigma_{001}^{(2)}}{\varsigma_{002}} w \right]^2 K(u,v)K(w)\,dudvdw \;+ \right.
$$

$$
\left. \int\int\int_{Q^{(2)}} \left[1 - \varsigma_{000}^{(2)} - \frac{\varsigma_{100}^{(2)}}{\varsigma_{200}} u - \frac{\varsigma_{010}^{(2)}}{\varsigma_{020}} v - \frac{\varsigma_{001}^{(2)}}{\varsigma_{002}} w \right]^2 K(u,v)K(w)\,dudvdw \right)^{1/2}.
$$

The main difference between this case and the previous case in the proof is in the derivation of the result of (A8). For $e(x,y;t)$, the corresponding result is

$$
A_3(x,y;t) = \frac{1}{NH} \sum_{(x_i,y_j;t_k)} \left[f(x_i,y_j;t_k) - \widehat{a}(x,y;t) - \widehat{b}(x,y;t)(x_i - x) - \right.
$$

$$
\left. \widehat{c}(x,y;t)(y_j - y) - \widehat{d}(x,y;t)(t_k - t) \right]^2 K_{ijk}
$$

$$
= \frac{1}{NH} \sum_{(x_i,y_j;t_k)} \left[f(x_i,y_j;t_k) - f_-(x_\tau,y_\tau;t_\tau) - d_\tau \varsigma_{000}^{(2)} - \frac{d_\tau \varsigma_{100}^{(2)}}{\varsigma_{200}} \frac{x_i - x}{h_x} - \right.
$$

$$
\left. \frac{d_\tau \varsigma_{010}^{(2)}}{\varsigma_{020}} \frac{y_j - y}{h_y} - \frac{d_\tau \varsigma_{001}^{(2)}}{\varsigma_{002}} \frac{t_k - t}{h_t} \right]^2 K_{ijk} + o_p(1)
$$

$$
= \frac{1}{NH} \left(\sum_{(x_i,y_j;t_k)\in I^1} + \sum_{(x_i,y_j;t_k)\in I^2} \right)
$$

$$
\left[f(x_i,y_j;t_k) - f_-(x_\tau,y_\tau;t_\tau) - d_\tau \zeta_{000}^{(2)} - \frac{d_\tau \zeta_{100}^{(2)}}{\zeta_{200}} \frac{x_i - x}{h_x} - \right.
$$

$$
\left. \frac{d_\tau \zeta_{010}^{(2)}}{\zeta_{020}} \frac{y_j - y}{h_y} - \frac{d_\tau \zeta_{001}^{(2)}}{\zeta_{002}} \frac{t_k - t}{h_t} \right]^2 K_{ijk} + o_p(1)
$$

$$
= \frac{1}{NH} \sum_{(x_i,y_j;t_k)\in I^1} \left[-d_\tau \zeta_{000}^{(2)} - \frac{d_\tau \zeta_{100}^{(2)}}{\zeta_{200}} \frac{x_i - x}{h_x} - \right.
$$

$$
\left. \frac{d_\tau \zeta_{010}^{(2)}}{\zeta_{020}} \frac{y_j - y}{h_y} - \frac{d_\tau \zeta_{001}^{(2)}}{\zeta_{002}} \frac{t_k - t}{h_t} \right]^2 K_{ijk} +
$$

$$
\frac{1}{NH} \sum_{(x_i,y_j;t_k)\in I^2} \left[d_\tau - d_\tau \zeta_{000}^{(2)} - \frac{d_\tau \zeta_{100}^{(2)}}{\zeta_{200}} \frac{x_i - x}{h_x} - \right.
$$

$$
\left. \frac{d_\tau \zeta_{010}^{(2)}}{\zeta_{020}} \frac{y_j - y}{h_y} - \frac{d_\tau \zeta_{001}^{(2)}}{\zeta_{002}} \frac{t_k - t}{h_t} \right]^2 K_{ijk} + o_p(1)
$$

$$
= d_\tau^2 \int\int\int_{Q^{(1)}} \left[\zeta_{000}^{(2)} + \frac{\zeta_{100}^{(2)}}{\zeta_{200}} u + \frac{\zeta_{010}^{(2)}}{\zeta_{020}} v + \frac{\zeta_{001}^{(2)}}{\zeta_{002}} w \right]^2 K(u,v)K(w)dudvdw +
$$

$$
d_\tau^2 \int\int\int_{Q^{(2)}} \left[1 - \zeta_{000}^{(2)} - \frac{\zeta_{100}^{(2)}}{\zeta_{200}} u - \frac{\zeta_{010}^{(2)}}{\zeta_{020}} v - \frac{\zeta_{001}^{(2)}}{\zeta_{002}} w \right]^2 K(u,v)K(w)dudvdw
$$

$$
+ o_p(1)
$$

$$
= d_\tau^2 (C_\tau)^2 + o_p(1).
$$

Appendix A.4. Proof of Theorem 3

For the case when $(x,y;t) \in \Omega_h \setminus J_h$, the estimator $\widehat{f}(x,y;t)$ is one of $\widehat{a}(x,y;t)$, $\widehat{a}^{(1)}(x,y;t)$, $\widehat{a}^{(2)}(x,y;t)$ and $(\widehat{a}^{(1)}(x,y;t) + \widehat{a}^{(2)}(x,y;t))/2$, all of which are consistent estimators of $f(x,y;t)$. So, we have the result in the theorem.

For the case when $(x,y;t) \in J_h \setminus S_h$, it is easy to see that we have either i) $e(x,y;t) = \sigma^2 + d_\tau^2(C_\tau)^2 + o_p(1)$, $e^{(1)}(x,y;t) = \sigma^2 + o_p(1)$, and $e^{(2)}(x,y;t) = \sigma^2 + d_\tau^2(C_\tau^{(2)})^2 + o_p(1)$, or ii) $e(x,y;t) = \sigma^2 + d_\tau^2(C_\tau)^2 + o_p(1)$, $e^{(1)}(x,y;t) = \sigma^2 + d_\tau^2(C_\tau^{(1)})^2 + o_p(1)$, and $e^{(2)}(x,y;t) = \sigma^2 + o_p(1)$. In both cases, we have $D(x,y;t) = d_\tau^2(C_\tau)^2 + o_p(1)$. Therefore, asymptotically $D(x,y;t) > u$. Since $e^{(1)}(x,y;t) < e^{(2)}(x,y;t)$ in i), the estimator $\widehat{f}(x,y;t)$ is $\widehat{a}^{(1)}(x,y;t)$ in this case, which is a consistent estimator of $f(x,y;t)$. A similar result follows in the case ii).

References

1. Zanter, K. *Landsat 8 (L8) Data Users Handbook*; Version 2; 2016; LSDS-1574; Department of the Interior, U.S. Geological Survey. Available online: https://landsat.usgs.gov/landsat-8-l8-data-users-handbook (accessed on 1 October 2020).
2. Qiu, P. Jump regression, image processing and quality control (with discussions). *Qual. Eng.* **2018**, *30*, 137–153. [CrossRef]
3. Gonzalez, R.C.; Woods, R.E. *Digital Image Processing*, 4th ed.; Pearson: New York, NY, USA, 2018.
4. Qiu, P. Jump surface estimation, edge detection, and image restoration. *J. Am. Stat. Assoc.* **2007**, *102*, 745–756. [CrossRef]
5. Geman, S.; Geman, D. Stochastic relaxation, Gibbs distributions and the Bayesian restoration of images. *IEEE Trans. Pattern Anal. Mach. Intell.* **1984**, *6*, 721–741. [CrossRef]
6. Besag, J. Spatial interaction and the statistical analysis of lattice systems (with discussions). *J. R. Stat. Soc. (Ser. B)* **1974**, *36*, 192–236.
7. Fessler, J.A.; Erdogan, H.; Wu, W.B. Exact distribution of edgepreserving MAP estimators for linear signal models with Gaussian measurement noise. *IEEE Trans. Image Process.* **2000**, *9*, 1049–1055. [CrossRef] [PubMed]

8. Perona, P.; Malik, J. Scale space and edge detection using anisotropic diffusion. *IEEE Trans. Pattern Anal. Mach. Intell.* **1990**, *12*, 629–639. [CrossRef]

9. Weickert, J. *Anisotropic Diffusion in Imaging Processing*; Teubner: Stuttgart, Germany, 1998.

10. Beck, A.; Teboulle, M. Fast gradient-based algorithms for constrained total variation image denoising and deblurring problems. *IEEE Trans. Image Process.* **2009**, *18*, 2419–2434. [CrossRef] [PubMed]

11. Rudin, L.; Osher, S.; Fatemi, E. Jump regression, Nonlinear total variation based noise removal algorithms. *Phys. D* **1992**, *60*, 259–268. [CrossRef]

12. Yuan, Q.; Zhang, L.; Shen, H. Hyperspectral Image Denoising Employing a Spectral–Spatial Adaptive Total Variation Model. *IEEE Trans. Geosci. Remote Sens.* **2012**, *50*, 3660–3677. [CrossRef]

13. Chang, G.S.; Yu, B.; Vetterli, M. Spatially adaptive wavelet thresholding with context modeling for image denoising. *IEEE Trans. Image Process.* **2000**, *9*, 1522–1531. [CrossRef]

14. Mrázek, P.; Weickert, J.; Steidl, G. Correspondences between wavelet shrinkage and nonlinear diffusion. In *Scale Space Methods in Computer Vision*; Griffin, L.D., Lillholm, M., Eds.; Springer: Berlin/Heidelberg, Germany, 2003.

15. Gijbels, I.; Lambert, A.; Qiu, P. Edge-preserving image denoising and estimation of discontinuous surfaces. *IEEE Trans. Pattern Anal. Mach. Intell.* **2006**, *28*, 1075–1087. [CrossRef]

16. Qiu, P. Discontinuous regression surfaces fitting. *Ann. Stat.* **1998**, *26*, 2218–2245. [CrossRef]

17. Qiu, P. Jump-preserving surface reconstruction from noisy data. *Ann. Inst. Stat. Math.* **2009**, *61*, 715–751. [CrossRef]

18. Qiu, P.; Mukherjee, P.S. Edge structure preserving 3-D image denoising by local surface approximation. *IEEE Trans. Pattern Anal. Mach. Intell.* **2012**, *34*, 1457–1468.

19. Polzehl, J.;Spokoiny, V.G. Adaptive weights smoothing with applications to image restoration. *J. R. Stat. Soc. (Ser. B)* **2000**, *62*, 335–354. [CrossRef]

20. Kervrann, C.; Boulanger, J. Optimal Spatial Adaptation for Patch-Based Image Denoising. *IEEE Trans. Image Process.* **2006**, *15*, 2866–2878. [CrossRef] [PubMed]

21. Jain, P.; Tyagi, V. A survey of edge-preserving image denoising methods. *Inf. Syst. Front.* **2016**, *18*, 159–170. [CrossRef]

22. Qiu, P. *Image Processing and Jump Regression Analysis*; John Wiley & Sons: New York, NY, USA, 2005.

23. Fan, J.; Gijbels, I. *Local Polynomial Modelling and Its Applications*; Chapman and Hall: New York, NY, USA, 1996.

24. Altman, N.S. Kernel smoothing of data with correlated errors. *J. Am. Stat. Assoc.* **1990**, *85*, 749–759. [CrossRef]

25. Opsomer, J.; Wang, Y.; Yang, Y. Nonparametric regression with correlated errors. *Stat. Sci.* **2001**, *16*, 134–153. [CrossRef]

26. Brabanter, K.D.; Brabanter, J.D.; Suykens, J.; Moor, B. Kernel regression in the presence of correlated errors. *J. Mach. Learn. Res.* **2011**, *12*, 1955–1976.

27. Rudin, L.; Osher, S.; Fatemi, E. **neuRosim:** An R package for generating fMRI data. *J. Stat. Softw.* **2011**, *44*, 1–18.

28. Hall, P.; Qiu, P. Blind deconvolution and deblurring in image analysis. *Stat. Sin.* **2007**, *17*, 1483–1509.

29. Svoboda, D.; Ulman, V.; Kováč, P.; Šalingová, B.; Tesařová, L.; Koutná, IK.; Matula, P. Vascular network formation in silico using the extended cellular potts model. *IEEE Int. Conf. Image Process.* **2016**, 3180–3183.

30. Shuford, W.D.; Warnock, N.; Molina, K.C.; Sturm, K. The Salton Sea as critical habitat to migratory and resident waterbirds. *Hydrobiologia* **2002**, *473*, 255–274. [CrossRef]

31. Davydov, Y.A. Convergence of Distributions Generated by Stationary Stochastic Process. *Theory Probab. Its Appl.* **1968**, *13*, 691–696. [CrossRef]

MDPI

St. Alban-Anlage 66

4052 Basel

Switzerland

Tel. +41 61 683 77 34

Fax +41 61 302 89 18

www.mdpi.com

Entropy Editorial Office

E-mail: entropy@mdpi.com

www.mdpi.com/journal/entropy

Contents

Editors
S. Ejaz Ahmed   Farouk Nathoo
Brock University   University of Victoria
Canada   Canada

Editorial Office
MDPI
St. Alban-Anlage 66
4052 Basel, Switzerland

This is a reprint of articles from the Special Issue published online in the open access journal *Entropy* (ISSN 1099-4300) (available at: https://www.mdpi.com/journal/entropy/special_issues/Big_Data_Biomedical_II).

For citation purposes, cite each article independently as indicated on the article page online and as indicated below:

LastName, A.A.; LastName, B.B.; LastName, C.C. Article Title. *Journal Name* **Year**, *Volume Number*, Page Range.

ISBN 978-3-0365-5549-2 (Hbk)
ISBN 978-3-0365-5550-8 (PDF)

Big Data Analytics and Information Science for Business and Biomedical Applications II

Editors

S. Ejaz Ahmed
Farouk Nathoo

MDPI • Basel • Beijing • Wuhan • Barcelona • Belgrade • Manchester • Tokyo • Cluj • Tianjin